The Dusky Dolphin
Master Acrobat off Different Shores

The Dusky Dolphin
Master Acrobat off Different Shores

Edited by Bernd Würsig and Melany Würsig

AMSTERDAM • BOSTON • HEIDELBERG • LONDON • NEW YORK • OXFORD
PARIS • SAN DIEGO • SAN FRANCISCO • SINGAPORE • SYDNEY • TOKYO

Academic Press is an Imprint of Elsevier

Academic Press is an imprint of Elsevier
32 Jamestown Road, London NW1 7BY, UK
30 Corporate Drive, Suite 400, Burlington, MA 01803, USA
525 B Street, Suite 1900, San Diego, CA 92101-4495, USA

First edition 2010

British Library Cataloguing in Publication Data
A catalogue record for this book is available from the British Library

Library of Congress Cataloging-in-Publication Data
A catalog record for this book is available from the Library of Congress

ISBN: 978-0-12-373723-6

For information on all Academic Press publications
visit our website at www. elsevierdirect.com

Typeset by Macmillan Publishing solution
www.macmillansolutions.com

This book is dedicated to Charlie Walcott, the finest of scientists, naturalists, mentors, and humanists. We thank you for making our journey possible, and for always—in direct fashion or in spirit—guiding us along the way. We owe you more than two life-times of gratitude.

Professor Charles Walcott, Punta Los Conos, Golfo San José, Argentina, 24 December 1973

Bernd and Melany Würsig,
Muritai Maui,
Kaikoura, New Zealand,
January 2009

Contents

Dusky dolphins (*Lagenorhynchus obscurus*) are indeed master acrobats, as the book title states, for they leap in a spectacular variety of ways. There are simple head-first re-entry leaps that make hardly a splash; side, belly, or backslaps that are structured to make noise; long distance in-air salmon leaps designed for the dolphin to travel more rapidly than through the dense medium of water; and acrobatic leaps, spins, and somersaults that seem to be created for the pure joy of creation. We call the latter acrobatic leaps "social," and believe that these indicate high levels of alertness and a "party-like" atmosphere that may very well be important in social mammals for reinforcing existing social bonds and forging new ones. Curiously, the structure of this book is such that we do not detail these different leap types in great detail, although leaps are mentioned in several contexts of behavior throughout. The interested reader may wish to visit Würsig and Whitehead (2009) for a more detailed discussion of aerial behavior.

Acrobatic leaping dusky dolphins leap in sequences, with from 1 or 2 leaps to as many as 36 such leaps during one bout. Whenever a dusky dolphin starts out with a particular type of acrobatic leap, say a forwards somersault with a half-body twist to the right, it will keep that same leap during the leap bout, with no change in repertoire. The 36th leap will be the same as the first, although one can tell that the dolphin is tiring by "the end," just before it quits leaping and slowly travels at the surface while breathing rapidly. This appears to be much like the athletic human who starts a sequence of "chin-ups" at the local gymnasium with vigor, but the final attempts to lift the chin above the exercise bar are arduous indeed, and the human then needs to take a rest. So it is for the dolphin; it rests, and several minutes later may begin a new leap sequence, which is hardly ever the same leap type as exhibited before. This time it may enter the air with a rapid forwards leap followed by a right-sided body slap onto the water, with final tail slap as it descends, and so on.

We have no (real) idea of what is going on when dusky dolphins engage in acrobatic "social" leaps, but these occur only at levels of high group alertness, and seem to signal (and perhaps incite) "excitement" in group members as a whole. We (Würsig and Würsig 1980) long-ago attempted to describe leap types relative to progressions in cooperative "bait-ball herding" foraging stages in Argentina: noiseless head-first re-entry leaps at the initial stages of foraging are designed for dolphins to come from depth, overshoot the surface, breathe, and descend to depth again; noisy side-slaps help to frighten and thereby tighten fish schools once they are at the surface, and possibly also signal other dolphins of the stage of feeding nearby; and acrobatic leaps are part of that important bonding "party" atmosphere towards the end of and after feeding. We are pretty sure that we were right in those generalities, but are also keenly aware that these "meanings" of leaps have never been proven, and that dolphins can leap in many different ways for many different reasons. For example, leaps are often parts of sexual displays, may signal changes in travel direction of a group (Markowitz 2004), and may also be important at times for dislodging ectoparasites. It is not impossible that leaps provide a "massage-like" therapy for feeling good, just as a red deer (*Cervus elephus*) or New World peccary (*Pecari angulatus*) may massage its sides and back by rubbing against trees, shrubs, and rocks.

We are tremendously pleased that many present and former students, and other fine colleagues, have heeded our request to contribute to this book. They are enriching our knowledge of dusky dolphins, other marine mammals, and the natural world. Each chapter was reviewed by a minimum of two but more often three to (at times) four peers, and we are particularly pleased that reviewers provided detailed suggestions and information for chapter improvements. Reviewers of entire or parts of chapters are A. Acevedo-Gutierrez, J. Alvarado-Bremer, A. Baker, C.S. Baker, L. Ballance, A. Baxter, R. Brownell, Jr., F. Cipriano, A. Dahood, S. Dawson, S. DuFresne, D. Fertl, A. Frankel, S. Gowans, W. Grant, A. Harlin-Cognato, S. Hung, T. Jefferson, L. Karczmarski, D. Lundquist, S. Lynn, T. Markowitz, J. Mann, L. Marino, K. McHugh, H. Pearson, D. Reiss, L. Slooten, R. Vaughn, J. Weir, R. Wells, and D. Weller.

The book begins with the taxonomic placement of dusky dolphins in the genus *Lagenorhynchus* and family Delphinidae. April Harlin-Cognato (Chapter 1) took on this daunting task, with a historical overview, and reminds us that the final chapter of dusky taxonomy is not yet written. Frank Cipriano and Marc Webber (Chapter 2) then give a description of dusky dolphin life history and demographic patterns, with old data gathered >20 years ago, and new information from around hot spots of dusky occurrence in the southern hemisphere. Silvana Dans and colleagues (Chapter 3) discuss duskies in terms of their roles in the food web, also with considerations of predation, competition, and parasitism. Theirs is a much welcome perspective, largely from South America, as much of the rest of the book concentrates on data from New Zealand.

The next several chapters deal with dusky dolphins interacting with their habitats, and we decided that it would be best to start it with a description of dusky dolphin sound production and probable perception, by Whit Au et al. (Chapter 4). Sound reception (and this is preceded by production during echolocation) is arguably their most important sensory modality, although we are firm believers that vision, a highly sensitive "tactile" skin, and taste are also important, perhaps critically so. Adrian Dahood and Kelly Benoit-Bird then tell us generally new things about foraging at night in deep waters (Chapter 5), and Robin Vaughn and colleagues tell us about foraging in daylight on "bait-ball" schooling fishes in shallow waters (Chapter 6). Nick Duprey and Griselda Garaffo (in a separate "box," embedded in Chapter 5) set the stage for the variability of dusky dolphin behavior by day and season, in anticipation of more detailed data to follow in subsequent chapters. We now know that dusky dolphins dive down to meet the rising deep scattering layer in the evening, but "prefer" to do so no deeper than about 130 m. If they find a less dense prey patch at 80 m depth than occurs at 120 m, they will dive only to 80 m (although we do not know the lower limit of density of prey to make feeding worthwhile). They are time-constrained, not energy-constrained, divers (see Acevedo-Gutiérrez et al. 2002 for differences in large lunge-feeding whales). When bait-balling, dolphins do indeed herd the fish school towards the surface, as was postulated about 30 years ago (Würsig and Würsig 1980), but not demonstrated until recently (Vaughn et al. 2008). The over-arching theme is that dusky dolphins are highly flexible creatures, forming party affiliations and a fission–fusion society in line with habitat and foraging strategy. Mridula Srinivasan and Tim Markowitz discuss dolphin responses to threats by predators (Chapter 7), and remind us that the mere possibility of predation threat can affect social structure and behavior of potential predators. Tim Markowitz et al. and Jody Weir et al. describe mating strategies and calf rearing (Chapters 8 and 9, respectively), both with much new syntheses of information. We round out this general "behavioral" section with new information on dolphin sexual segregation and hints at genetic relationship, by Deborah Shelton and colleagues (Chapter 10), with the understanding that much more data are needed to help us describe true social structure in variable habitats.

The next several chapters describe duskies in interaction with humans. Tim Markowitz and colleagues summarize generally negative (but not totally so) human interactions (Chapter 11), and Simon Childerhouse and Andrew Baxter summarize management ("regulation") perspectives, from New Zealand, Australia, both sides of South America, and southern Africa (Chapter 12). Overall, dusky dolphins are doing "alright" in most parts of their ranges, but there are problem areas, especially in South America. An inserted "box" by Maurice Manawatu on an indigenous people's perspective is embedded in Chapter 12. Then (Chapter 13), Dennis Buurman of Kaikoura's dolphin-watching operation gives a flavor of what it is like to be both a tour operator who cares about giving an excellent experience to tourists, and what a tourist might expect to encounter with duskies at sea.

The final section might be considered a "grab-bag" of not totally related topics, but we hope that here we cover things not addressed elsewhere, and wrap up with more general thoughts. Peter Best and Michael Meÿer give us a summary of duskies off southern Africa, covering a gamut of information from life history strategies to human interactions (Chapter 14). This is very welcome information, as not much has been published on dusky dolphins from that part of the world. Sonja Heinrich and colleagues address the oft-overlapping habitats of the genera *Lagenorhynchus* and *Cephalorhynchus,* and discuss how it is that there is generally little interaction among the species (Chapter 15). Heidi Pearson and Deborah Shelton then take us on a consideration of dusky dolphins as large-brained intelligent social mammals, with interesting comparisons among dolphins and great apes (Chapter 16), and Bernd Würsig (Chapter 17) harks back to some things learned throughout the book, makes guesses about dolphin culture, and discusses worries and conservation needs of what on the face of it seems like a quite "robust" species in terms of prospect for survival.

There is some overlap of information, and quite a few chapters "again describe" the basic differences of night-time deep water feeding and daytime shallow water bait-ball herding, for example. While we have tried to reduce the most glaring duplications of information, we have also given the authors some latitude in their concepts of what needs to be described for *their* chapter to be a stand-alone contribution, so that the interested reader can pick up the book and read just one or two chapters without the need to go to earlier ones for clarification. We realize that such duplication may be somewhat distracting for the reader who dedicates several hours to reading the entire book (a rare reader, we guess!), and apologize for the duplication.

We asked each senior author to come up with glossary terms and definitions, and most did so. These terms range from the "quite obvious" to the "quite esoteric," but we hope that the inclusion of definitions will speed the reader through topics with which she or he may not be as familiar. We hope that the cross-referenced index will also help to find topics. After much debate, we decided to place all references at the end of the book, not particular to each chapter. There are advantages and disadvantages to both methods for such a work as ours, but since the main topic is only ONE dolphin species, we were able to limit quite a few often-cited references to one instead of the one-half dozen or more times they were cited by individual chapters. We justify this by claiming that we are saving paper and therefore trees.

We two editors have been working with dusky dolphins since 1972, although with several long-term hiatuses while parenting, research elsewhere, and school endeavors took us away from these little acrobats. It is understandable that there are many (many!) people to thank. We can do so only partially here, and do not repeat all acknowledgments made by authors at the end of their chapters. If we accidentally missed you, we apologize and meant no harm. Our special thanks to Charlie Walcott and Roger Payne who started us on this journey, and the authors and reviewers who helped us to this juncture some 37 years later.

We further thank, in (very) rough chronological order of help: George C. Williams, Doug Smith, Larry Slobodkin, Peter Tyack, Russ Charif, Marty Hyatt, Des and Jen and Les Bartlett, Juancito Olazabal, Katy and John and Holly and Laura and Sam Payne, Hugo Callejas, Carlos Garcia, Juan Carlos and Diana Lopez, Chris Clark and Janie Moon (now Clark), Lisa Leyland (now Struhsaker), Jan Wolitzky, Ken and Phylly Norris, Guy Oliver, Bill Evans, Steven Ferraro, Matt Lamishaw, Kim Wuersig (now Kirkhart) and Paul Wuersig, Mari Smultea, Thomas Henningsen, Anna Forest, Dagmar Fertl, Steve Leatherwood, Janet Doherty, Nanette Parker, Elizabeth Zuñiga, Stefan Bräger, Jack and Betty van Berkel, Dennis and Lynnette Buurman, Rik and Anja Buurman, Ian Bradshaw and Jackie Wadsworth, Mike Morrissey, Andrew Baxter, Nikki Brown, Kirsty Barr, Amy Beier (now Engelhaupt), Dan Engelhaupt, Olga Sychenko, Glenn Gailey, Cynthia McFadden, Rochelle

"Brave" Constantine, Wang Ding, Rachel Smolker, Ron Peterson, Suse Shane, Sheila and Alan Baldridge, Mike Donoghue, Koen Van Waerebeek, Bill Perrin, Tom Jefferson, Dave Weller, Whitlow Au, Richard Connor, Ginger Barnett, Mark Webber, Bob Brownell, Shannon Gowans, Peter Simard, Becky Simard, Nathalie Patenaude, Liz Slooten, Steve Dawson, Stacie Arms, Bailey Arms, Nicole Luukko, Gabriella Fox, a long list of several hundred Earthwatch volunteers who helped in the late 1980s, and mid 1990s through 2006, and a long list of interns who graced Bernd's lab throughout the years at Moss Landing Marine Laboratories and Texas A&M University. Robin Vaughn, Nikki Hughes, Sarah Piwetz, Mridula Srinivasan, Heidi Pearson, Adrian Dahood, and Emily Gillespie gave up parts of their 2008 end-of-year holidays to help with final editing, organization, and other bits for the book. We thank you all.

Funding for parts of our work through the years was provided by the National Geographic Society, Earthwatch, New York Zoological Society, World Wildlife Fund, Neurobiology and Behavior Program of Stony Brook University, National Institutes of Health, National Science Foundation, National Marine Fisheries Service, Cetacean Society International, Moss Landing Marine Laboratories, the Packard Foundation, American Cetacean Society, Cetacean Society International, Gruppe 5 Film Company, MacGillivrey-Freeman "IMAX," Fulbright International, New Zealand Department of Conservation (with special thanks to Andrew Baxter), Marlborough District Council (New Zealand), Research and Graduate Studies Office of Texas A&M University at Galveston (with special thanks to Bill Seitz and Tammy Holliday), Regents of Texas A&M University, Department of Marine Biology of Texas A&M University at Galveston, The Erma Lee and Luke Mooney Graduate Student Travel Grants and Fellowships, and multiple other grants and fellowships to graduate students. We are grateful to all.

Mel and Bernd, "Whale Camp", Península Valdés Chubut Argentina, Winter 1973

Bernd and Melany Würsig,
El Deseado, Southeastern Sonoran Desert,
Arizona, December 2008

Bernd Würsig is regents professor at Texas A&M University, and chair of the Marine Biology Graduate Program. Melany Würsig is curriculum instruction specialist at Cloverleaf Elementary School in Houston, Texas. Bernd has been senior advisor to 65 graduate students since 1981, and authored or co-authored approximately 150 research papers and six books, including the second edition of the *Encyclopedia of Marine Mammals* (2009, with Bill Perrin and Hans Thewissen). Bernd and Melany also co-authored *The Hawaiian Spinner Dolphin* (1994, with Ken Norris and Randy Wells). They tend to tropical plants in south Texas, and are visited by raccoons every night.

Chapter 1

The Dusky Dolphins' Place in the Delphinid Family Tree

April D. Harlin-Cognato

Department of Zoology, Michigan State University, East Lansing, Michigan, USA

It is a cool autumn Saturday in Michigan, and I am sitting in my office with a cup of coffee, scouring dusty volumes in search of yet another piece of the taxonomic puzzle of the dusky dolphin. My imagination wanders back in time to the British Museum of Natural History in the 1800s—smack dab in the middle of a period of discovery in biology. I wonder, what would it have been like to be a scientist during the great period of exploration in the 1700s and 1800s? Each new shipment that arrived at the British Museum from voyages of exploration contained an abundance of plants, animals, and geologic samples that had never been described before. I imagine myself as John Edward Gray, a preeminent taxonomist at the British Museum of Natural History in 1846, the day he received from the voyage of the HMS Erebus *and* Terror, *a ship commissioned by Queen Victoria to explore Antarctic seas, the crate that contained the type specimen for the genus* Lagenorhynchus. *The excitement! As I lift the skull from its surrounding packaging materials, my eyes search for those features that I know from experience should identify the species. Aha! The head is rather convex, gradually sloping into the beak in front, and the beak is very short relative to its braincase. I reference my memory of the current specimens at the Museum that I have previously described, and do a quick mental comparison with this new specimen. I am certain it is unique! Once again, I am holding in my hand the representative of a species of marine mammal, amazing creatures in their own right for having adapted to life at sea, from a far off, exotic region of the Earth's oceans few have ever traversed. My imagination races! What was the bearer of this skull like when alive? What were its patterns of coloration, its behaviors? Did it live in groups, or alone? What was the nature of the habitat in which it lived? I must take this amazing specimen to my lab immediately to make a careful comparison with the collection . . .*

The tools of taxonomy have changed drastically over the past 160 years since Gray first described the genus Lagenorhynchus, *but the excitement of discovery, I imagine, is not all that different today. As scientists, we are driven to understand the origin of living things. It is this inherent curiosity that fuels my search to answer the questions about the evolutionary history of the dusky dolphin. Now my tools are not the features of bony remains, but those of individual pieces of DNA—the code that determines the physical characteristics of species that make them different from each other. In this sense, each chain of nucleic acids represents a unique snapshot of evolutionary time—the culmination of millions of years of natural selection that has shaped the amazing diversity of life on Earth. I never cease to be amazed by this process. I imagine Gray looking over my shoulder as I use DNA sequences from multiple species to produce a phylogeny, equally eager to discover where the newly added individuals will end up in the dolphin family tree. Decades of scientific advances lie between Gray and me, but the passion for the question, regardless of the approach taken to answer it, remains the same.*

April Harlin-Cognato
Michigan State University, October 2008

A STATE OF CONFUSION

There has also been a great amount of imperfect and hasty compilation, and attempts at systematizing, based upon erroneous conceptions of affinities and imperfect anatomical knowledge, which have thrown a haze over the subject, often most difficult to penetrate.

William Henry Flower, On the Characters and
Divisions of the Family Delphinidae, 1883

Sir William Henry Flower (1831–1899; Figure 1.1), a prominent surgeon and keen student of mammalian osteology, assumed in the early 1880s the directorship of the Natural History Departments at the British Museum. Of particular interest to Flower were the cetaceans, and as a trained comparative anatomist, he was pre-equipped to tackle the issue of cetacean taxonomy via the study of osteological characters; however, upon initiating his studies of the Cetacea at the Museum, he was stymied by the poor organization of the collection:

I carefully examined all the Dolphins' skulls in the collection …A difficulty at once arose from the fact that none of the skulls have any number or mark upon them by which their history could be traced with certainty … the names and indications of origin are written on cards fixed on the stands, and there is unfortunately no guarantee that the latter may not have been changed, as in some cases it is quite evident has been done. (Flower 1883, p. 492)

The state of the cetacean osteological collection prompted Flower, a disciplined man with attention to detail (Lydekker 1906), to take on the task of reorganization, and in 1883 and 1885, within the first years of his directorship, he published two important works in which he presents a revision of the Museum's collection of Cetacea. In the first paper he attempts to verify provenance, and re-examines, and in many cases redefines, taxonomy based on his anatomical expertise (Flower 1883). In the second, he provides a revised taxonomy of the Cetacea and a list of the specimens held at the British Museum (Flower 1885). These two publications are landmarks in the field of cetacean taxonomy, as they mark the end of an era plagued with ambiguity and confusion, no doubt attributed to the spurious, and sometimes arbitrary, assignment of names based on incomplete specimens or poorly defined anatomical characters (Flower 1883).

When Flower assumed the directorship, it had been almost 10 years since the death of John Edward Gray (1800–1875; Figure 1.2), who, during his 45-year

Sir William Henry Flower; K.C.B.

FIGURE 1.1 Sir William Henry Flower (1831–1899) was the Director of the British Museum of Natural History from 1884 to 1898, during which time he made significant contributions to the taxonomy of the Cetacea, including the genus *Lagenorhynchus* (photo courtesy of Wikimedia Commons, www.commons.wikimedia.org).

tenure at the Museum, became probably the greatest contributor to the study of cetacean taxonomy. Much of the success of Gray can be attributed to a fortuitous convergence of events, most of which were related to the rapid discovery of new and unusual forms of animals during the great exploratory voyages of Great Britain and other nations between the years 1820 and 1870, during which time Gray served as Keeper of Zoology at the Museum. It is clear from Museum records (e.g., Flower 1885) that it was during this period that the majority of cetacean osteological specimens were acquired for the collection, and Gray, a very prolific zoologist, was often the first to describe these novelties. As a result, his contributions to cetacean taxonomy were extensive—nearly half of the original descriptions of currently recognized dolphin genera and a third of species are attributed to Gray (Mead and Brownell 2005).

Gray's industrious nature was considered one of his greatest assets; however his taxonomic expertise was not. His colleagues criticized him for his lack of expertise in anatomy, which in some cases "distorted his view of the relative importance of character based only on osteological features" (Anonymous 1875, p. 368). Given that the majority of taxonomic classification was based on osteological remains, Gray's lack of anatomical expertise profoundly influenced

FIGURE 1.2   John Edward Gray, F.R.S. (1800–1875) was the Keeper of Zoology at the British Museum in London from 1840 to 1874. During this period he was responsible for describing almost half of currently recognized genera of cetaceans (photo courtesy of Wikimedia Commons, www.commons.wikimedia.org).

the history of cetacean taxonomy. It was not until after his death in 1875, just 4 years after he published his final synopsis of the Cetacea (Gray 1871), that Gray was publicly criticized for his frequent mischaracterization of osteological features. It was Flower, likely frustrated with the confused state of the cetacean osteological collection left by Gray, who perhaps was his harshest critic:

> [Gray's] tendency to multiply divisions and impose names almost at random, his want of accuracy in description, and his defective anatomical knowledge, are exhibited in his writing on [Cetacea] in their fullest development ... [Gray's] names and views of affinity are becoming so deeply rooted in zoological literature, it appears time that an attempt should be made to supply something upon a more scientific basis. (Flower 1883, p. 467)

Gray's "defective anatomical knowledge" is apparent, according to Flower, in the choice of characters Gray used to categorize dolphin taxa. These traits were predominantly based on the relative proportion of skull features and the number of teeth, which were known by Flower (1883) to be unreliable because they can vary with age and gender:

> Thus the proportions of length and width of beak, and number of teeth in a given space (so much used by Gray to distinguish species) cannot be relied upon, except in comparing *perfectly adult animals.* ... Sex also appears to exercise an important influence upon the form of the skull. ... Such differences as these, it will be observed, quite as great as many upon which Dr. Gray has founded distinct species. (emphasis Flower, pp. 469–470)

In addition, Flower disapproved of what he considered "the rule with some zoologists," including Gray, of assuming a new individual is a different species if found in a different locality (Flower 1883):

I have abandoned the old assumption, upon which so many new species were founded, which limited the geographical area of each species to a small and circumscribed portion of the ocean, and placed imaginary barriers to its distribution where none really existed. (Flower 1883, p. 469)

Given what we now know about sex- and age-specific trait variation and the wide geographical distribution of many dolphin species, it is not surprising that the use of such traits to diagnose species threw "a haze over the subject [of dolphin taxonomy], often most difficult to penetrate" (Flower 1883, p. 466).

A HISTORICAL OVERVIEW OF THE GENUS *LAGENORHYNCHUS*

Perhaps no group of cetaceans epitomizes the confused state of cetacean taxonomy at the end of the nineteenth century more than the genus *Lagenorhynchus*. There are currently six recognized species (Figure 1.3), all of which are distributed in temperate to cool waters in the North Pacific (*L. obliquidens*), North Atlantic (*L. albirostris*, *L. acutus*), and Southern Pacific and Indian Oceans (*L. obscurus*, *L. australis*, *L. cruciger*; Fraser 1966; Figure 1.4). All species of the genus share a general body plan defined by a relatively short rostrum, a prominent falcate dorsal fin, a stocky robust body shape, and a complex pattern of body coloration, dominated by bands, stripes, and flares of multiple hues (Figure 1.3). It was this suite of traits that Gray first used to define the genus in 1846 based on a skull sent to him from a "Mr. Brightwell," who had initially identified it as *Delphinus tursio*. Gray found it distinguishable from *Delphinus* by the white color of the beak, the smaller size of the teeth, and the short rostrum compared to the length of the braincase (Gray 1846a). Later that same year, Gray (1846b) published one of his most famous works, *The Zoology of the Voyage of the HMS Erebus and Terror*, in which he not only puts "into scientific order the materials brought home by this Expedition," he takes the opportunity to arrange and name "the extensive collection of specimens of these animals, and their osseous remains, in the British Museum ... and throw the result of my labors into a synoptic revision of the species of the entire family" (p. 13). In this synopsis, Gray presents a more detailed description of the diagnostic characteristics of the genus *Lagenorhynchus*, which include a short beak, high falcate dorsal fin, and a front of face "short, broad, flat above and rather narrowed in front, and scarcely longer than the length of the brain-cavity" (p. 34).

This description was based on a new species, *Lagenorhynchus albirostris*, which was, and still is, recognized as the type species for the genus. Unfortunately, however, Gray had chosen, as was his tendency, traits with high levels of intraspecific variation to define the genus. In Gray's defense, the traits he chose were based on the limited knowledge of dolphin morphology at the time. Nevertheless, his choice

FIGURE 1.3   Currently recognized members of the genus *Lagenorhynchus*. Left to right from the top: white-beaked dolphin, *L. albirostris* (photo © Rudolf Svensen, www.UMPhoto.no); dusky dolphin, *L. obscurus* (photo by April Harlin-Cognato); Atlantic white-sided dolphin, *L. acutus* (photo: MarineBio, www.marinebio.org); Pacific white-sided dolphin, *L. obliquidens* (photo: NOAA, SWFSC, www.photolib.noaa.gov); hourglass dolphin, *L. cruciger* (photo © David Walsh, www.davidwalshphotography.com), Peale's dolphin, *L. australis* (photo courtesy of Wikimedia Commons, www.commons.wikimedia.org).

of characters hindered at the outset the ability to properly define new members of the genus, which was worsened by his repeated redefinition of *Lagenorhynchus* spp. based upon increasingly slight variations in the characteristics of the skull (e.g., Gray 1866a, 1866b, 1868a, 1868b, 1871, Flower 1883; Table 1.1).

Gray's taxonomic heritage had a widespread and long-term impact on the taxonomy of *Lagenorhynchus*. From the mid 1800s to the middle part of the twentieth century scientists such as Cope (1866, 1876, Scammon and Cope 1869), True (1889, 1903), and Kellogg (1941), the latter from the National Museum of the United States (part of the Smithsonian Institution), made multiple contributions

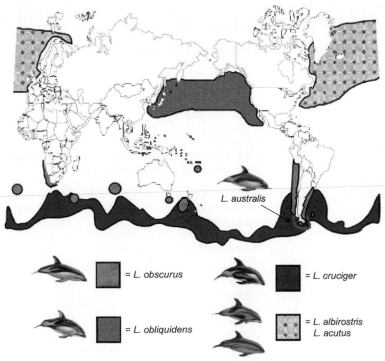

FIGURE 1.4 Distribution of the six currently recognized species of *Lagenorhynchus* (Jefferson et al. 2008).

TABLE 1.1 An abridged summary of the taxonomic history of the genus *Lagenorhynchus* between the years 1846–2008

Gray 1846b	*Lagenorhynchus leucopleurus (= L. acutus*)* *L. albirostris** *L. acutus** *L. electra* *L. asia* **Delphinus obscurus (= L. obscurus*)**
Gray 1850	*L. leucopleurus (= L. acutus*)* *L. albirostris** *L. acutus** *L. electra* *L. asia* *L. caeruleo-albus* *L. clanculus (= L. cruciger*)* *L. thicolea* **D. obscurus (= L. obscurus*)**

(Continued)

TABLE 1.1   (Continued)

Gray 1866a	*L. leucopleurus (= L. acutus*)* *L. albirostris** *L. acutus** *L. electra* *L. asia* *L. caeruleo-albus* *L. clanculus (= L. cruciger*)* *L. thicolea* *L. breviceps (= L. obscurus*)* **Tursio obscurus (= L. obscurus*)**
Gray 1868a	*Electra acuta (= L. acutus*)* *Leucopleurus arcticus (= L. leucopleurus = L. acutus*)* *L. albirostris* **Clymenia similis (= L. obscurus*)** **Clymenia obscura (= L. obscurus*)**
Flower 1885	*L. albirostris** *L. acutus** *L. electra* *L. clanculus (= L. cruciger*)* *L. fitzroyi*[a] *(=* **L. obscurus***)* *L. (Delphinus) cruciger*[*a] *Prodelphinus obscurus =* **(L. obscurus*)**
True 1889	*L. albirostris** *L. acutus** *L. electra* *L. thicolea* *L. obliquidens** *L. cruciger** *L. superciliosus (=* **L. obscurus***)* *L. fitzroyi (=* **L. obscurus***)* **L. obscurus***
Norman & Fraser 1948	*L. acutus** *L. australis** *L. obliquidens** *L. albirostris** *L. cruciger** *L. fitzroyi (=* **L. obscurus***)* **L. obscurus***
Hershkovitz 1966	*L. acutus** *L. albirostris** *L. cruciger** *(includes L. australis,* **L. obscurus***)* *L. electra* *L. [cruciger] obliquidens** *L. thicolea*

(Continued)

TABLE 1.1   (Continued)

Perrin 1989; Jefferson, et al. 1993	*L. acutus*[*] *L. obliquidens*[*] *L. albirostris*[*] *L. cruciger*[*] *L. australis*[*] **L. obscurus**[*]
Rice 1998; Jefferson et al. 2008	*L. acutus*[b] *L. albirostris*[b] *L. obliquidens* *L. cruciger* *L. australis* **L. obscurus** *L. o. obscurus (South Africa)* *L. o. fitzroyi (South America)* *L. o. ? (New Zealand)*
Harlin-Cognato 2008	*L. acutus* *L. albirostris* *L. obliquidens* *L. obscurus* *L. o. obscurus (South Africa)* *L. o. fitzroyi (Argentina)* *L. o. posidonia (Peru/Chile)* *L. o. superciliosus (New Zealand)* *Sagmatias australis* *Sagmatias cruciger*

*Each list contains all of the members of the genus Lagenorhynchus *recognized by the author, plus members of other genera that were later synonymized under* Lagenorhynchus *as one of the six currently recognized species in the genus (indicated by asterisk). Note that this is not a complete taxonomy; please reference the original publications for a complete list of synonyms and the citations for their provenance.*

[a]Flower (1883) suggests D. cruciger and D. fitzroyi are probably members of Lagenorhynchus, *but is uncertain due to limited evidence; Flower (1885) places D. fitzroyi in* Lagenorhynchus.

[b]Jefferson et al. (2008) recognize recent genetic studies that suggest these species are not members of Lagenorhynchus.

to *Lagenorhynchus* taxonomy using the same problematic characters and general tendency to name new species as had Gray. The result was the addition of more confusion to an already tangled taxonomic mess. From 1828 to 1885, the taxonomy of *Lagenorhynchus* was particularly labile, with the assignment (and often re-assignment) of *Lagenorhynchus* spp. into at least nine different genera during this period (e.g., *Delphinus*, Gray 1828; *Electra*, Gray 1871; *Phocoena*, Peale 1848; *Tursio*, Gray 1866a; *Clymenia*, Gray 1868a; *Clymene*, Gray 1868b; *Leucopleurus*, Gray 1871; *Sagmatias*, Cope 1866; *Prodelphinus*, Flower 1885; Table 1.1).

TAXONOMIC HISTORY OF THE DUSKY DOLPHIN

Gray first described the dusky dolphin as *Delphinus obscurus*, subgenus *Grampus*, in his 1828 *Specilegia Zoologica, or Original Figures and Short Systematic Description of New and Unfigured Animals*. He reports that the specimen, captured in the vicinity of the Cape of Good Hope by a "Captain Heaviside"[1], came to the British Museum via the Royal College of Surgeons in 1827.

However, Gray was not actually the first person to describe a specimen of the dusky dolphin. According to publications by Gray and others (e.g., Gray 1846a, 1846b; Schlegel 1841) the dusky dolphin was first described as *Delphinus superciliosus* from a specimen collected off the coast of Tasmania 2 years earlier by René-Primevère Lesson and Prosper Garnot, pre-eminent French surgeons and naturalists aboard Captain Duperrey's round-the-world voyage of *La Coquille* (1822–1825; Lesson and Garnot 1826; Figure 1.5). Although Lesson and Garnot captured the dolphin and brought it on board, only an illustration and a description were retained. The first code for systematic nomenclature adopted by the British Association for the Advancement of Science in 1846 (Darwin et al. 1843a, 1843b) allowed recognition of a species if its original definition was published as a "distinct exposition of essential characters" (Darwin et al. 1843a, p. 114)[2]. Gray (1846b) considered

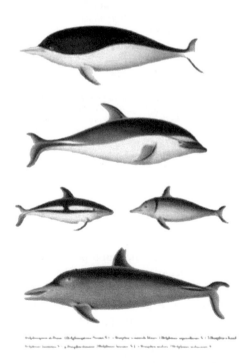

FIGURE 1.5   Illustrations of dolphins described by Lesson and Garnot (1826) from the voyage of *La Coquille. Delphinus superciliosus*, now recognized as a synonym of *Lagenorhynchus obscurus*, is the second from the top.

D. superciliosus as a junior synonym of his *D. obscurus* in his 1846 synopsis of the cetaceans at the British Museum (Figure 1.6), but gives Lesson and Garnot (1826) credit for their original description. The implication is that Gray did not consider Lesson and Garnot's illustration of *D. superciliosus* "essential" enough to justify it as a type specimen for a new species.

Hermann Schlegel (1804–1884), future director of the Leiden Museum, published another description of *D. superciliosus* in 1841 from a specimen collected from the Cape of Good Hope. Interestingly, in future publications, Schlegel (1841) alone is credited with *D. superciliosus*, and Lesson and Garnot's contribution is not acknowledged, even though Schlegel himself acknowledges their original description (Schlegel 1841, p. 22). For example, True (1889) and Kellogg (1941) resurrect *D. superciliosus* as *Lagenorhynchus superciliosus* (Figure 1.6), but give precedence to Schlegel's description over that of Lesson and Garnot. In this case, there is direct evidence that this was due to a resistance to describe a new species without a specimen "in hand." True (1889), Kellogg (1941), and Fraser (1966) clearly indicate their position against such practices:

I share the hesitancy of Kellogg about the specific reference of *Delphinus superciliosus* Lesson and Garnot 1827 ... and agree with him that sufficient material is not yet available for a decision about [its] status. (Fraser 1966, p. 17)

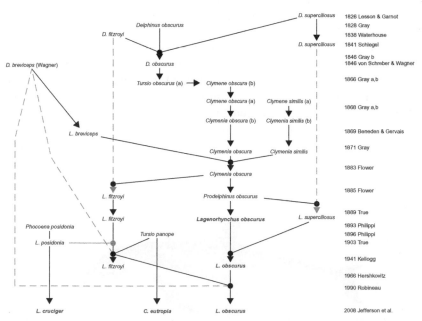

FIGURE 1.6 Overview of the major events in the taxonomic history of the dusky dolphin, *Lagenorhynchus obscurus*, synthesized from landmark publications from 1826 to present. Synonymy is represented by joining lines, and dashed gray lines indicate the resurrection of a taxon under a different genus; arrows indicate direction of changes in nomenclature. Currently recognized species are in red type. In cases where the same author had more than one publication in the same year, the letter next to the species name indicates in which publication it appeared.

In 1966, Hershkovitz made *L. superciliosus* a synonym of *L. cruciger*, but more recent reviews of dusky dolphin taxonomy consider "*superciliosus*" as a junior synonym of *L. obscurus* (e.g., Brownell and Cipriano 1999, Mead and Brownell 2005).

A third early representative of the dusky dolphin was collected sometime between 1831 and 1836 from the HMS *Beagle* during the voyage that sparked the beginnings of Darwin's theory of natural selection. George Waterhouse (1838a) in *The Zoology of the HMS Beagle* described *Delphinus fitzroyi*, named in honor of Robert Fitzroy, Captain of the HMS *Beagle*, who made the first illustration of the captured dolphin (Figure 1.7):

> This porpoise, which was a female, was harpooned from the Beagle in the Bay of St. Joseph, out of several, in a large troop, which were sporting round the ship. I am indebted to Captain Fitzroy for having made an excellent coloured drawing of it, when fresh killed. (C. Darwin in Waterhouse 1838a, p. 26)

Gray (1846a) recognized the "Fitzroy" dolphin as a junior synonym of *D. obscurus* in his synopsis in the *Zoology of the HMS Erebus & Terror.* Yet "*fitzroyi*" is resurrected multiple times as *L. fitzroyi* (e.g., Flower 1885, True 1889, Fraser 1948; Table 1.1, Figure 1.6) and persists in the literature as late as 2008 as a South American subspecies of *L. obscurus* (*L. o. fitzroyi*, Rice 1998, Jefferson et al. 2008; Table 1.1, Figure 1.6).

It is in 1866 that Gray's taxonomic legacy of more than 40 years begins to take a noticeable turn towards repeated, and what often appears spurious, reclassification of genera and species, including the dusky dolphin. What brought on this sudden and precipitous decline in the clarity of his taxonomic assessments is unclear, but his taxonomic expertise appeared to decline, and there was an increasing tendency to make mistakes in names, dates, and other facts related to previous taxonomic work, including his own[3]. With respect to the taxonomy of the dusky dolphin, between 1864 and 1868, Gray changes the generic classification of *D. obscurus* three times, sometimes more than once in the same

FIGURE 1.7 Reproduction of the original drawing of *Delphinus fitzroyi* (= *L. obscurus*) made by Captain Fitzroy during the voyage of the HMS *Beagle* (in Waterhouse 1838a).

year (Figure 1.6). In 1866, Gray moves *D. obscurus* into the genus *Tursio* as
T. obscurus (Gray 1866a). Later that year he makes *T. obscurus* a junior synonym of *Clymene obscura* (Gray 1866b), which he changes to *Clymenia obscura*
in 1868 (Gray 1868a). Again in 1868, Gray defines a new species, *C. similis*,
based on slight differences in the pterygoid bone discovered while re-examining
a specimen formerly classified as *D. obscurus* (Gray 1868b; Figure 1.6). Gray's
C. obscura and *C. similis* persisted until after his death in 1875. Flower reclas-
sified *C. similis* as a junior synonym of *C. obscura* in 1883, and then in 1885
moved *C. obscura* to a new genus, *Prodelphinus*, as *P. obscurus*. Frederick True
(1889) at the United States National Museum first recognizes the dusky dolphin
as a member of *Lagenorhynchus*, and makes *Clymenia obscura* and *Prodelphinus
obscurus* junior synonyms of *L. obscurus*:

> It has been customary among authors since 1868 to refer [*L. obscurus*] to *Clymenia*
> (=*Prodelphinus*) … the form of the head in the type specimen is unlike that of any
> species of *Prodelphinus* … there is no real beak … the color seems rather that of a
> *Lagenorhynchus* than a *Prodelphinus*. (True 1889, p. 104)

Dr. Rodolpho A. Philippi (1808–1904), Director of El Museo Nacional de
Chile, published two papers in which he described two new species, *Phocoena
posidonia*, from a specimen collected off the coast of Chile (Philippi 1893), and
Tursio panope, from a skull in the Chilean museum (Philippi 1896). Based on
the illustration published by Philippi (1893; Figure 1.8), True (1903) proposed
P. posidonia was very similar to *L. fitzroyi* and thus a member of *Lagenorhynchus*
(i.e., *L. posidonia*). Similarly, Kellogg (1941) considered *P. posidonia* "doubtfully
distinct from *L. obscurus*," but because Philippi published only an illustration
of external features, neither he nor True (1903) were confident in their species
assignments. Hershkovitz (1966) made *P. posidonia* a synonym of *L. cruciger*,
but Brownell and Cipriano (1999) confirmed synonymy with *L. obscurus*.

   Tursio panope presented a greater problem for True (1903) than *P. posidonia*; he was "unable to determine even the genus to which this singular species
belongs," and concluded, "Further study of the type specimen can alone resolve
the problem." Brownell and Mead (1989, in Brownell and Cipriano 1999) re-
examined the holotype and concluded it was a junior synonym of *L. obscurus*;
however, recent analysis of mitochondrial DNA sequence obtained from the

FIGURE 1.8 Scanned reproduction of an illustration of *P. posidonia* (= *L. obscurus*) from Philippi
(1893). Despite the poor quality of the image, the close resemblance to *L. obscurus* is apparent.

holotype identified *T. panope* as *C. eutropia* and not *L. obscurus* (Pichler and Olavarria 2002). From the late 1940s to mid 1960s, several authors proposed a single southern hemisphere species, with *L. obscurus* and *L. australis* as junior synonyms of *L. cruciger* (Bierman and Slipjer 1948, Hershkovitz 1966). It was Fraser (1966) who solidified the taxonomy of the southern hemisphere *Lagenorhynchus* with three distinct species, *L. obscurus*, *L. australis*, and *L. cruciger*, all of which are currently recognized as valid (Rice 1998).

In 1998, Rice proposed three subspecies of the dusky dolphin based on geographical distribution: *L. obscurus fitzroyi* (South America), *L. o. obscurus* (South Africa), and an unnamed subspecies from New Zealand (Rice 1998, Jefferson et al. 2008). The division of *L. obscurus* into subspecies reflected the morphometric study done by Van Waerebeek in the early 1990s in which he quantitatively compared a suite of cranial measurements between groups of skulls collected from Peru, Chile, South Africa, and New Zealand (Van Waerebeek 1993a). The results of this study revealed a striking degree of variation in skull size and tooth number between populations; 29 of 31 morphometric characters were geographically different. With regard to overall skull length, New Zealand + South Africa and Chile + Peru formed two groups with a 3 cm difference between them. Furthermore, dusky dolphins from New Zealand had 1.5–1.8 more teeth per tooth row than those from South Africa. Although the subspecies classification of the dusky dolphin is rarely used, recent phylogeographic studies using nuclear and mitochondrial genes suggest genetic divergence between New Zealand, South Africa, and South America (Harlin-Cognato et al. 2007), which agrees with the results of Van Waerebeek's morphometric study and the subspecies designations of Rice (1998).

Van Waerebeek (1993a) did not have a large enough sample of dusky dolphins from Argentina to make morphometric inference, but a recent study of population genetics suggests there is an additional genetic division within South America between Argentina and Peruvian populations (Cassens et al. 2005). It is clear from the molecular phylogeography that regional dusky dolphin populations are genetically isolated from each other, and for management purposes should be considered as different stocks, if not different subspecies (*sensu* Rice 1998).

EVOLUTION, MOLECULAR SYSTEMATICS, AND A NEW TAXONOMY

It is impossible to speak of the objects of any study, or to think lucidly about them, unless they are named. It is impossible to examine their relationships to each other and their places among the vast, incredibly complex phenomena of the universe, in short to treat them scientifically, without putting them into some sort of formal order.

George Gaylord Simpson, *The Principles of Classification and A Classification of Mammals*, 1945

Prior to the Darwinian revolution, species were considered static, unchangeable entities, and it was the goal of the taxonomist to identify, name, and organize this diversity. After Darwin published his theory of natural selection in 1859, it was no longer satisfactory to simply group species according to similarity of features—it became important that organisms be classified into "natural groups" to reflect their shared evolutionary history. This paradigm shift was apparent in the study of cetacean evolution, which began to use terms like "adapted," "modification," and "vestiges," and recognized that the ancestors of living whales and dolphins were land mammals (Kellogg 1928). Yet in the late 1800s and early 1900s, modern quantitative methods for examining evolutionary history were not available, and the definition of a "natural group" often was still defined by the number of traits shared in common (Simpson 1945). This was based on the assumption that similarity served as a reliable indicator of shared evolutionary history.

We now understand that some traits are more informative about shared ancestry than others. It was Willi Hennig in the 1960s who first proposed that taxonomy should be derived from testing a series of hypotheses with carefully defined rules based on the philosophical premise of parsimony. It was the birth of Hennig's methodology, called "cladistics" (Hennig 1966), which revolutionized the science of modern taxonomy.

During this volatile period in the history of biology the taxonomy of *Lagenorhynchus* was experiencing its own revolution (Table 1.1). By the late 1800s the taxonomy of the North Atlantic species, *L. acutus* and *L. albirostris*, was well-established (Flower 1885, True 1889; Table 1.1), but it was not until the 1960s that the taxonomy of the southern hemisphere species stabilized (e.g., Fraser 1966; Table 1.1, Figure 1.6). Fraser's study of *Lagenorhynchus* in 1966 marked the end of a period of uncertainty and volatility in the taxonomic history of the genus that was to persist nearly 40 years (Table 1.1).

In the late 1990s, systematic biology again underwent a second major revolution that would have a profound effect on our understanding of cetacean evolution. The invention of new technology such as the polymerase chain reaction (PCR), automated DNA sequencing, and new computer-based quantitative methods transformed the way in which taxonomic hypotheses were tested. It was the implementation of modern phylogenetic analysis of DNA sequence data that once again challenged the taxonomy of *Lagenorhynchus*. These early analyses of mitochondrial DNA sequences suggested that the genus *Lagenorhynchus* was not a natural group, but rather a polyphyletic assemblage of antitropical and disjunctly distributed species (Cipriano 1997, LeDuc et al. 1999; Figure 1.9). For example, LeDuc et al. (1999) published a phylogeny of the entire family Delphinidae that indicated strong support for a more recent common ancestry between *L. obliquidens*, *L. obscurus*, *L. australis*, *L. cruciger*, and the genera *Cephalorhynchus* and *Lissodelphis*, than to *L. albirostris* and *L. acutus* (Figure 1.9). Based on these findings, LeDuc et al. (1999) proposed a revised taxonomy of the genus with *Lagenorhynchus albirostris* and *Leucopleurus acutus* representing the North Atlantic species and *Sagmatias,* from *Sagmatias amblodon,* a junior synonym of *L. australis* (Cope 1866) for species in the North Pacific

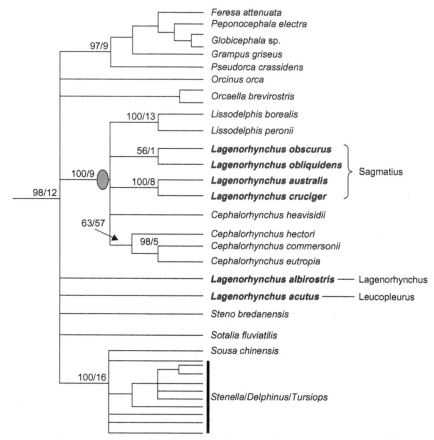

FIGURE 1.9 Evolutionary relationships of the family Delphinidae from LeDuc et al. (1999), derived from cytochrome *b* sequences. Numbers above branches are bootstrap and decay indices, respectively, for adjacent clades. These results clearly support that the six currently recognized species of *Lagenorhynchus* (bold text) are not monophyletic, but the relationships between species within and outside of the well-supported subfamily Lissodelphininae (shaded circle) are not well resolved. Generic names suggested by the taxonomic revision of LeDuc et al. (1999) are indicated to the right of the phylogeny (modified from LeDuc et al. 1999).

and southern hemisphere (Figure 1.9). However, their taxonomic revisions were based on inconclusive evidence, and the phylogenetic relationships of the *Lagenorhynchus* spp. within the Lissodelphininae and between the North Atlantic species remained unresolved (Figure 1.9).

A more recent Bayesian analysis of mitochondrial cytochrome *b* sequences by May-Callado and Agnarsson (2006) supported the paraphyly of the genus *Lagenorhynchus* as suggested by LeDuc et al. (1999), but improved the phylogenetic resolution within the Lissodelphininae (Figure 1.10). Because of the sister-clade relationship between *L. cruciger/L. australis* and the monophyletic *Cephalorhynchus* (Figure 1.10), May-Callado and Agnarsson (2006) suggested

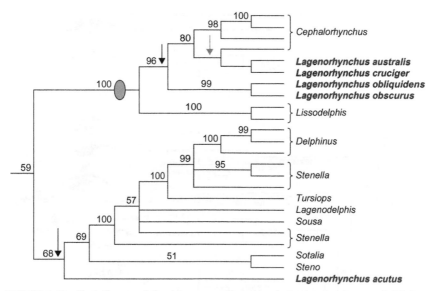

FIGURE 1.10 Evolutionary relationships among five of the six *Lagenorhynchus* spp. (bold) from the study of May-Collado and Agnarsson (2006), derived from Bayesian analysis of cytochrome *b* sequences. Numbers on branches are posterior probabilities for each clade. These results indicate a paraphyletic *Lagenorhynchus*, with at least two (black arrows), and perhaps three (gray arrow), independent origins, and support the monophyly of Lissodelphininae (shaded circle) as first proposed by LeDuc et al. (1999). Note that *Lagenorhynchus albirostris* was not represented in this analysis (adapted from May-Collado and Agnarsson 2006).

it would be easiest to move *L. cruciger* and *L. australis* to *Cephalorhynchus,* instead of *Sagmatius.* However, although *Cephalorhynchus* and *Lagenorhynchus* were both first described by Gray (1846b) in the same publication, the phylogenetic position of the two genera would give precedence to *Lagenorhynchus* (i.e., *L. obscurus* and *L. obliquidens* are basal; Figure 1.10). Furthermore, as with LeDuc et al. (1999), some of the relationships among species of *Lagenorhynchus* and *Cephalorhynchus* within the Lissodelphininae remained unresolved (Figure 1.10); thus taxonomic revision based on this hypothesis alone may not be warranted.

Harlin-Cognato and Honeycutt (2006) used multiple genetic markers to examine the evolutionary history of the genus *Lagenorhynchus* and its relationship to other members of the family Delphinidae. This study included complete taxonomic sampling from all currently recognized members of the subfamily Lissodelphininae (LeDuc et al. 1999), including the genera *Lagenorhynchus,* *Cephalorhynchus,* and *Lissodelphis,* plus a number of other delphinids (Harlin-Cognato and Honeycutt 2006). Similar to previous studies, the results of this study supported the monophyly of the Lissodelphininae, and therefore concur with LeDuc et al. (1999) and May-Callado and Agnarsson (2006) that the genus *Lagenorhynchus* is polyphyletic (Figure 1.11). In contrast to LeDuc et al. (1999), the parsimony and Bayesian analyses in this study recovered weak support for

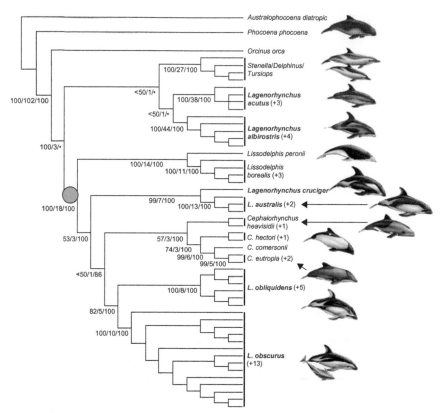

FIGURE 1.11 Evolutionary relationships among all six of the currently recognized *Lagenorhynchus* spp. (bold text) from the study of Harlin-Cognato and Honeycutt (2006). The phylogeny was derived from the simultaneous parsimony analysis of actin, RAG2, cytochrome *b*, and control region data. Numbers to the left of each node are bootstrap proportions, decay indices, and posterior probabilities, respectively; those in parentheses indicate the increased sample size for each species compared to LeDuc et al. (1999). A circle denotes the Lissodelphininae clade (modified from Harlin-Cognato and Honeycutt 2006).

the monophyly of *L. acutus* and *L. albirostris* (Figure 1.11), but stronger support for a close relationship of these species with the *Delphinus/Stenella/Tursiops* clade (Figure 1.11). We therefore propose that both *L. acutus* and *L. albirostris* remain as *Lagenorhynchus* until further analyses are performed with better taxonomic sampling outside of the Lissodelphininae.

In contrast to the phylogeny of May-Callado and Agnarsson (2006), Harlin-Cognato and Honeycutt (2006) recovered a monophyletic *Cephalorhynchus*, and two monophyletic clades of *Lagenorhynchus* within the Lissodelphininae (Figure 1.11). Their evidence supports LeDuc et al. (1999) and suggests that *L. cruciger/L. australis* are monophyletic and should be placed into a separate genus, *Sagmatius,* first described from *Sagmatius amblodon* (Peale 1848) and later synonymized with *L. australis* (Cope 1866).

The monophyly of *L. obliquidens* and *L. obscurus* is now well established from morphological (e.g., Webber 1987) and molecular studies (Harlin-Cognato and Honeycutt 2006, May-Callado and Agnarsson 2006), with the molecular studies suggesting that these two species share a separate evolutionary history from *Sagmatius* spp. and *Lagenorhynchus* spp. (Figures 1.10 and 1.11). Thus, given the phylogenetic impetus of taxonomy for monophyletic taxa (i.e., evolutionary units), both *L. obliquidens* and *L. obscurus* warrant recognition as a new genus. However, all generic synonyms of *Lagenorhynchus* are preoccupied (e.g., Hershkovitz 1966). For example, *Clymene* (= *Clymenia*; Gray 1864) and *Prodelphinus* (Flower 1885), both obvious candidates (Figure 1.6), were later synonymized with *Stenella* (Hershkovitz 1966, Mead and Brownell 2005). Therefore, it is my opinion that a description of a new genus is required.

This new description might include a combination of morphology and DNA sequence characters in a simultaneous analysis. For example, Fraser and Purves (1960) found distinct differences among genera of dolphins, including *Lagenorhynchus*, based on a suite of characters in the sinuses related to the structure and function of hearing. A re-evaluation of these characters and additional molecular loci and individuals, especially *L. acutus* and *L. albirostris*, has the greatest potential to resolve remaining taxonomic issues.

ACKNOWLEDGMENTS

I thank Bernd and Melany Würsig for giving me the opportunity to make a contribution to this delightful collection of work on the dusky dolphin. This chapter would never have been finished without the help of my husband, Anthony, who took care of our little boy while I pushed hard to make (and often break) deadlines. Robert Brownell provided helpful suggestions to improve the manuscript. A portion of this research was made possible from a Doctoral Dissertation Improvement Grant, U.S. National Science Foundation. The majority of this work was made possible by the legacy of the many excellent scientists who came before. Thank you.

NOTES

1. The specimen of *D. obscurus* is described by Gray (1828) in the "Specilegia Zoologica" as having come to the Museum via the Royal College of Surgeons who obtained it from a "Captain Heaviside." However, "Heaviside" was a misspelling of Captain Thomas Haviside who collected the specimen (and those of *D. heavisidii* and *D. capensis*) during one of his many voyages past the Cape of Good Hope while employed by the East India Company. Captain John Heaviside (1748–1828) was a prominent surgeon, a lecturer at the prestigious Royal College of Surgeons, a Fellow of the Royal Society, and served as the personal surgeon to King George III. Although known for his anatomical collection of Cetacea, he did not collect this particular specimen (see Fraser 1966).

2. One of the most important of the early attempts to regulate zoological nomenclature was that by Hugh Strickland. The rules proposed by Strickland and his colleagues developed into

what has since been called the British Association Code or the Stricklandian Code; its official title was Series of Propositions for Rendering the Nomenclature of Zoology Uniform and Permanent. Following its presentation at the British Association for the Advancement of Science in 1842, by a committee that included such distinguished zoologists as Charles Darwin, Richard Owen, and John Westwood, that Code was translated and circulated widely and had great influence. It was published in France, Italy, and the USA. It was received by the Scientific Congress at Padua in 1843, by the American Society of Geologists and Naturalists in 1845, and was adopted by the British Association for the Advancement of Science in 1846 (excerpt from the ICZN, http://www.iczn.org).

3. For example, in his 1871 "Supplement to the Catalogue" Gray incorrectly references Cope as the author of the original description of *L. obliquidens*, when it is in fact Gill (1865). Also in this publication he incorrectly lists his *D. obscurus* from the *Voyage of the HMS Erebus and Terror* (Gray 1846b) as *Tursio obscurus*.

Chapter 2

Dusky Dolphin Life History and Demography

Frank Cipriano[a] and Marc Webber[b]

[a]*Department of Biology, San Francisco State University, San Francisco, California, USA*
[b]*Columbus, Indiana, USA*

In December 1983 two enthusiastic young biologists boarded a huge container ship in Long Beach, California and began a month-long journey across the Pacific. They had heard that they could make the passage for free, and were excited to spend it observing the marine mammals and birds of the Pacific; they had even prepared sighting data sheets, an observation protocol, and had their binoculars and cameras at the ready. Alas, instead they were told to spend daylight hours mainly in the 110°F engine room wiping pipes, sweeping, and painting in slow motion. But, the journey WAS free, and once in New Zealand, they could begin studying dusky dolphins along the spectacular Kaikoura coast of New Zealand.

But back to the trip and the idea to get there: After stops in Sydney and Melbourne, they finally arrived in Auckland and offloaded bicycles, backpacks, binoculars, camera gear, an Apple IIe computer, and sundry pieces of field gear and sent it ahead by truck, following along by bus. Some eight months earlier, Frank had gone to see Bernd Würsig, the new faculty member at Moss Landing Marine Laboratories, with a copy of Gaskin's 1968 paper "The New Zealand Cetacea" in hand, impressed with the variety of cetacean life to be found in those waters. "New Zealand, New Zealand" Bernd exclaimed— "Marc Webber knows about New Zealand!" and indeed Steve Leatherwood had introduced Marc (also a Moss Landing student) to Kaikoura a few years earlier, during a stopover on the way from Antarctica. From that beginning we knew that "duskies" were to be found with some regularity in the area, and the presence of a field station operated by the University of Canterbury, run by the unflappable Jack van Berkel, gave us a home base and an inflatable boat to use (with the Apple IIe as our down payment).

From a very slow start, pedaling up and down the coastal road on our 10-speed bicycles to the delight of the indulgent but bemused locals, a pattern started to emerge: although dusky dolphins could be found anywhere off the coast, they were often seen to the south of the Kaikoura Peninsula, in the 22 km open embayment flanked at the south end by Haumuri Bluffs. After some false starts, we eventually launched the mighty 3-m-long dinghy Arancia from the South Bay ramp with Jack's help, and headed out to find ourselves for the first time surrounded by a large group of dolphins—Marc snapping dorsal fin photos, Frank steering and barely able to speak past a big grin. After some time spent chatting up the locals and distributing "Wanted" posters, specimens started to come in—incidentally netted dolphins brought into the field station by fishermen, or mysteriously appearing on the beach nearby. We eventually scoped out a series of elevated observation stations, from the water tank atop the peninsula, to the hill above the Kahutara River, to Ota Matu point above Goose Bay, to the railroad tracks rising to the tunnel at Haumuri, and the bicycles were replaced with a motorbike and then a big yellow van.

Over the intervening 25 years many more students under the skilled tutelage of Bernd and Melany Würsig have continued to study dusky dolphins from the same vantage points and have extended their observations north to Admiralty Bay, where some Kaikoura dolphins are seen in winter months, foraging on different prey and in different ways than seen in summer. New studies and new students continue, and Kaikoura is now the whale and dolphin-watching capital of New Zealand, with much of the local economy fueled by eco-tourism, and many foreign tourists and New Zealanders alike dazzled by the leaps and flips and somersaults that captivated two no longer young but still enthusiastic biologists from California.

Frank Cipriano, San Francisco, October 2008

DUSKY DOLPHIN TAXONOMY

Dusky dolphins (*Lagenorhynchus obscurus*; Gray 1828), along with five other dolphin species, are currently classified in the genus *Lagenorhynchus*, but that is likely to change in the near future once the details are better understood. Analysis of full-length cytochrome *b* DNA sequences by LeDuc and co-authors (LeDuc et al. 1999) gave the first evidence that the current naming scheme, which implies that all six species are closely related by placing them in a single genus, is incorrect. Two currently recognized *Lagenorhynchus* species, Atlantic white-sided dolphins (*L. acutus*) and white-beaked dolphins (*L. albirostris*), do not appear closely related to the other four, or to each other. As the earliest-named member of the genus, white-beaked dolphins will probably remain classified in *Lagenorhynchus* because of the rules of zoological nomenclature, and the others may be assigned to at least two other genera (LeDuc et al. 1999). Dusky dolphins and the remaining three species—one found only in the North Pacific, the others found only in the southern oceans, do appear to be closely related, but are also related to dolphins currently placed in two other genera—two species in genus *Lissodelphis* (*L. borealis* or northern right whale dolphin in the North Pacific, *L. peronii* or southern right whale dolphin in the southern ocean) and four species in genus *Cephalorhynchus* (*C. commersonii* and *C. eutropia* in South America, *C. heavisidii* in South Africa, and *C. hectori* in New Zealand; see also Chapter 15). Although the final assignment of genus names to the species awaits detailed study and formal taxonomic review and publication, the new (if incomplete) understanding of evolutionary relationships within this group has been recognized in several recent papers (e.g., LeDuc et al.

1999, Pichler et al. 2001, May-Collado and Agnarsson 2006, Harlin-Cognato et al. 2007, see also Chapter 1), which place all but the two North Atlantic species into subfamily Lissodelphininae.

Morphological evidence suggests differentiation, possibly at the subspecies level, between geographically distinct dusky dolphin populations, including the observations that skulls from New Zealand and southwest Africa are on average 3.1 cm (8.5% of condylobasal length) shorter than those from Peru and Chile (concordant with differences in body length), and New Zealand specimens also have smaller teeth and a larger number of teeth than African specimens (Van Waerebeek 1993a,b). Molecular data provide further evidence for significant differentiation; for example, overall φ_{st} for comparisons between New Zealand, Argentina, South Africa, and Peru specimens were 0.49 (cytochrome b) and 0.74 (control region), and all pairwise comparisons between regions were significant (Harlin-Cognato et al. 2007). Although patterns of differentiation are complex and interpretation of details (such as identifying the source population for the founding of contemporary populations) difficult, the bulk of evidence suggests little or no contemporary gene flow between New Zealand, Argentina, South Africa, and Peru. There is tantalizing evidence that dusky dolphins from Argentina and southern Africa recently (that is, in evolutionary terms) separated from an ancestral Atlantic population (Cassens et al. 2005), although not all agree with this conclusion and instead suggest a Pacific/Indian Ocean source population (Harlin-Cognato et al. 2007). Formal taxonomic consideration of subspecies status awaits concentrated study utilizing both morphological and molecular analysis, as has been accomplished recently for North Island Hector's dolphins (*Cephalorhyncus hectori mauii*; Baker et al. 2002) and Kerguelen Island Commerson's dolphins (*Cephalorhynchus commersonii kerguelenensis*; Robineau et al. 2007).

DISTRIBUTION AND ABUNDANCE

The distribution of the three northern hemisphere *Lagenorhynchus* species, the white-beaked dolphins, Atlantic white-sided dolphins, and Pacific white-sided dolphins (*L. obliquidens*), are well known (e.g., Rice 1998, Jefferson et al. 2008). The distributions of the three southern hemisphere species, dusky dolphins, Peale's dolphins (*L. australis*), and hour-glass dolphins (*L. cruciger*), are sparsely documented and less clearly understood (Brownell 1974), although more information is now available, as summarized below. No member of *Lagenorhynchus* or indeed of the Lissodelphininae is normally found in warm tropical waters—all seem to be found regularly only in cold or cold-temperate oceans. The northernmost extent of dusky dolphin sightings anywhere is off the coast of central Peru at about 8° S (e.g., Van Waerebeek 1992a), in the region also associated with a cyclically shifting boundary between the cold north-flowing waters of the Humboldt current and the intrusion of warm tropical waters into the coastal zone known as the "El Niño Southern Oscillation" (ENSO; Taylor and Wolff 2007). This region is one of the most productive marine ecosystems known, and is also characterized by a cyclical change in dominance of anchovy in coastal waters (cold periods) and the movement of sardines inshore (warmer periods; Ñiquen and Bouchon 2004). Reports of dusky dolphins in warmer waters (e.g., Kasamatsu et al. 1990) have been rejected as unsubstantiated or erroneous (e.g., Van Waerebeek et al. 1995). The latitudinally disjunct distribution of the northern and southern hemisphere species of *Lagenorhynchus* was described as "antitropical" by Davies (1963a), and this was followed by Brownell (1974), and Gaskin (1982; Figure 2.1).

Dusky dolphins are widely distributed in the southern hemisphere, where they are found in coastal and continental shelf waters off both coasts of South America, southwest South Africa, most of New Zealand, sporadically off southern Australia and around certain oceanic island groups including Tristan da Cunha and Gough Island, Prince Edward and Marion Island, Amsterdam and St. Paul Islands (Van Waerebeek et al. 1995). Little information is available on the abundance of dusky dolphins throughout their range, although recent dedicated line-transect surveys in southern Argentina provided the first detailed evidence on abundance levels in that area (Schiavini et al. 1999). The continental and New Zealand populations all appear to be discontinuous (Figure 2.1), although over evolutionary time, rare instances of interchange are possible and there must have been some dispersal in the past leading to the current distribution. There is potential for genetic divergence between such widely separated populations and some suggestion that dusky dolphins found in these areas may be distinct enough to deserve subspecific status, and should definitely be considered different management units (e.g., Van Waerebeek 1993b, Cassens et al. 2003).

New Zealand

The presence of dusky dolphins in New Zealand's waters was first documented more than a century ago, when they were identified as *Clymenia obscura* (Hector 1873, 1878). Since then, dusky dolphins have been reported as common in many locations around the South Island and the southern and central regions of the North Island (Lillie 1915, Oliver 1922, Gaskin 1968a,b, 1972, Baker 1972b, 1983). In New Zealand waters, the stronghold of dusky dolphins is along the North Island's east coast from East Cape (in winter) and Cape Palliser (in summer) to Timaru or Oamaru on the South Island in the cold waters of the north-flowing Southland and Canterbury Currents (Gaskin 1972). This species may be present continuously in the inshore waters of the North Canterbury coast of the South Island, and a long series of observations have shown that this is particularly true in the Kaikoura area (e.g., Stonehouse 1965, Gaskin 1972, Webber 1987, Cipriano 1992, Markowitz 2004).

Gaskin (1972) reported that the winter and spring distribution of dusky dolphins off the North Island are from East Cape south along the Pacific coast to Cook Strait, with groups frequently seen south of Cape Kidnappers in Hawke Bay, and infrequently north of there. Personnel at Marineland of New Zealand, Napier

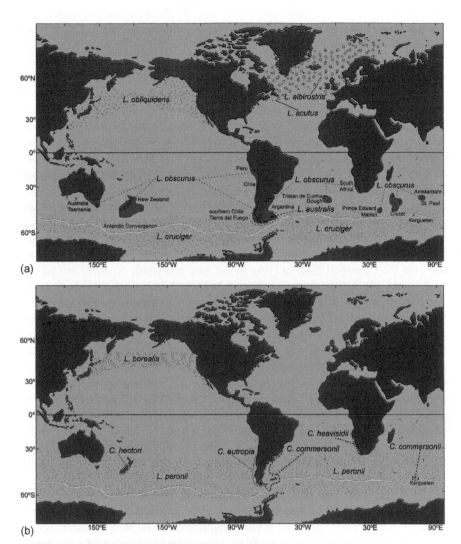

FIGURE 2.1 Worldwide distribution of the Lissodelphininae. (a) Distribution of dusky dolphins (pink shaded areas) and other species within the genus *Lagenorhynchus*, including the two North Atlantic species that do not appear to be closely related to the other four (after Webber 1987). (b) Distribution of species within genera *Lissodelphis* and *Cephalorhynchus*. Note that the distributions of *Lagenorhynchus cruciger* and *Lissodelphis peronii* are at present poorly understood, so the distributions shown here are broad approximations.

reported that dusky dolphins were found in Hawke Bay mainly in the winter (Gaskin 1968a). The northern limit of dusky dolphins on the North Island east coast at the time seemed to be Cape Palliser (Webber 1987). Along the west coast, dusky dolphins are present in winter in the Taranaki Bight and off Cape Egmont, where they are only "very occasionally" sighted in the summer (Gaskin 1972).

On the eastern side of the North Island, dusky dolphins are rare north of East Cape at 37°40'S (Baker 1983).

There are a limited number of sightings from the rest of New Zealand including: the south and west coasts of the South Island, Stewart Island, and out to the offshore islands on the Campbell Plateau. Gaskin (1972) reported that dusky dolphins had been recorded from among the islands of the Foveaux Strait. He stated they occurred sporadically off the South Island's west coast south to Greymouth, with no records from there to West Cape in Fiordland. Webb (1973a) added sightings of two schools west of Stewart Island and south of Solander Island, one of which was of a very large group reported to number more than 5000, and also recorded an important sighting for the west coast, approximately 50 miles southwest of Greymouth, thus extending the range given earlier by Gaskin (1972).

Although primarily a coastal species, dusky dolphins have been recorded from far offshore in the New Zealand region. Sightings have come primarily from the Campbell Plateau lying to the east and south of the South Island (Gaskin 1968b, Leatherwood et al. 1982, M.A. Webber, unpublished data). Records from New Zealand's outlying and sub-Antarctic islands have come primarily from strandings (Webber 1987). The southernmost records from New Zealand waters are three stranded dusky dolphins from Campbell Island (52°30'S). Baker (1977) reported the collection of three skeletons from Garden Cove, Campbell Island in 1974 by the staff of the meteorological station (Webber 1987: table 2, record nos. 80, 82, 83). An adult male dusky dolphin was collected at sea east northeast of the Auckland Islands at 49°06'S, 166°29'E during an International Whaling Commission cruise on December 19, 1980 (Webber 1987: table 2, record no. 97, N. Miyazaki personal communication to M.P. Francis, MAF-NZ). Lillie (1915) reports what is probably a reliable sighting of this species on February 7, 1913 at 51°56'S, 168°02'E between Auckland and Campbell Islands.

To the east and southeast, there are several records of sightings and strandings from the Chatham Islands, but none from the Antipodes or Bounty Islands. Gaskin (1968b) reports the most easterly sighting from the New Zealand region, southeast of the Chatham Islands. There have been two records of strandings, both involving groups of animals. In October 1966, a group of six dusky dolphins were found stranded together (Gaskin 1968b, 1972). On December 20, 1986 four dusky dolphins were trapped in a tidal pool (A.N. Baker personal communication, Webber 1987: table 2, record nos. 76 and 119).

From mark–recapture analyses of naturally marked and photo-identified dolphins, Markowitz (2004) estimated a mean of just under 2000 dusky dolphins in the Kaikoura area at any one time, from a population of over 12 000 individuals. Harlin et al. (2003) used diversity indices and rate of substitution among sites to estimate effective female population size of dusky dolphins in New Zealand waters, and a lineages-through-time analysis was used to test hypotheses of population growth. No significant partitioning of variance among the four geographic regions of the South Island were detected, and the structure of the neighbor-joining phylogeny, the nested haplotype network, and results of the lineages-through-time analysis suggested that the New Zealand dusky dolphin population has undergone at least one, and possibly two, historical population expansions (Harlin et al. 2003).

Australia/Tasmania

Although not commonly found in south Australia and Tasmanian waters, the availability of some sightings and specimen records prompted Davies (1963b), Brownell (1974), Baker (1983), and Bannister et al. (1996), to include dusky dolphins in the region's fauna. Some reports from the region are problematic and some species identifications have been revised; for example, an undocumented skull in the Tasmanian Museum, often referred to as dusky dolphin, may not actually represent a collection from Tasmania, and at any rate has been re-identified as southern right whale dolphin (Van Waerebeek 1993c). The specimen of "*Delphinus superciliosus*" (Lesson and Garnot 1826) collected off South Cape, Tasmania (43°38'S, 146°44'E) in January 1824 and examined on deck seems to be the first documentation of dusky dolphins in these waters (Gill et al. 2000). Additional sightings and confirmed strandings reviewed by Gill and colleagues (Gill et al. 2000) include a total of 12 records collected over 175 years. All these reports are limited to times between October/November–March/April, suggesting that changes in oceanographic conditions (possibly a seasonal or ENSO-associated northward shift of the subtropical convergence) allow occasional movement into Australian waters from some other area, most likely New Zealand (Gill et al. 2000). There are no abundance estimates for dusky dolphins in Australian/Tasmanian waters, but they are likely rare transients in this region.

South Africa and Namibia

Dusky dolphins are known only from a narrow portion of the western side of coastal southern Africa, particularly in the Cape of Good Hope region (Sclater 1901, Barnard 1954, Brownell 1974, Best 2007). The normal range of dusky dolphins off the west coast of southern Africa extends to Luderitz (26°59'S, 14°31'E) and probably as far north as Walvis Bay, Namibia and possibly farther. The southern limit extends to at least Hout Bay (34°04'S, 18°21'E) in the Cape region, but rarely farther east than False Bay (Findlay et al. 1992, Best 2007). There have been no dedicated abundance surveys and there are no overall estimates for the total number of dusky dolphins in the region, but the species appears to be relatively common in South African inshore waters (see also Chapter 14). Incidental sightings (which could easily have included duplicate sightings of the same individuals) during an aerial survey between approximately 31° and 34°S included a total count of 1144 dusky dolphins (Best and

Ross 1984). Between January 21 and April 8, 2001, the Whale Unit of the Mammal Research Institute (MRI Whale Unit), University of Pretoria surveyed the west coast of South Africa using an inflatable boat, and counted 814 dusky dolphins in the area between Rooiduinpunt, Lamberts Bay, and Bok Point (Oosthuizen 2001). Between July 24 and December 20, 2001, the MRI Whale Unit maintained a shore-based watch; in 681.39 hours of watch on 102 days there were five sightings of dusky dolphins, including a total of 235 individuals (Oosthuizen 2001). Again, these surveys were not designed to estimate abundance and could have included duplicate sightings of the same individuals.

South America

Early reports on the distribution of dusky dolphins in Chilean waters are limited: Aguayo (1975) noted a few records and Brownell (1974) listed only two specimens and two sightings. Van Waerebeek (1992a) authenticated 47 locality records from the west coast of South America using original data from artisanal fisheries landings, museum specimens and the literature, and confirmed the distribution limits between Chimbote (09°05'S) in north-central Peru and Isla Treble (55°07'S, 71°02'W), Magallanes, in southern Chile. Two skulls found by A. García-Godos and J. Alfaro in Salaverry (08°14'S) represent the most northerly records for the species (Van Waerebeek et al. 1997). Earlier studies suggested a lack of dusky dolphin records from a roughly 1000 km coastal strip between 36°30'S and 46°S, suggesting a discontinuous distribution (Van Waerebeek 1992a), but more recent records from this region (Van Waerebeek, personal communication) may indicate a continuous range from southern and central Peru into Chile. A large number of dusky dolphin specimens have been obtained from coastal net fisheries off southern and central Peru over decades of study (e.g., Read et al. 1988, Van Waerebeek and Reyes 1990, Van Waerebeek et al. 1997). Although abundance is difficult to estimate, fishery mortality data indicated that dusky dolphins were the most common cetacean off central Peru and, based on limited sightings data, possibly also off northern Chile (Van Waerebeek 1992a).

On the east coast of South America, dusky dolphins are regularly found from Peninsula Valdés in the Province of Chubut, Argentina to Mar del Plata (e.g., Gallardo 1912, Brownell 1974, Lichter and Hooper 1983, Würsig and Bastida 1986, Bastida and Rodriguez 2005). Dusky dolphins are relatively uncommon (four specimens available, about one dozen sightings from 1972 to 2008) in the Beagle Channel and inshore waters of the Tierra del Fuego region (Goodall 1978, 1989, Goodall et al. 1997). Historically, dusky dolphins are assumed to have been present in the Beagle Channel, Tierra del Fuego because of the presence of remains in aboriginal kitchen middens from approximately 6000 years ago (Piana et al. 1985). Observations compiled by Natalie Goodall suggest that Peale's dolphins are more common in the inshore waters of Tierra del Fuego, while dusky dolphins are more often found along the outer coast and off Cape Horn (Natalie Goodall, personal communication). Records from around the Malvinas (Falkland) Islands were reviewed by Van Waerebeek et al. (1995); although some reports of dusky dolphins from this area are erroneous or unsubstantiated, sightings by reliable observers and at least three skulls in the British Museum of Natural History collected by J.E. Hamilton and identified by Van Waerebeek (Van Waerebeek et al. 1995) are conclusive evidence that the species does occur there.

Along the coast of Argentina during the early 1980s, population abundance was roughly estimated at "several thousand (exact number unknown)" (Würsig and Bastida 1986). The total number of dusky dolphins in the area subjected to the highest shrimp trawler fishing effort between Peninsula Valdés and Puerto Deseado, Argentina (42°S to 46°S), off the Patagonian coast, was more recently estimated to be 6628 dolphins (95% confidence interval = 4039–10877; Schiavini et al. 1999).

Southern Indian and Atlantic Oceans

Van Waerebeek et al. (1995) examined a specimen collected by S. Hits near Amsterdam Island (37°55′S, 77°40′E) and described by C.F. Lutken in 1889 as *Prodelphinus persersii*, and concluded that it was in fact a dusky dolphin, and thus the first specimen documenting occurrence of the species in this area. Van Waerebeek et al. (1995) reviewed other records of sightings from the vicinity of oceanic island groups in the southern Indian and Atlantic oceans and concluded that there was good evidence for the presence of dusky dolphins in the area north of the Prince Edwards Islands, north of the Crozet Archipelago, and possibly between the Crozet Archipelago, Kerguelen Island, and Gough Island, but that reported sightings to the south of Madagascar (in much warmer water) were unlikely to be dusky dolphins.

In 1951, a dolphin skull was retrieved from a badly decomposed carcass found on Guite Beach, Morbihan Bay, Kerguelen Island in the south central Indian Ocean and the specimen was identified as dusky dolphin by Fraser (Paulian 1953). Robineau (1989) re-examined this specimen (MNHN-AC:n 1952–197) and concluded that it was a young specimen of Commerson's dolphin. There are thus no confirmed records of dusky dolphins from Kerguelen waters or from Heard Island, also located in the adjacent relatively shallow waters of the Kerguelen Plateau.

HABITAT CHARACTERISTICS

Dusky dolphins are associated with cool water, upwelling areas, and cold currents throughout their circumpolar southern hemisphere range, and appear to be an inshore species chiefly inhabiting waters to the outer continental shelf and similar zones around offshore islands (Gaskin 1968b, 1972, Brownell 1974, Würsig and Würsig 1979, 1980, Leatherwood and Reeves 1983, Würsig and Bastida 1986). From radio-tracking studies in Argentina and New Zealand, both long- and short-range movements have been reported, with most evidence indicating that animals remained within the coastal zone (Würsig 1982, Würsig and Bastida 1986, Würsig et al. 1991). Sightings of this species from the open ocean areas between the major landfalls of the temperate southern ocean are virtually nonexistent, leading to earlier suggestions that the distribution of this species is not only coastal, but discontinuous (e.g., Gaskin 1968b, 1972, Brownell 1974). However, seasonal and occasional sightings at locations such as off southern Australia and the oceanic island groups closest to continents in the South Atlantic and southern Indian oceans suggest that there might be infrequent but regular wanderings of this species from continental and New Zealand coastal areas, or as yet unrecognized offshore populations.

Dusky dolphins found off the west coast of southern Africa and both coasts of South America are associated with continental shelves and the cool waters of the Benguela, Humboldt and Falkland Currents, and in New Zealand these dolphins are also associated with the cold Southland and Canterbury Currents (Gaskin 1972). Van Waerebeek et al. (1995) suggested that this species may be primarily limited to water shallower than 200 m, and water temperature also appears to be a major influence on distributional bounds, probably through restrictions on the range of associated prey species. Dusky dolphins are not common in water deeper than 2000 m, and sightings in deeper water are generally adjacent to abrupt continental shelf or island drop-offs (Jefferson et al. 1993). Seasonal changes in distribution observed in Argentina (e.g., Würsig and Würsig 1980) and New Zealand (e.g., Würsig et al. 1991, Harlin et al. 2003) suggest that prey distribution and/or abundance rather than water temperature or depth *per se*, may be the important factor controlling or limiting distribution.

Würsig and Würsig (1980) observed dusky dolphins in Golfo San José, Argentina in waters that varied seasonally between 11°C and 17°C. Kasamatsu et al. (1990) reported sightings of dusky dolphins in waters with sea surface temperatures between 10°C and 16°C, except for an anomalous sighting at 31°46'S at a surface temperature of 25°C, which (based on photographs from this sighting) Van Waerebeek et al. (1995) believe to be an erroneous identification. Gaskin (1968a) reported sightings of this species offshore from New Zealand at surface temperatures between 14°C and 15°C. Cipriano (1992) regularly observed dusky dolphins off the Kaikoura coast of South Island, New Zealand in waters that varied seasonally between 10°C and 18°C and suggested that sightings were concentrated over the flanks of the Kaikoura Canyon.

Gaskin (1968b) suggested that the distribution of dusky dolphins in the New Zealand region is linked to sea surface temperature, and associated the species' distribution closely with the Subtropical Convergence—the region of the southern oceans at about 42°S latitude where the surface temperature of the sea drops sharply from about 18°C to 10°C. Gaskin (1968b) also noted that he did not find this species east of the Chatham Island coastal shelf, providing the first evidence that the distribution of dusky dolphins, while circumpolar in the southern hemisphere, is discontinuous. The irregular movement of water masses in the Cook Strait area and the associated variations in temperature probably account for sporadic sightings there throughout the year (Gaskin 1968b). Webb (1973b) found dusky dolphins associated with waters between 9.8 and 16.0°C in this area.

The presence of dusky dolphins far from continental landmasses in the southern Indian and South Atlantic Oceans is somewhat paradoxical, given their association with relatively shallow nearshore waters off the coasts of South America, South Africa, and New Zealand. An oceanographic feature which appears to mark the southern distributional boundary of dusky dolphins in the oceanic island areas is the Antarctic Convergence (sometimes referred to as the "Antarctic Polar Frontal Zone")—a 32–48 km band encircling Antarctica, where cold, northward-flowing Antarctic waters meet and sink beneath sub-Antarctic waters, associated with a sudden 2.8–5.5°C average drop in temperature, to below 2°C. The position of the Antarctic Convergence is circumpolar across the southern oceans between 48 and 61°S, and varies in latitude seasonally, but usually does not stray more than one-half degree of latitude from its mean position. A number of islands and island groups occur along the circumpolar path of the Antarctic Convergence: the South Shetland, South Orkney, South Georgia, South Sandwich, Bouvet, Heard, and McDonald islands all lie south of the Antarctic Convergence; the Malvinas (Falkland) Islands, Prince Edward and Marion Islands, Crozet Island, Amsterdam and St. Paul Islands, Tristan de Cunha and Gough Island, Campbell Island and Macquarie Island all lie north of the Convergence (Figure 2.1a). Dusky dolphins are known from all of the oceanic island groups that lie north of the convergence, except Macquarie Island (which lies about halfway between Tasmania and Antarctica and is regularly observed from ship traffic supplying the Australian Antarctic Division station there), but are not known from any of the islands that lie south of the Convergence, or from Kerguelen Island, which lies almost directly on the Convergence.

There is no information regarding prey types utilized by dusky dolphins associated with island areas of the southern Indian and South Atlantic Oceans, so stomach contents recovered from even a single stranding on one of these remote islands would be invaluable, and tissue samples for genetic analysis and comparison to dusky dolphins from other regions would be similarly of great interest.

EXTERNAL CHARACTERISTICS

Color pattern

Most of the features of coloration pattern discussed here use the terminology of Mitchell (1970). One exception to this is made when discussing the observations of Fraser (1966), where both Fraser's terms, and the equivalent names from Mitchell (in parentheses) are provided. All six species currently classified within the genus *Lagenorhynchus* have in common the presence of a dark blaze or line of variable intensity originating from the gape or eye and extending to the flipper. All have a light-colored lateral flank patch (terminology of Fraser 1966 and Mitchell 1970). Mitchell (1970) does not refer to the zone of light coloration found on the dorsal fin of some *Lagenorhynchus*, which is described here as a fin patch. For five of the six species, a dorsal flank blaze reaches anterodorsally from the area of the fin to as far forward as the melon, where it can form a blowhole chevron (Mitchell 1970: figures 12–14). Atlantic white-sided dolphin appears to be the only exception to this pattern, since it "does not show the unifying feature of a dorsal flank blaze and its anterior expansion, the blowhole chevron" (Mitchell 1970, p. 727). Mitchell goes on to relate Atlantic white-sided dolphins to the rest of the genus *Lagenorhynchus* through a similarity with hourglass dolphin flank patch pigmentation patterns, and considers Atlantic white-sided dolphins to have the same general degree of pigmentation specialization as hourglass dolphin. However, recent genetic evidence (e.g., LeDuc et al. 1999) suggests instead that similarities between Atlantic white-sided dolphins, white-beaked dolphins and the remaining four species currently classified within *Lagenorhynchus* are due to convergent evolution or retention of ancestral characters, and the genus as currently described is paraphyletic.

The pigmentation of the dusky dolphin is similar to that of the North Pacific white-sided dolphin. In lateral view, the pattern (Figure 2.2) can be divided into four parts: the spinal field, the flank patch, the thoracic patch, and the white abdominal field (terminology from Mitchell, 1970). The most conspicuous features of the coloration are flank blazes, thoracic patch, and rostrum color. Detailed descriptions of the pigmentation are found in Fraser (1966), Mitchell (1970), and Van Waerebeek (1993a). Also see figure 77 in Gaskin (1968b). Both dusky dolphins and Pacific white-sided dolphins show a light gray flank patch with associated dorsal and ventral flank blazes, a peduncular saddle, bicolored dorsal fin (dark with a light fin patch), a light gray thoracic patch, a dark beak blaze and lip patch, a dark eye patch, and a white abdominal field (Webber 1987).

Van Waerebeek (1993b) provided an excellent review of coloration and variation in dusky dolphins. No significant variation in coloration was found between the New Zealand, South American, or South African populations, or between the sexes, although some age-related differences were seen in animals examined from Peru. The lip patch and flipper blaze are typically lighter in juveniles, and darken with maturity in Peruvian dusky dolphins. Figure 2.3 shows photographs of the typical adult color pattern as observed in South Africa, Argentina, Peru, and New Zealand.

A darker, heavily melanized form with black upper and lower lip and eye patches, a nearly all black dorsal fin, and yellowish-brown at the interface of dark and light areas from Peninsula Valdés, Argentina is described from the work of Gallardo (1912) and Würsig and Würsig (1979). Van Waerebeek (1993b) discusses a darker form of dusky dolphin reported as *L. fitzroyi* in the historical literature, and concludes that the "Fitzroy form" is very similar to the

(a)

(b)

FIGURE 2.2 (a) Lateral view of a dusky dolphin, showing color pattern terminology as used in the text (following Fraser 1966, Mitchell 1970, Webber 1987) (photo by Michael Richlen). (b) Lateral view of a Pacific white-sided dolphin (photo by Nancy Black).

heavily melanized dusky dolphin form, and that the name could be used if this phenotype is ever identified as a discrete form. The type specimen of "*L. fitzroyi*" was collected by Charles Darwin in Golfo San José, Argentina and named "*Delphinus fitzroyi*" (Waterhouse 1838b) after the captain of the Beagle, Robert Fitzroy; Kellogg (1941) declared *Delphinus fitzroyi* a junior synonym of dusky dolphin. In contrast to the heavily melanized form are observations of a lighter phenotype in which darker areas such as the flipper stripe, eye and lip patches, the dark dorsal surface of the flipper, and the dark portion of the dorsal fin are all so pale as to be indistinct (Van Waerebeek 1993a). Dusky dolphins are also known in an anomalous, highly variable partial albino or piebald form from Peru and South Africa—these piebald animals have various amounts of white blotching and spotting on various parts of the body including sides, face, fin,

FIGURE 2.3 Comparison of color pattern of adult dusky dolphins from four regions. (a) Peru (photo by Anne Douglas), (b) South Africa (photo by Paula Olson), (c) New Zealand (photo by Sarah Piwetz), (d) Argentina (photo by Mariana Degrati).

flippers, and flukes, and some normal color fields can be obscured or absent from these individuals (Van Waerebeek 1993a).

A photograph in Bastida and Rodriguez (2005, p. 149) shows a very unusual looking animal reported by Bastida as a dusky dolphin × southern right whale dolphin hybrid; exhibiting features of both species, including what appears to be a flank blaze far back on the tailstock, and hints of other pigmentation resembling dusky dolphins (Figure 2.4). Documentation of additional potential hybrids and an analysis of the coloration of hybrids would be of interest (see also Yazdi 2002 for the article regarding this potential hybrid, and Reyes 1996 for possible dusky × common dolphin hybridization).

Dusky dolphins and Pacific white-sided dolphins are the only two *Lagenorhynchus* dolphins that have bicolored dorsal fins (Webber 1987). Both species show light gray patches of variable intensity, shape, size, and extent of coverage on a dark fin. In Pacific white-sided dolphin, the fin patch can cover most of the fin, reaching the animal's back at the insertion of the fin (see figures 214 and 216 in Leatherwood et al. 1982), or it can form just a small crescent along the trailing edge (Walker et al. 1986). Although the fin patch of dusky dolphins from New Zealand can extend onto the back, it is consistently more restricted, rarely covering more than one-half of the fin, as opposed to more than

FIGURE 2.4 Anomalous dolphin photographed off the Argentine coast by Gabriel Rojo, and suspected by him to be a hybrid between *Lagenorhynchus obscurus* and *Lissodelphis peronii.*

two-thirds coverage typical in Pacific white-sided dolphins, and the fin patch in Pacific white-sided dolphins is consistently more sharply demarcated from the dark body color. However, none of the characteristics of the fin patch was found to show a difference between these two species in all cases (Webber 1987).

The most striking differences between dusky dolphins from New Zealand and Pacific white-sided dolphins are the lack of a distinct well-developed eye to anus stripe in the former, particularly the flipper to anus portion, which is entirely absent in dusky dolphins, as noted by Fraser (1966). The differences between length of dorsal flank blaze, the presence or absence of a blowhole chevron, and presence or absence of a well-developed eye to anus stripe, especially in the flipper to anus portion, appear to be the most reliable features for separating dusky dolphins and Pacific white-sided dolphins on the basis of coloration pattern (Webber 1987).

The utility of coloration pattern in describing species of delphinid cetaceans is unclear (Gaskin 1982). Numerous authors have discussed delphinid coloration patterns from a taxonomic, evolutionary, and ecological perspective (e.g., Mitchell 1970, Perrin 1972, 1975; Evans 1975, Madsen and Herman 1980, Gaskin 1982, Ross 1984). As yet, no criteria have been put forward to distinguish the degree or number of differences in coloration patterns or how to identify an adequate sample size to determine the differences useful in separating species. Many species of delphinids are polymorphic in their external appearance. For example, the magnitude of the differences between dusky dolphins and Pacific white-sided dolphins does not seem to exceed either the differences in color pattern between populations of spinner dolphins (*Stenella longirostris*) deemed subspecifically distinct by Perrin (1975, note that additional anatomical information was also used to make these distinctions for spinner dolphins), or the regional and worldwide differences within *D. delphis* noted by Evans (1975) and Perrin (1984).

SIZE, SHAPE, AND SEXUAL DIMORPHISM

The dolphins of the genus *Lagenorhynchus* have long been noted for their distinctive coloration and unique shape (Gray 1866a, True 1889), and they share several features of external appearance. They are sometimes referred to as the ploughshare-headed dolphins, undoubtedly a reference to the abbreviated beak and wedge-shaped head of all six species. The dorsal fin is tall, usually pointed, and, in dusky dolphins and Pacific white-sided dolphins, bicolored. Like other species of *Lagenorhynchus*, dusky dolphins have virtually no beak and the head slopes evenly down from the blowhole to the rostrum. The dorsal fin is "falcate"—tall with a concave posterior border (Figure 2.2).

Lagenorhynchus dolphins are small to medium size relative to most other dolphin species. The smallest adult female and male from Peru were 173 cm and 176 cm, respectively (Manzanilla 1989). The largest female dusky dolphin reported from Peru was 205 cm, while the largest male from Peru, at 210 cm, represents the maximum size record for the species (Van Waerebeek 1993a). The largest dusky dolphin reported from South Africa was 191 cm ($N = 58$, data from P. Best summarized in Van Waerebeek 1992b). The largest of 19 adult dusky dolphins (4 females, 15 males) collected in New Zealand was only 186 cm, somewhat smaller than the maximum size of specimens from South Africa and much smaller than those from Peru (Cipriano 1992); on average, adult dusky dolphins from both South Africa and New Zealand are 8–10 cm shorter than Peruvian specimens (Van Waerebeek 1993b).

No significant differences between sexes in body length were found by Cipriano (1992) or Van Waerebeek (1993a) for New Zealand and Peruvian dusky dolphins, respectively. Both studies found that males had slightly larger dorsal fins, with a broader base and greater surface area than females, and no sexual dimorphism in color pattern was noted (Cipriano 1992, Van Waerebeek 1993a).

Weight

The 15 adult dusky dolphins from New Zealand reported by Cipriano (1992) weighed 69–78 kg for females (total lengths 167–178 cm) and 70–85 kg for males (total lengths 165–175 cm). Best (1976) reported that two captive females from South Africa weighed 53.6 kg (191 cm) and 69.9 kg (184 cm), but suggested that these two individuals were below normal weight at the time of their death. Length/weight relationships for 63 individuals from southern Africa are given in Best (2007).

INTERNAL ANATOMY

Skull and teeth

Van Waerebeek (1993a) compared skulls of mature dusky dolphins from Chile ($N = 22$), South Africa ($N = 40$), New Zealand ($N = 47$)—and a large sample from central Peru ($N = 189$) was analyzed separately by sex—but found little evidence for sexual dimorphism in overall size, although males tended to have a wider rostrum and a longer temporal fossa than females. Skulls from New Zealand and South Africa were on average 3.1 cm shorter than those from Peru and Chile, while New Zealand specimens had smaller tooth size and a slightly larger number of teeth than those from South Africa (Van Waerebeek 1993a). Mean tooth counts per row only varied slightly between regions: Peru females 30.8–32.0/row, Peru males 31.0–31.8/row, Chile 30.2–30.9/row, South Africa 29.6–30.4/row, New Zealand 31.4–32.1/row; upper jaw tooth rows tended to have on average one more tooth than lower rows (Van Waerebeek 1993a).

Dusky dolphins have a narrower skull and rostrum, throughout their length, than Pacific white-sided dolphins, and the rostrum of dusky dolphins is proportionately longer than that of Pacific white-sided dolphins (Figure 2.5, reviewed more thoroughly in Webber 1987). The mean rostrum width (at all points measured along the length of the rostrum) of Pacific white-sided dolphins is greater than the mean rostrum width of dusky dolphins; however, the rostrum base width to condylobasal length relationship suggests that while dusky dolphins are consistently smaller than most Pacific white-sided dolphins in this relationship, it is nearly the same proportionately (Webber 1987). Skulls of dusky dolphins are distinct from those of hourglass dolphin and Peale's dolphin—dusky dolphin skulls have a longer, narrower rostrum than either hourglass dolphin or Peale's dolphin of comparable length and maturity, and the pre- and postorbital and zygomatic widths of dusky dolphin skulls are smaller than in the other two species (Fraser 1966). Mean condylobasal length (CBL) in 16 specimens from Peru (all in the USNM) of dusky dolphins is 373 mm (range 333–421 mm; Brownell and Mead 1989). Mean CBL of 25 specimens from New Zealand waters is 369.5 mm, with a range of 347–387 mm (Webber 1987).

Postcranial skeleton

The vertebral count for dusky dolphins from Chile reported by Cárdenas et al. (1986) was C_7, T_{15}, L_{23}, Ca_{36}, giving a total of 81 vertebrae. Best (2007) reported vertebral counts of 49 dusky dolphins from South Africa as C_7, T_{12-15}, L_{16-26}, Ca_{28-36}, with a maximum per individual of 71–75. The difference between these counts may be population specific, or it may be that Cárdenas et al. (1986) reported the maximum number of vertebrae found in all individuals analyzed. For comparison with other lissodelphines, vertebral count for a single individual Pacific white-sided dolphin (USNM 504851) was C_7, T_{12}, L_{24}, Ca_{32} (75 total), for Northern right whale dolphin (USNM 504350) C_7, T_{14}, L_{33}, Ca_{31} (85 total), and for Commerson's dolphin (USNM 550156) C_7, T_{12}, L_{13}, Ca_{31} (63 total; Buchholtz et al. 2005).

Organs and tissues

Gallardo (1913) provided some preliminary notes on the nasal tract complex of a dusky dolphin, and Schenkkan (1973) gave a more detailed description of this complex. Hearts of six adult dusky dolphins collected by Cipriano (1992) in New Zealand weighed 595–657 g (2 females) and 571–773 g (4 males). Livers of

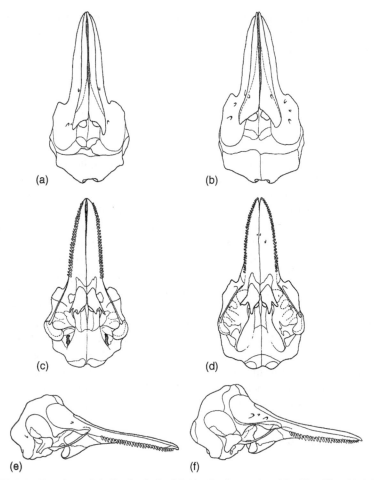

FIGURE 2.5   Drawings of skulls of a dusky dolphin (*L. obscurus*) and Pacific white-sided dolphin (*L. obliquidens*), to scale. (a) Dorsal, (c) ventral, and (e) lateral aspects of the skull of
L. obscurus. (b) Dorsal, (d) ventral, and (f) lateral aspects of the skull of *L. obliquidens* (original
drawings by Claudia Stevens, used by permission).

these same specimens weighed 2068–2663 g (females) and 1347–2160 g (males);
kidneys of the specimens weighed 339–425 g (females) and 191–417 g (males);
right and left lungs of the two females weighed 646/663 g and 1054/1000 g,
respectively; the lungs of a single male weighed 722/779 g (Cipriano 1992).

GROWTH AND REPRODUCTION

Age, growth and reproductive status

Manzanilla (1989) reported on the reproductive status of dusky dolphins (26
females, 14 males) landed in the Peruvian ports of Pucusana and Cerro Azul

between 1985 and 1986. Eight immature females ranged in total length from 127 cm to 179.5 cm; the three largest immature females (174, 177, 179.5 cm) had developing follicles in their ovaries; the remaining 18 females were all mature and ranged in total length from 173 cm to 193 cm (Manzanilla 1989). Age (based on dentine layers) for all adult females was 4.0–11.5 years (Manzanilla 1989). Two immature males measured 141 cm and 172 cm; 12 mature males, with seminal fluid in the epididymis, ranged in total length from 176 cm to 206 cm (Manzanilla 1989). The largest known male from Peru was 210 cm (Van Waerebeek 1993a), while the largest of 19 adult dusky dolphins analyzed by Cipriano (1992) was 186 cm and weighed 77 kg. The maximum age of 32 New Zealand dusky dolphins analyzed by Webber (1987) was 26–30 years old.

Kastelein et al. (2000) monitored the growth and food consumption of a female dusky dolphin that was captured on February 24, 1981 and maintained for 13 years at Sea World, Durban, South Africa—the dolphin's weight increased gradually from approximately 52 kg at the age of 2 years to approximately 66 kg at 6 years, and then fluctuated between 61 and 74 kg. Total length at 3 years of age (measured along the body contour) was approximately 170 cm; at age 15, total length (measured in a straight line parallel with the body axis) was 162 cm (Kastelein et al. 2000). The dolphin was fed five to six times per day on a diet that was composed of, on average by weight, 30% hake (*Merluccius* spp.), 45% cephalopod, and 25% miscellaneous teleosts depending on availability. Annual food consumption increased from 1870 kg at the age of 3 years to 2170 kg when she was 5 years old; during her sixth year, her annual food intake increased to approximately 2900 kg (coincident with the installation of a cooling system that was used in the summer in the years thereafter); after her sixth year, her food consumption fluctuated between 2400 and 2800 kg per year (Kastelein et al. 2000).

Age at first reproduction

Van Waerebeek and Read (1994) analyzed growth and reproductive parameters of 852 dusky dolphins (522 females, 330 males) captured in a directed fishery off the central coast of Peru during 1985–1990. Another study of the same specimens indicated that female and male dolphins reached sexual maturity at 72.1 kg/174.9 cm and 75.0 kg/175.2 cm, respectively, at about 4–6 years of age (Van Waerebeek 1992b).

Age of first reproduction of dusky dolphins from New Zealand was estimated as very approximately 7–8 years at 160–165 cm total length (Cipriano 1992).

Gestation and lactation

The reproductive cycle of dusky dolphins from Peru totaled about 28.6 months, including 12.9 months' gestation, 12.0 months' lactation, and a resting period of 3.7 months (Van Waerebeek and Read 1994). Cipriano (1992) roughly estimated a gestation period of about 11 months for New Zealand dusky dolphin, and lactation was estimated to last around 18 months (Leatherwood and Reeves 1983). A new analysis of reproductive intervals in southern African dusky dolphins is presented in Chapter 14.

Calving

The main calving season in Peru extended from August to October, with some neonates collected in February; mean size at birth was 91.2 cm with a weight of 9.6 kg (Van Waerebeek and Read 1994). In New Zealand, small calves were found stranded or sighted during boating surveys only between November and January, suggesting a limited summer calving period which also coincided with the months that dolphin groups were found closest to shore (Cipriano 1992, Markowitz 2004, see also Chapter 9). In Golfo San José, Argentina calves of dusky dolphins were also born mainly during the austral summer (Würsig and Würsig 1980).

Testis weight and seasonal recrudescence

Five of the nine adult males collected by Cipriano (1992) in New Zealand had semen present in the epididymis and enlarged testes (995–2143 g each, including epididymis), and were collected in spring or summer months. The other four adult males collected in fall or winter (165–175 cm, minimum age 12–26 years) had smaller (252–479 g) and apparently inactive testes (Cipriano 1992). Testis weight of juvenile and subadult males (130–160 cm) from New Zealand was 12–663 g (Cipriano 1992). Best (1976) reported that a 173 cm subadult male dusky dolphin from South African waters had a combined testes weight of only 134 g, and that adult males have combined testes weights up to 1.6 kg. Testis weight (including epididymis) of 125 dusky dolphins of all ages from Peru, analyzed by Van Waerebeek and Read (1994), covered a huge range, from 53 to 5120 g (left) and 53–4930 g (right), with a maximum combined weight of up to 9730 g. Testis weight in these Peruvian dolphins increased in August, peaked in September–October, and then decreased again in November, in synchrony with the timing of ovulation and conception (Van Waerebeek and Read 1994).

Such a huge increase in the testis size of adult males (up to 8% of body weight) during the reproductive season—followed by a decrease in size or "recrudescence" in the non-breeding season—is an indication that sperm competition and promiscuous breeding characterize the breeding system of dusky dolphins. The relative size of the testes (in relation to body size) has been extensively studied in primates, and is closely linked to breeding systems in other groups of mammals as well (Gomenido et al. 1998).

ASSOCIATIONS WITH OTHER SPECIES

Würsig and Würsig (1980) observed dusky dolphins in Golfo San José, Argentina associating closely with southern right whales (*Eubalaena australis*) and South

American sea lions (*Otaria flavescens*), actively avoiding killer whales (*Orcinus orca*), close to but not apparently interacting with bottlenose dolphins (*Tursiops truncatus*) and potentially feeding in the same area as Risso's dolphins (*Grampus griseus*). In this area, a wide variety of marine birds associate with surface-feeding dusky dolphins, and the dolphins in this region may be unique in apparently using bird aggregations over surface-feeding groups to coordinate foraging efforts and increase efficiency in corralling prey (Würsig and Würsig 1980, but see also Chapter 6).

Dusky dolphins have also been observed in mixed cetacean schools with southern right whale dolphins off Namibia (Cruickshank and Brown 1981, Rose and Payne 1991). In summer, dusky dolphin groups off Kaikoura, New Zealand were occasionally accompanied by small groups of common dolphins (*Delphinus delphis*), which traveled as a cohesive subgroup within the larger dusky dolphin group (Cipriano 1992). Dusky dolphins were also observed with pilot whales (*Globicephala* spp.) off southwest Africa and the Prince Edward Islands (Cruickshank and Brown 1981) and over the Campbell Plateau in New Zealand waters (M.A. Webber, personal observation). Dusky dolphins have been observed to react strongly to the presence of killer whales—moving rapidly away and/or into very shallow water (Würsig and Würsig 1980, Cipriano 1992, see also Chapter 7)—and killer whales have been directly observed to prey on dusky dolphins (Constantine et al. 1998). As with other small cetaceans, dusky dolphins are also likely subject to predation by large sharks, and their use of shallow water habitats during resting periods and especially when small calves are present has been suggested to be related to avoidance of such predators (e.g., Würsig and Würsig 1980, Würsig et al. 1991, Cipriano 1992, see also Chapter 7). Markowitz (2004) presents a detailed account of all other marine mammal species known to interact with dusky dolphins off Kaikoura, New Zealand (see also Chapter 11).

PARASITES, COMMENSALS, AND DISEASE

Endoparasites from stomachs of dusky dolphins were reported by Gibson and Harris (1979) from specimens in the British Museum, including the cestodes *Trigonocotyle prudhoei* and *Phyllobothrium delphini*, and nematodes *Anisakis typica* and *Anisakis* sp. Nematode, cestode, and trematode parasites from a sample of two dozen dusky dolphins collected in New Zealand have been reported (Cipriano et al. 1985, Cipriano 1992). Nematodes were found in dusky dolphin intestines and stomachs, and other types (*Stenurus* spp.) in upper respiratory tracts, ear capsules, pterygoid sinuses, and lung tissue; cestodes were found in stomachs and heart tissue, and pleurocercoid cysts (probably *Phyllobothrium* spp.) were common in the blubber of most adult specimens; trematodes were found in intestines, stomachs, and liver tissue; *Braunina cordiformis* was often attached to the lining of the second stomach (Cipriano et al. 1985, Cipriano 1992).

A large number of fresh carcasses from Peru and northern Chile were examined for phoronts and parasites by Van Waerebeek and colleagues; 275

of 707 animals examined had the sessile barnacle *Xenobalanus globicipitis* attached to flukes, flippers or dorsal fin (but no other body parts); none of these specimens were found to carry cyamid amphipods (Van Waerebeek et al. 1993). Dusky dolphin skulls from Peru ($N = 240$) and northern Chile ($N = 26$) were inspected, but only a single individual showed evidence of bone lesions associated with *Crassicauda* roundworm infestation; the digenetic trematode *Nasitrema* was found in 76 of 97 fresh dolphin heads in which the pterygoid and maxillary sinuses were examined (Van Waerebeek et al. 1993). *Anisakis* spp. nematodes were detected in 87 of 218 dusky dolphin stomachs examined; the stomach fluke *Braunina cordiformis* was found in 18 of 212 digestive tracts examined, usually in the pyloric stomach and with no evidence for a difference in infestation rate between males and females; neither *Anisakis* spp. nor *Braunina cordiformis* were found in juveniles examined, and likely are obtained through predation on a primary cephalopod or fish host after weaning (Van Waerebeek et al. 1993). *Halocercus* (Pseudaliidae) lungworms were found in the stomach of a single animal (although lungs were not systematically surveyed for such parasites); the trematode *Pholeter gastrophylus* was found in 17 of 212 stomachs, also mainly in the pylorus; the cestode *Phyllobothrium delphini* was found in 14 of 20 animals in which the ventral blubber was surveyed; adult tapeworms were found in a single individual female of 27 specimens in which the intestines were flushed and examined (Van Waerebeek et al. 1993).

The stomachs and intestines of 23 dusky dolphins incidentally captured in the shrimp trawl fishery off the coast of Patagonia between 1990 and 1995 included five parasite species; all the dolphins were parasitized, and more than half of them harbored all five parasites (Dans et al. 1999). Parasite species included the nematode *Anisakis simplex* found mainly in the stomach, the acanthocephalan *Corynosoma australe* found only in the stomach and at much lower prevalence, and three digenetic trematodes: *Braunina cordiformis* (Brauninidae) found mostly in the duodenal ampulla, *Pholeter gastrophilus* (Troglotrematidae) found only in the stomach and at low prevalence, and *Hadwenius* spp. (Campulidae) found mostly in the stomach (Dans et al. 1999).

Skin lesions associated with herpesvirus-like particles were documented by Van Bressem et al. (1994) in four dusky dolphins, all sexually immature individuals with standard lengths between 150 and 157 cm, which were incidentally captured in the artisanal fishery off central Peru in 1991. Herpesvirus infection in these individuals seemed only mildly pathogenic since there was no evidence of ill-health, and as they were all apparently captured together it is possible that the young individuals were members of a group in which infection was enzootic (Van Bressem et al. 1994).

Van Bressem and Van Waerebeek (1996) found poxvirus infection in 68 (34.7%) of 196 dusky dolphins captured in gill net fisheries off coastal Peru between March 1993 and July 1994; infection was independent of sex but strongly correlated with body length; tattoo lesions associated with poxvirus infection were absent in neonates and some juveniles, but prevalence peaked in the next larger size class and gradually declined again in the largest animals.

Van Bressem et al. (2000) found ovarian cysts and one ovarian tumor, uterine tumors, vaginal calculi, and abscesses of the broad ligament or undetermined testicular lesions in 25 of 502 female and male dusky dolphins of all ages caught in coastal waters of central Peru during the periods 1985–1987 and 1993–1994. Tentative or definitive diagnoses included Graafian follicle cysts, luteinized cysts, ovarian parasitic granulomatous inflammation, dysgerminoma, leiomyoma, fibroleiomyoma, and chronic fibrino-suppurative inflammation of the broad ligament; all lesions described represented first reports for dusky dolphins (Van Bressem et al. 2000).

DIET AND FORAGING ECOLOGY

Studies in the Valdés Peninsula region of Argentina (Würsig and Würsig 1980) showed that during daylight hours the dusky dolphins in that area regularly utilized cooperative surface feeding on southern anchovy (*Engraulis anchoita*) to locate, herd and tightly contain concentrated schools of fish near the surface. Cooperative surface feeding occurred mainly in the interior portions of the Golfo San José during spring and summer months; in late summer and fall surface feeding was more often concentrated near the mouth of the bay; during winter months when most southern anchovy move out of the bay, dolphin group size was much smaller and groups often moved in shallower water close to shore, likely feeding on other prey type(s) (Würsig and Würsig 1980).

Cipriano (1992) analyzed the stomach contents of 24 dusky dolphins from the Kaikoura area of New Zealand (some returned by fishermen and others found beachcast but showing evidence of incidental capture in gill nets), and identified a total of 17 735 prey pieces. Deep scattering layer (DSL)-associated species included squid (especially *Nototodarus/Todaroides* spp.) and lanternfish (Myctophidae), such as *Symbolophorus* spp., *Diaphus* spp., *Myctophum* spp., and *Hintonia* spp. Hoki (*Macruronus novaezelandiae*) was found in the analysis of some dolphins in each season, whereas red cod (*Pseudophycis bachus*), hake (*Merluccius australis*), hatchetfish (Gonostomatidae/Sternoptychidae), carapids (Carapidae), *Bathylagus* spp., and some unidentified fishes appeared to be taken only occasionally. Although myctophids are smaller than the other main food types, they are the most energy-rich (Cipriano 1992). Theodolite and radio-tracking of dusky dolphins in the Kaikoura area showed both seasonal and diurnal patterns: during spring, summer, and early fall months dolphins were found very close to shore in early morning hours and gradually moved farther from shore in late afternoon; during these months, dive times increased markedly at twilight, suggesting that the dolphins were moving into deeper water to meet DSL organisms as they moved into surface waters; during winter months the dusky dolphins almost always remained much farther from shore and appeared to traverse through the area much more rapidly than in summer (Cipriano et al. 1989, Würsig et al. 1991, Cipriano 1992, Benoit-Bird et al. 2004).

Dusky dolphins that feed in Admiralty Bay in the Marlborough Sounds region of South Island, New Zealand in winter eat pilchard (*Sardinops neopilchardus*), yellow-eyed mullet (*Aldrichetta forsteri*), and sprat (*Sprattus antipodum;* Markowitz et al. 2004, see also Chapter 6). In the Marlborough Sounds region, approximately 250 km north of Kaikoura, dusky dolphins have been observed to forage cooperatively on schooling fishes during the daytime (Markowitz 2004, Markowitz et al. 2004, see also Chapter 6), perhaps similar to the cooperative feeding behavior observed in Argentina. Photographic records indicate that some of these dolphins also frequent Kaikoura at other times of the year, showing that individual dolphins alter their foraging strategies and prey types on a seasonal/regional basis (Markowitz 2004, Markowitz et al. 2004). Ecological differences between the deep waters of Kaikoura Canyon and the shallow nearshore waters of Admiralty Bay appear to influence how, when, and in what social groupings dusky dolphins forage (Benoit-Bird et al. 2004).

McKinnon (1994) analyzed contents from 136 stomach samples containing recognizable hard parts, 130 from dusky dolphins taken in the artisanal gill net fishery of central Peru and six additional stomachs from dolphins taken by a purse seiner in the same area, collected in the austral summers (January 1 through March 31) and austral winters (July 1 through September 30) of 1985 and 1986. The anchoveta (*Engraulis ringens*), the most abundant vertebrate in Peruvian coastal waters, was the principal prey of dusky dolphins, regardless of reproductive class, in both seasons and in both years (McKinnon 1994). Other prey species commonly found in dusky dolphin stomachs were horse mackerel (*Trachurus symmetricus*), hake (*Merluccius grayi*), sardine (*Sardinops sagax*), Patagonian squid (*Loligo gahi*), and jumbo flying squid (*Dosidicas gigas*); all prey species averaging less than 30 cm in estimated length and 300 g in estimated weight, consistently smaller than those caught by the fishery (McKinnon 1994).

Sekiguchi et al. (1992) analyzed stomach contents of 37 dusky dolphins from southern Africa, collected between 1969 and 1990, and including 17 individuals from fisheries by-catch, 10 from a scientific catch, and 10 stranded individuals. Dusky dolphins from this region had frequently fed on pelagic schooling prey such as horse mackerel (*Trachurus trachurus capensis*) and hake (*Merluccius* spp.), but also some deep-water squid (e.g., *Todarodes angolensis*), lanternfish (*Lampanyctodes hectoris*), and lightfish (*Mauloicus muelleri*), suggesting possible foraging in the DSL at night (Sekiguchi et al. 1992). A full summary of new information on prey items identified from stomach contents of dusky dolphins in southern African waters is given in Chapter 14.

ANTHROPOGENIC THREATS

South Africa

There is little published information on the magnitude of by-catch or other anthropogenic threats to dusky dolphins in South Africa (but see Best and Ross 1977, and Chapter 14, for updated information). Dusky dolphins have been captured for oceanarium display in Argentina, New Zealand, and South Africa;

37 individuals were captured in South Africa 1961–1981; 13 of these were immediately released and the remainder died either at capture or during captivity (Best and Ross 1984).

Argentina

Early reports suggested that catches of dusky dolphins in Patagonian waters of Argentina were mostly rare and accidental (Crespo et al. 1997, Dans et al. 1997a), but continued studies showed that by-catch in some years may approach critical levels (Dans et al. 2003a, 2003b). Mid-water trawling for Argentine red shrimp produced the highest by-catch reported to date (Crespo et al. 1997, 2000) with a maximum of 560 dolphins in 1984, and 852–1377 dolphins killed in the period 1982–1995, while catch levels estimated for 1984 (442–560 dusky dolphins) exceeded all critical values (Dans et al. 1997a). Comparisons between annual by-catch rates and critical values suggest that dusky dolphin catches off the Patagonian coast may be low in absolute numbers, but are close to or exceed replacement thresholds with the potential for population decline (Dans et al. 1997a, 2003a, 2003b). Annual by-catch was estimated at 1–3% of the Argentine population (estimated at 7000 total), and these fisheries operate at depths, on prey, in particular areas, and at night when catches of young but mature female dusky dolphins make up the majority (70%) of mortality (Dans et al. 2003b). An annual removal of 70 or more dolphins from this regional population always produced a significant negative trend in population dynamic models, and the authors estimated that if this continued, the population would decline by approximately 20% from current levels in about 50 years (Dans et al. 2003b).

Peru

Catches of dusky dolphins in Chile and Peru were documented in the late 1980s (e.g., Guerra et al. 1987, Van Waerebeek et al. 1988). Early records indicated that small cetaceans were taken incidental to other fishing operations (Mitchell 1975), but after the decline of the anchoveta fishery in 1972, the majority of dusky dolphins landed were intentionally captured in a drift net fishery targeting this species for human consumption (Read et al. 1988). Large catches (approximately 10 000) of small cetaceans were reported from the coastal waters of central Peru in 1985 (Read et al. 1988); in the 1991–1993 period an estimated 7000 dusky dolphins were captured per year, and it appears that this level of exploitation has diminished since dolphin hunting was banned by law in 1996 due to depletion of the population (Van Waerebeek and Reyes 1990, Van Waerebeek and Würsig 2002); however, the clandestine taking of dolphins and selling of dolphin meat continues (Van Waerebeek and Würsig 2009).

Cassens et al. (2005) pointed out that Peruvian dusky dolphins might have experienced loss of genetic variability because small cetaceans off Peru have been severely affected by direct takes for three decades (Van Waerebeek and Reyes

1990, 1994, Van Waerebeek et al. 1997), and furthermore, Peruvian dusky dolphins live in an extremely unstable environment where El Niño events recurrently cause increased rates of mortality among marine organisms, with the most severe oscillations of the last century having occurred in 1982–1983 and 1997–1998 (Berta and Sumich 1999). Cassens et al. (2005) used nine nuclear species-specific microsatellite loci and two mitochondrial gene fragments (cytochrome *b* and control region) to investigate processes that have shaped the geographical distribution of genetic diversity exhibited by contemporary dusky dolphin populations, from a sample of 221 individuals from four locations (Peru, Argentina, southern Africa, and New Zealand), covering most of the species' distribution range. Results from this study are compatible with a more recent demographic decline in the Peruvian dusky dolphin stock, but comparison between the different molecular markers used suggested an ancient bottleneck predating both recent El Niño oscillations and human exploitation (Cassens et al. 2005). See Chapter 14 for a further discussion of population differences relative to ecology.

New Zealand

By-catch of dusky dolphins in New Zealand waters does not appear to be high, relative to the apparently large population size in the region. M. A. Webber and F. Cipriano (unpublished data) estimated that approximately 200 dusky dolphins off Kaikoura, New Zealand were incidentally captured in 1984, but in recent years it appears that incidental catches associated with gill net fisheries have declined as the local economy has become more tied to eco-tourism rather than fishing (B. Würsig, personal communication).

Markowitz et al. (2004) identified an increase in aquaculture activities as a new potential threat to dusky dolphins wintering in the Marlborough Sounds region of the South Island, New Zealand, due to displacement and interference with feeding activities in areas where mussel farms and dolphin habitat overlap. Overlap of dusky dolphin habitat use with proposed marine farms is high, and dolphins rarely used areas within the existing farms—during observations over five winter seasons, only eight of 621 dusky dolphin groups monitored in Admiralty Bay were observed to enter the boundaries of mussel farms at any point (Markowitz et al. 2004).

Another potential threat to dusky dolphins is disturbance to critical activities or displacement from areas of critical habitat by eco-tourism and especially "swim with dolphins" activities (Samuels et al. 2003). In New Zealand, such programs have been going on since 1989 and year-round activities have continued since 1994—with humans interacting with the dolphins in some areas for up to 70% of daylight hours (Samuels et al. 2003). At Kaikoura, 83% of vessel approaches resulted in behavioral change with interruptions to feeding and resting behavior; departure of boats did not always lead to resumption of previous activities; when boats were present, dolphins formed more compact groups, changed direction of travel, and exhibited higher levels of activities during periods normally characterized by lower level "resting" activities (Samuels et al. 2003, but see also Chapters 11 and 13 for more positive interpretations of dusky dolphins and human tourism activities).

Heavy metals and organic pollutants tend to concentrate in upper level predators, with the potential for decreased reproductive potential and injury to developing young. Pollutant levels in odontocete cetaceans have been recognized as a serious potential threat, but few such studies have been carried out on dusky dolphins. The maximum concentration of DDT reported in the blubber of dusky dolphins from New Zealand waters was 175 ppm wet weight (Koeman 1973).

SOME ANSWERS AND MORE QUESTIONS

Over the past 25 years, much has been added to our knowledge of the biology of dusky dolphins from studies in South America, New Zealand, Australia, and southern Africa, resulting in a great increase in breadth and depth of our understanding of these fascinating and photogenic animals. After long-term observations by a series of dedicated researchers in several locations, dusky dolphins now arguably rank as one of the best-known dolphin species, making the remaining mysteries all the more tantalizing in the context of what we think we know about their basic life history. Dusky dolphins occur relatively close to shore off the coasts of Peru and Chile, Argentina, New Zealand, and southern Africa, but are occasionally seen in the waters off southern Australia and Tasmania. Given the restriction of sightings to certain periods of the year, it seems likely that the dolphins off south Australian coasts are seasonal vagrants from New Zealand, but what about the ones seen thousands of kilometers from continental margins, around oceanic islands of the South Atlantic and southern Indian Ocean—are they vagrants from continental populations, or resident in those waters?

Dusky dolphins often feed on schooling fishes at the surface in Argentine and Peruvian waters, and seem to feed primarily on DSL organisms in the Kaikoura region of New Zealand and in southern African waters. However, some Kaikoura dolphins occur in Admiralty Bay, foraging on schooling fishes in winter—why do some individuals switch location and foraging strategy, while others move somewhat offshore in winter and maintain a diet of DSL prey? Sightings of dusky dolphins around South Atlantic and southern Indian Ocean island groups seem to be limited to those island groups north of the Antarctic Convergence—is it a barrier, and if so, is it limiting the dolphins or their prey, and what are they eating there anyway?

Molecular studies have revolutionized our understanding of relationships among dusky dolphins and their relatives, indicating that two species once considered congeners are only distantly related. Now it seems that four "*Lagenorhynchus*" species plus *Cephalorhynchus* and *Lissodelphis* are closely related—when and why did such physically, ecologically, and behaviorally different groups diverge, and what was their common ancestor like?

Additional questions include concerns for our coexistence with dusky dolphins in places where human activities influence their behavior, deplete or pollute their prey, and exclude them from or alter their (and our) environment, especially the giant unplanned experimental manipulation of global temperature. Continued longitudinal studies in established study sites, observations in and samples from additional areas—especially remote locations where the species is poorly known—and the use of new technology, such as sophisticated dive recorders and satellite tracking systems will help, but we still have a lot to learn.

ACKNOWLEDGMENTS

We thank Bernd and Melany Würsig and Hal Markowitz for getting us (and so many others) started on our professional careers with knowledge, inspiration, humor and great kindness; and Steve Leatherwood for recognizing the potential for dusky dolphin research at Kaikoura and for sharing that insight with us. We are very grateful to Bernd, Melany, Hal, Steve, Alan N. Baker and Martin Cawthorn, and many others for providing encouragement and support for our work in Kaikoura. Marc specifically and gratefully thanks Al Myrick for teaching him how to process teeth for aging, and Nina Lisowski, Isidore Szczepaniak, and Bill Keener for help, friendship, and encouragement along the way. We also gratefully acknowledge the assistance of Koen Van Waerebeek, Natalie Goodall, Peter B. Best, and Alan N. Baker for tracking down new information and presenting it (we hope) clearly; and Tom Jefferson, Bob Pitman, Alan Jackson, and Ricardo Bastida for helping us find pictures, Jack van Berkel for making us welcome in Kaikoura and providing logistical and technical help when it all began, and all the photographers and illustrators who provided material for inclusion in this chapter—Michael Richlen, Anne Douglas, Mariana Degrati, Sarah Piwetz, Nancy Black, Paula Olson, Gabriel Rojo, and Claudia Stevens.

Chapter 3

Dusky Dolphin Trophic Ecology: Their Role in the Food Web

Silvana L. Dans[a], Enrique A. Crespo[b], Mariano Koen-Alonso[c], Tim M. Markowitz[d], Bárbara Berón Vera[a], and Adrian D. Dahood[e]

[a]Centro Nacional Patagónico, CONICET, Chubut, Argentina
[b]Facultad de Ciencias Naturales, Universidad Nacional de la Patagonia, Chubut, Argentina
[c]NW Atlantic Fisheries Center, Fisheries and Oceans, St. John's, Newfoundland & Labrador, Canada
[d]School of Biological Sciences, University of Canterbury, Kaikoura, New Zealand
[e]Marine Mammal Research Program, Texas A&M University, Galveston, Texas, USA

Have you ever seen a full net on the deck of a hake trawler from Argentina? It is an impressive sight. A massive mesh bag, several meters long and way above the heads of fishers looking at it with smiles on their faces and a readiness to get to work. Even though this net was "searching" and "hunting" hakes, many other unfortunate species were also caught. Sharks, skates, fishes, sea lions, and dolphins all lie altogether with the same destiny: to be part of the by-catch.

How can this by-catch be reduced? We can try to develop better fishing techniques, improve the selectivity of the gear; in essence, use our technological capabilities to minimize this "collateral damage." However, regardless of how much we improve our technology, it is virtually impossible to eliminate by-catch issues. What solves one problem often enough creates another.

Why is that? As soon as we pose the question, we realize that all these species are part of by-catch because they are all together at the same place and time. They are sharing much more than their fates in human hands. Like the net, some of them are searching and hunting prey, while others are swimming away or trying to hide to avoid being eaten. Despite their current and ephemeral roles—someone's prey is also someone else's predator—all of them are sharing a common habitat and through the relationships that link them, they are helping to define the ecosystem where they live, reproduce, and die.

Our question of how to reduce by-catch evolves. We can gain new insights about how to deal with human influences if we understand how the species are interrelated and how they manage to coexist. Now we find ourselves immersed in the construction of food web diagrams, drawing boxes with species names, connecting them with arrows, trying to come up, in the best case scenario, with meaningful numbers to attach to a box or an arrow. But, we often end up holding a species name in our hands without the slightest idea where to place it in the big and somewhat messy picture of trophic roles we are putting together.

This is hard work, and is it worth it? Sometimes the answer presents itself even before the question takes shape in our minds. Nowadays, the catch on the deck often is only several meters long and several heads BELOW, not above, the fishers looking at it with sad faces and ready to work. The present catch has declined over the years, and yet many unfortunate by-caught species are still there, sharing way more than their fate, as always.

Dusky dolphins are part of these marine ecosystems, and as any other species, they have a place in them and a role to play. As well, they are also subject to human impacts, and constitute an integral part of the by-catch of fisheries. In some cases, their populations may be at risk as a consequence of these impacts, but we still do not know what the repercussions of these impacts could be on the rest of the ecosystem. This same reasoning applies to many other species. The problem extends far beyond dusky dolphins, but they—as well as other marine mammals—act like the tip of the iceberg. We can see them, but we also know that they are representative of so much that goes on below, unseen.

Marine ecosystems worldwide are subject to high levels of extraction by fisheries, and it is clear by now that there is no such a thing as a pristine ecosystem. If we want that dusky dolphins remain on our Earth in a healthy state, if we want our children to have the opportunity to enjoy them as we have the privilege today, then we need to face the challenges and continue the sometimes frustrating and painstaking job of trying to understand how marine ecosystems work, and what is the role that dusky dolphins, along with other top predators, play in their structure and functioning. Substantial work has been done, but we are barely scratching the surface of knowledge. We need to coordinate efforts and integrate different approaches, but most of all, we need to be creative and put our ingenuity to the test. Maybe, just maybe, answers lie in the wake that dusky dolphins leave behind as they pass so effortlessly through the placid waters.

Silvana Dans, Puerto Madryn, Chubut, Argentina, November 2008

INTRODUCTION

Dusky dolphins (*Lagenorhynchus obscurus*) are an integral part of several marine ecosystems of the southern hemisphere. Among other cetaceans, pinnipeds, seabirds, and sharks, dusky dolphins represent the near-terminal links in the food webs in which they are embedded. This set of species is referred to as the "top predators" or "upper trophic level predators" (Boyd et al. 2006) and the study of the trophic role dusky dolphins play in marine ecosystems will be driven mostly by the same interests we have in top predators. In an ecological sense, the term "role" implies something about the functional significance of a species. In the case of top predators, this importance mainly involves the tropho-dynamic function they have as consumers and the effects of this consumption on species interactions and community structure (Bowen 1997).

Analysis of the trophic role of a top predator provides a context for evaluating its predation impact on prey populations, as well as understanding how variation in prey populations, both as a result of harvesting by humans and due to environmental changes, can affect predator population dynamics (Bowen 1997). For example Mohn and Bowen (1996) modeled the impact of gray seal (*Halichoerus grypus*) predation mortality on Atlantic cod (*Gadus morhua*) of the Eastern Scotian shelf off Nova Scotia. In addition to direct effects on prey abundances, top predators can also indirectly affect the abundance of other species by out-competing them or consuming species that prey upon them (Trites 2002).

A major hypothesis is that predators control community structure and function, producing forces that cascade across successively lower trophic levels, sometimes reaching the base of the food web, in a process known as top-down control. One of the best examples of such control is the effect of killer whale (*Orcinus orca*) predation on sea otters (*Enhydra lutris*) in the Aleutian Islands, Alaska, which sets in motion a host of effects in this coastal ecosystem (Estes et al. 1998). Estes et al. (1998) proposed that killer whale predation has caused a decline of sea otters; this in turn released sea urchin populations from the limiting influence of sea otter predation, and the urchins increased rapidly and overgrazed the kelp forests.

Most top predators are generalist and opportunistic foragers, preying on a wide variety of alternative prey according to availability (Trites 2002). This mechanism is an important aspect for the persistence of top predator populations. Most top predators have diets dominated by only several main prey species, but including a large number of minor prey species. As a result, a richer community will allow the predator to change from the main prey if it becomes scarce (Sergio et al. 2006). However, the precise way in which top predators modify their diet as the abundances of alternative prey change requires further experimental and observational studies. For example, more than 40 taxa of fish and invertebrates were found in 682 gray seal stomachs from eastern Canada, although relatively few species account for most of the energy ingested (Bowen and Siniff 1999). This example and others suggest that seals would show some degree of preference.

The opportunistic feeding behavior of most top predators in marine ecosystems generates complex trophic webs, since many of them consume prey at different trophic levels. Moreover, they can prey on pelagic and benthic domains, integrating different energy channels in their ecosystems. This type of integration has recently been identified as one of the mechanisms that could provide stability to natural ecosystems (Rooney et al. 2006).

Another aspect not deeply addressed is the effect a top predator may have through parasitism relationships. Parasitism can be viewed as another trophic relationship, since it involves some degree of energy transfer between two partners—the parasite, who takes some benefit, and the host, who receives some form of disadvantage or harm, but not necessarily by dying (Bush et al. 2001). Such interaction is well represented by *Crassicauda* spp. infection, which causes cranial damage in spotted dolphins (*Stenella attenuata*) from the eastern tropical

Pacific (Raga et al. 2002). But even when parasitism does not represent a significant source of mortality for the host, the indirect effects may be important. Most parasites require a number of species linked mainly by predator–prey relations, such as when an organism eats an infected host, in order to be transmitted and achieve the adult stage. Therefore, the definitive host may have an impact on adult parasite abundances, and in turn on larvae abundances.

The interactions between host and parasite abundances have been studied thoroughly in economically important species. In North Atlantic waters, seals are an important phase in the life cycle of *Pseudoterranova decipiens* (Bonner 1982), whose larvae cause heavy losses in the cod industry (Bowen 1990). However, in the past few decades, the study of parasites has shifted from an economic view to a more ecological one, and their potential to structure marine communities (Poulin 1999).

The role that top predators may have on a marine ecosystem gains relevance when fisheries also exploit the ecosystem, mainly because they share prey items such as fish and squid species, which may result in competitive interactions (Pauly et al. 1998). However, since they are both part of the same food webs, ultimately they would be competing for the primary productivity of the region where they occur (Trites et al. 1997). For example, it is generally believed that today's competition between marine mammals and fisheries is probably more significant than much of former direct exploitation was (Crespo and Hall 2001). Several hypotheses concerning fisheries removal effects were proposed to explain the decline of the western Steller sea lion (*Eumetopias jubatus*) population, such as changes in food availability (localized prey depletion close to rookeries) or changes in food quality (a less diverse diet and a replacement of herring by pollock, which is a low nutritional quality prey; Committee on the Alaska Groundfish Fishery and Steller Sea Lions, National Research Council 2003). There is also an increased awareness about the effects of fisheries on entire marine food webs and predators' responses to these potential changes. Landings from global fisheries have shifted in the past 45 years from large piscivorous fishes toward smaller invertebrates and planktivorous fishes, especially in the northern hemisphere, implying major changes in marine food webs (Pauly et al. 1998).

Top predators play a role in a conservation and management context because they can be used as indicators of ecosystem shifts. Their demography and abundances can be directly influenced by the availability of their prey, which are themselves often driven by abiotic influences. However, the choice of top predators as proxies for ecosystem status depends upon a range of factors that include their functional or structural importance, their amenability to measurement, and the extent to which they integrate variability at other levels in the ecosystem (Reid et al. 2005). In the Southern Ocean, krill predominate in the diet of most top predators, and seabirds and seals are reasonable indicators of krill availability (Reid and Croxal 2001, Weimerskirch et al. 2003). However, they are generalists, and the manner in which the availability of alternative prey may affect the consumption of krill must be understood (Asseburg et al. 2006).

Like most top predators, dusky dolphins may play different roles in marine ecosystems, although they are still poorly known. The goal of this chapter is to briefly summarize what we know about their trophic ecology and to attempt to develop, within the limitations of the available data, an initial description of the trophic role of dusky dolphins in their ecosystems. To achieve this goal, several topics will be presented, including diet and feeding habits, relationships with other top predators, competition and kleptoparasitism, the role in the transmission of parasites, and finally their trophic relationships with fisheries.

DUSKY DOLPHINS AS PREDATORS

The feeding habits of dusky dolphins have been described for New Zealand (Gaskin 1972), and Antarctic and sub-Antarctic waters (Goodall and Galeazzi 1985). Würsig and Würsig (1980) studied duskies in a closed bay off the Patagonian coast of Argentina and gave the first detailed description of their cooperative foraging behavior when feeding on southern anchovies (*Engraulis anchoita*).

The 1990s saw a worldwide increase in research activities directed towards understanding interactions between marine mammals and fisheries. One of the results of these efforts was the recovery of carcasses of dusky dolphins incidentally caught in fishing gear off the Peruvian, Argentine, and New Zealand coasts. This source of samples allowed more detailed studies of feeding habits through the analysis of stomach contents (Cipriano 1992, McKinnon 1994, Koen-Alonso et al. 1998). These analyses rely on the identification of prey species using the remains found in dolphin stomachs, such as fish otoliths and bones, squid beaks, and fragments of crustacean exoskeletons.

In the following section we summarize dusky dolphin feeding habits in different regions, mainly based on diet composition. Although foraging behavior is another part of the predation process, it is presented elsewhere in this book (see especially Chapters 5 and 6).

Diet composition of dusky dolphins in Peru, eastern South Pacific

Samples of stomach contents have been analyzed from entangled animals in an artisanal fishery, mainly in gill nets set from dusk to dawn within the coastal upwelling zone off the Peruvian coast (McKinnon 1994). In addition, the stomach contents of six individuals landed by a purse seiner have been included (Table 3.1). Samples were collected during the summers and winters of 1985 and 1986, in several ports near Lima, including Pucusana, Ancon, and Cerro Azul, where the entangled dolphins were landed after being caught. The stomach contents were collected between 6 and 48 hours after capture. The relative importance of each prey species was calculated as percentage frequency of occurrence (%FO), percentage total number (%N), percentage of total wet weight (%W), and percentage of gross energy contribution (%GE) (see Table 3.2 for definitions).

TABLE 3.1 Summary of stomach content analyses from dusky dolphins by region

	Peru	Argentina	New Zealand
Sampling interval	1985–1986	1989–1994	1984–1988
Number of stomachs	142	26	26
Full	136	24	13
Partially full	0	1	4
Empty	6	1 (newborn calf)	9
Number of individual prey	9137	3702	17735
Number of prey species identified	10	8	8
Number of prey species unidentified	8 (fishes)	1 (fish)	At least 5 Myctophidae, 1 Carapidae
Number of species per stomach	No data	3.40 ± 1.19	No data
Number of items per stomach	No data	148.08 ± 137.91	845.24 ± 247.21
Estimated total prey weight per stomach (g)		2675.63 ± 1848.44	1124.96 ± 216.61
More frequent prey species %FO	E. ringens	L. gahi	Myctophids
More important prey species %N	E. ringens	E. anchoita	Myctophids
More important prey species %W	E. ringens	E. anchoita	No data

Anchoveta (*Engraulis ringens*), horse mackerel (*Trachurus symmetricus*), hake (*Merluccius gayi*), and Pacific sardine (*Sardinops sagax*) among fishes, and the Patagonian squid (*Loligo gahi*), as well as *Dosidicas gigas*, constituted the bulk of the prey consumed by Peruvian dusky dolphins (McKinnon 1994). Of these, anchoveta was by far the most important and dominant prey species in the diet, accounting for 92.5%N and 83.8%W (Figure 3.1a). No other prey species accounted for more than 5.1%W (*M. gayi*) or 2.5%N (*T. symmetricus*; McKinnon 1994). Anchoveta was also the most important prey species by gross energy contribution, with 87.3%, followed by *M. gayi* with only 4%. In terms of habitat utilization, several demersal–pelagic[1] species such as hake and squid were found in the stomachs. However, the high importance of anchoveta clearly indicates a diet dominated by a pelagic species.

TABLE 3.2   Classical indices used to describe and characterize the diet of a predator

Index name	Abbreviation	Description
Frequency of occurrence	%FO	This index provides a measure of how often a given prey species is consumed by a predator. It is calculated as 100 multiplied by the number of stomachs in the sample that contained a given prey species divided by the total number of stomachs analyzed
Percent number	%N	This index provides a measure of how abundant a given prey species is in the diet of a predator. It is calculated as 100 multiplied by the number of individual prey belonging to a given prey species divided by the total number of individual prey found in all the stomachs analyzed
Percent weight	%W	This index provides a measure of how dominant by weight a given prey species is in the diet of a predator. It is calculated as 100 multiplied by the total weight estimated for a given prey species divided by the estimated total weight of prey in all the stomachs analyzed
Percent gross energy	%GE	This index provides an estimation of the relative energetic contribution to the diet by a prey species. It is typically calculated like %W but where the weight contribution of each prey species is multiplied by its average energy density (energy per unit of weight)
Percent index of relative importance	%IRI	This is a composite index that combines three of the above indices. The IRI is initially calculated for each prey species as IRI = %FO(%N + %W). Once these IRIs for all prey species are obtained, the percent IRI (%IRI) is calculated as 100 multiplied by the IRI of a given species divided by the summation of the IRIs from all prey species

Diet composition showed little temporal variation. *T. symmetricus* only showed differences in the %FO between years, and the squid *L. gahi* showed differences between years and seasons (summer and winter) but did not exhibit consistent seasonal or annual patterns. Differences in diet between lactating and

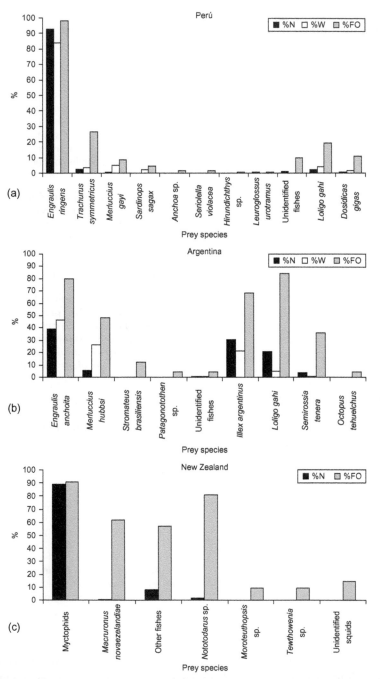

FIGURE 3.1 Relative importance of prey species in the diet of the dusky dolphin in different regions: (a) Peru, (b) Argentina, and (c) New Zealand. Importance was measured as %FO: percent frequency of occurrence, %N: percent total number, %W: percent total weight (see text for more explanation). %W is not available for New Zealand.

non-lactating mature females were not found, and the anchoveta was almost the exclusive prey species in both sexes (McKinnon 1994).

Individual prey were always below 30 cm in length and 300 g in weight. Mean total lengths for anchovetas consumed by dusky dolphins were 1.4–2.0 cm smaller than anchovetas taken by the fishery. However, the sizes caught by the fishery are possibly biased towards larger fish due to gear characteristics. Therefore, dusky dolphins in Peruvian waters appear to be opportunistic in their feeding habits, preying upon the size classes of anchoveta that are more abundant (McKinnon 1994).

Diet composition of dusky dolphins in Argentina, western South Atlantic

Samples were obtained from dolphins incidentally caught in the trawling fishery operating over the Argentine Patagonian Continental Shelf (Koen-Alonso et al. 1998). All samples were taken in the area between 42°00'S and 46°30'S and between the coastline and the 100 m isobath between 1989 and 1994 (Table 3.1). Dolphins were mainly caught in mid-water hauls for shrimp *Pleoticus muelleri* during the night (22 dolphins), as well as in bottom hauls for shrimp during the day (2 dolphins), but also by trawlers targeting Argentine hake *Merluccius hubbsi* (1 dolphin). Another dusky dolphin was of unknown origin. The sample was female-biased (7 males and 19 females) and included immature as well as mature individuals, although only one newborn calf (Dans et al. 1997b). Dolphins were frozen to −20°C after landing. Once landed, they were dissected and stomach contents were collected and stored frozen or in 70% alcohol. The measures of relative importance (%FO, %N, and %W) were calculated as described for the Peruvian samples. The index of relative importance (IRI) was also calculated, but the volumetric percentage was replaced by the wet weight percentage (Castley et al. 1991), and the index expressed as IRI = (%N + %W)%FO (Pinkas et al. 1971; see Table 3.2 for definitions).

The fish species identified in the diet were southern anchovy, Argentine hake, "pampanito" *(Stromateus brasiliensis)* and southern cod *(Patagonotothen ramsayi*; named *Nothotenia* spp. in Koen-Alonso et al. 1998). The cephalopod species were Argentine shortfin squid *(Illex argentinus)*, Patagonian squid, sepiolid *(Semirossia tenera)*, and octopus *(Octopus tehuelchus)*. Southern anchovy was the most important prey species (Figure 3.1b). It represented 39% in number and 46.4% in weight. However, the southern anchovy was not as dominant as the anchoveta in Peru, since hake and shortfin squid accounted for 26% and 21% by weight, respectively, while shortfin squid also accounted for 30% in number (Figure 3.1b). The most frequent prey species was the Patagonian squid, which occurred in 84% of stomachs (Figure 3.1b). The %IRI still indicated that the southern anchovy was the most important prey species, but both squids were also important items, followed by hake. For this reason, the diet of dusky dolphins in Argentina showed a higher contribution of demersal–pelagic items.

Mean length of every prey species was below 20 cm. Dusky dolphins preyed mainly upon pre-adult and adult southern anchovies, juvenile shortfin squid, and immature Patagonian squid (most of them smaller than 8 cm). The hake showed the widest range of individual sizes from juveniles of age 0 (8 cm) to adults (36 cm, ages 2–5). Mean individual wet weight of the important prey species ranged from 4 g to 92 g. The two most important prey species (southern anchovy and shortfin squid) had a mean individual wet weight of 21 g and 19 g, respectively. The heaviest individual prey estimated was a hake of 531 g.

Diet composition of dusky dolphins in New Zealand, western South Pacific

Samples came from gut contents of 24 dolphins caught incidentally in gill nets or found dead on shore, possibly as a result of entanglement in gill nets near Kaikoura, New Zealand (Cipriano 1992). Samples were obtained in fall, winter, and summer months of 1984–1988. The samples included males ($n = 11$), females ($n = 7$), nursing infants ($n = 5$) of both sexes, and one infant sized dolphin with fish lenses in its stomach (Table 3.1). The minimum number of individuals of each species was used to calculate the percentage contribution to total gut contents of each dolphin stomach (Cipriano 1992). For the purpose of comparing to Souh American data, %OF and %N were also calculated according to definitions in Table 3.2.

The most common prey was lanternfishes (family Myctophidae), which occurred in 90% of the stomachs, followed by squid (*Nototodarus* spp.) (88%) and hoki (*Macruronus novaezelandiae*) (Figure 3.1c). Other squid species, such as *Moroteuthopsis* spp. and *Teuthowenia* spp., were less frequent. The "other fish" category covered a wide range of species including: red cod (*Physiculus bacchus*), hake (*Merluccius australis*), hatchetfish (*Maurolicus muelleri*), *Bathylagus* sp., and unidentified fishes. Lanternfishes accounted for 88%N (14–100% of identifiable prey items in each stomach reported by Cipriano 1992). Squid only represented 2%N (0.2–100% of the identifiable prey in each stomach reported by Cipriano 1992). Hoki accounted for less than 1%N (0.2–21.3%) of prey items in each stomach. Lanternfishes are small (dry weight range 0.4–1.2 g), but contain a high concentration of energy (26.6 + kJ/g dry weight; Cipriano 1992).

Comparison among dusky dolphin populations

Among the three regions for which detailed stomach content analyses are available, dusky dolphins from Argentina showed the least diverse diet (9 prey species), while duskies from Peru had the highest number of species in the diet (18 species). See also data from southern Africa, by Best and Meÿer (Chapter 14). Even though sample size could have an effect on the total number of prey species identified, it is still interesting to note that the diet of dusky dolphins in New Zealand had a larger number of prey species than in Argentina (Table 3.1). In contrast with this higher diet richness, dusky dolphins from New Zealand had a higher number of prey items per stomach and a lower total weight of prey per stomach in comparison with Argentine dolphins (Table 3.1). This indicates that in Argentina, dusky dolphins tend to feed on larger prey items than in New Zealand.

The food of dusky dolphins in Peruvian as well as in Argentine coastal waters can be characterized as schooling, small to medium size, pelagic species. However, the importance of ecological groups of prey in the diet differed between these regions. In Peru, even though semipelagic or demersal species (such as hake (*M. gayi*) and squid) were found, the diet was strongly dominated by a pelagic species. In contrast, demersal components, such as squid and Argentine hake, represented 50% of total prey weight in the diet of dusky dolphins from Argentina. However, dolphins in Argentine waters mainly fed upon size groups of squid and hake that are expected to show pelagic behavior (Koen-Alonso et al. 1998).

These differences reflect differences in what is available in the marine community in each region, particularly in the pelagic domain, and not necessarily a selective behavior. In the western South Atlantic, dusky dolphins forage in shallow waters, where sea bottom depth is less than 200 m, due to the wide continental shelf where they occur. On the other hand, the continental shelf in the South Pacific, off Peru and Chile as well as New Zealand coasts, is narrower and then dolphins may be foraging in deeper waters, where sea bottom is at several thousand meters. Although dusky dolphins hardly ever seem to forage below 130 m (Benoit-Bird et al. 2004), species associated with the deep scattering layer become available to them at night in deep waters.

Due to the contribution of different species and sizes of fishes and squid to their diet, dusky dolphins prey on different trophic levels, from the typical pelagic and lower trophic level, such as engraulids, to large and upper trophic level squid (Pauly et al. 1998). The variation found in the contribution of these components in the diet from different regions suggests that different trophic levels for dusky dolphins could be expected. This is especially important for comparisons between dusky dolphins from New Zealand, which have a higher contribution of myctophids and other groups of the deep scattering layer, with dusky dolphins from Argentine or Peruvian waters, which have a very large contribution of pelagic fishes to their diet. These prey types correspond to different trophic levels (Pauly et al. 1998), but this interesting issue still remains to be explored because it would require better and comparable quantifications of dusky dolphin diet from different ecosystems.

DUSKY DOLPHINS AS COMPETITORS FOR FOOD RESOURCES

Competition requires that members of at least two populations rely on a common resource, and that the availability of this common resource is scarce enough that both (or more) populations cannot fulfill all their requirements at the same time. However, competitive relationships usually are far more complex than this, since populations are embedded in food webs, and most species have a diversity of food resources that they utilize. Therefore, a more complete assessment of the effects of one species on another would require the effects from direct as well as indirect interactions in the food web to be evaluated (Yodzis 1988, 1998, 2001).

The scientific assessment of competitive interactions is not an easy task. It requires confirmation that both populations actually share a critical resource, that the availability of that resource is a limiting factor for population growth, and that the utilization of that resource that competitors engage in controls resource availability. The first step towards a full assessment of competition is a comparative study of the use of food resources. This kind of approach would provide part of the puzzle that is necessary to complete an understanding of the relative role of each predator in the community, and to determine which predation links and species need to be considered for developing multispecies models.

A detailed study of the use of common resources among dusky dolphins and other top predators and potential competitors was carried out in the marine community of northern and central Patagonia. Information about diet and feeding habits of components of this marine community was mostly available for target species of the fishery (Angelescu 1982, Angelescu and Anganuzzi 1986, Menni 1986, Menni et al. 1986, Angelescu and Prenski 1987, Goldstein 1988, Prenski and Angelescu 1993, Ivanovic and Brunetti 1994, Crespo et al. 1997, García de la Rosa and Sánchez 1997, García de la Rosa 1998). This community is organized around the trophic system composed of southern anchovy, Argentine shortfin squid, and Argentine hake; these three species are key components in the diet of most analyzed predators, and at the same time are core fishery resources in the region (Angelescu and Prenski 1987).

More recently, a long-term research effort has described the feeding habits of several top predators of marine mammals and non-commercial fish species (Koen-Alonso et al. 1998, 2000a, 2001, 2002; Koen-Alonso 1999). Along with dusky dolphins, other top predators considered were the South American sea lion (*Otaria flavescens*), Commerson's dolphin (*Cephalorhynchus commersonii*), beaked skate (*Dipturus chilensis*), spiny dogfish (*Squalus acanthias*), and school shark (*Galeorhinus galeus*). All these species are part of the by-catch of the trawling fishery that operates on the Argentine continental shelf (Figure 3.2).

Diet studies of each predator species allowed for some distinguishing of intraspecific differences in food habits (Koen-Alonso 1999), such as differences between sexes, size groups, and maturity stage within a given species. For this reason, the comparative analysis among predators was carried out by considering each one of the identified subgroups within each biological species as a "trophic" species, or trophospecies. These trophospecies were males (OF_m) and females (OF_f) of the South American sea lion, dusky dolphin (LO), Commerson's dolphin (CC), immature (DC_{im}) and mature (DC_m) beaked skates, immature (SA_{im}), mature males (SA_{mm}), and mature females (SA_{mf}) of the spiny dogfish, and school sharks (GG; Table 3.3; Koen-Alonso 1999).

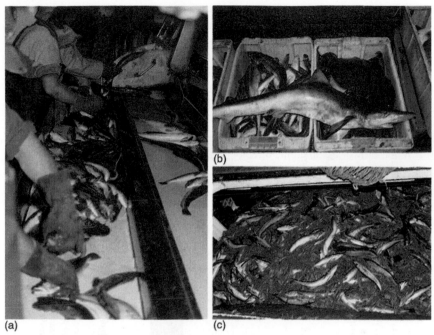

FIGURE 3.2   Sharks, skates, and other teleost fishes are part of the by-catch of hake and shrimp fisheries operating off Patagonia, Argentina, in the southwestern South Atlantic. (a) Selecting hakes for production on the right and discarding all other species including small hakes on the left, (b) sharks and skates by-caught in the hake fishery, (c) hakes and other fishes by-caught in the shrimp fishery.

Prey importance as well as the importance of ecological groups in the diet of each trophospecies were evaluated using a percentage index of relative importance, IRI (see Table 3.2 for definitions). Seventy prey taxa were identified in the diet of the entire sample set of trophospecies, but only nine showed IRI >5% in at least one trophospecies. These nine prey taxa were Argentine hake, Argentine shortfin squid, Patagonian squid, southern cod, southern anchovy, "raneya" (*Raneya brasiliensis*), red octopus (*Enteroctopus megalocyathus*), isopods, and ctenophores (Ctenophora).

Population trophic diversity (Hp) for each trophospecies was estimated by the Pielou's pooled quadrat method and based on the Brillouin diversity index. Population trophic diversity (Hp) showed a large range of variation among trophospecies (Table 3.3), although not all these diversities were different from a statistical perspective (Kruskal–Wallis test and paired comparisons; Siegel and Castellan 1995). Nonetheless, the analysis allowed for detection of a gradient of trophic diversity, where female sea lions showed the maximum and dusky dolphins showed the minimum values for Hp, respectively (Table 3.3). Possibly, Commerson's dolphin also has a low trophic diversity, but sample size was small and probably not representative of the population (Table 3.3). The gradient

TABLE 3.3 Population trophic diversity (Hp) of each trophospecies considered in the comparative analysis of the diet of several top predators of the Northern Patagonia marine community, Argentina

| | | | Trophospecies | | | | | | |
	OF Females	OF Males	DC Immature	DC Mature	SA Mature males	GG	SA Mature females	SA Immature	LO	CC
Hp	5.00	3.82	3.75	3.44	3.24	3.22	3.12	2.91	2.26	–
S	30	32	28	34	15	18	15	10	9	9
N	26	22	161	96	36	23	39	45	25	9

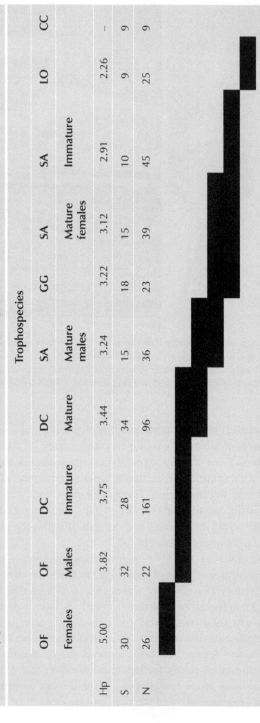

Predators are arranged by decreasing Hp and solid boxes link those predators whose Hp did not differ significantly.

S, number of prey species found in the predator; N, number of stomachs analyzed.

OF, Otaria flavescens; DC, Dipturus chilensis; SA, Squalus acanthias; GG, Galeorhinus galeus; LO, Lagenorhynchus obscurus; CC, Cephalorhynchus commersonii.

The population trophic diversity (Hp) was estimated using the pooled quadrat method of Pielou (1966) and based on the Brillouin diversity index. Before carrying out these estimations, cumulative diversity curves were constructed to determine if the sample size of each trophospecies was large enough. CC Hp could not be estimated due to sample size. The statistical comparison among Hp was done using a Kruskal–Wallis test and pairwise comparisons (Siegel and Castellan 1995).

observed among all top predators was positively correlated with the importance of benthic and demersal–benthic prey (Koen-Alonso 1999).

Trophic overlap among trophospecies was also quantified (Koen-Alonso 1999). Although there are many indices available, the specific overlap index (SO_{ij}) (Petraitis 1979, Ludwig and Reynolds 1988) was used in this case. This index is pair-wise (it calculates the overlap between two species) and asymmetric. This means that the overlap calculated from the perspective of one of the predators is different from the one calculated from the perspective of the other. For example, take two predators, one feeding on ten prey species and the other one on two, but only one of these two is part of the diet of the first predator. From the perspective of the first predator, its overlap onto the second one will be low because only one of its ten prey species overlaps with the second predator. However, from the point of view of the second predator, its overlap onto the first one would be expected to be much higher because half of its prey base is also used by the first predator (see also Chapter 15).

Dusky dolphin diet overlapped to some degree with male and female South American sea lions (SO_{ij} 0.619 and 0.543, respectively), mature female spiny dogfish, and school shark, although all these overlaps were significantly different from a complete overlap (Figure 3.3). From the perspective of the other predators, their overlaps onto dusky dolphin diet were negligible (Figure 3.3). These results suggest that the dusky dolphin trophic niche width is narrower than those of the other predators considered, and these results complement the low trophic diversity described previously.

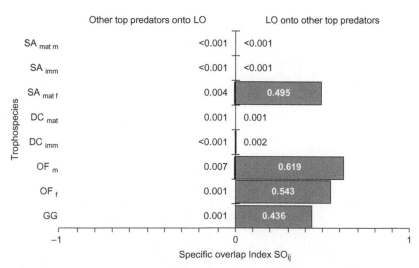

FIGURE 3.3 Specific overlap between the dusky dolphin and the other trophospecies considered top predators in the marine community of northern and central Patagonia, western South Atlantic. LO, *Lagenorhynchus obscurus*; SA_{mm}, mature male *Squalus acanthias*; SA_{im}, immature *Squalus acanthias*; SA_{mf}, mature female *Squalus acanthias*; DC_m, mature *Dipturus chilensis*; DC_{im}, immature *Dipturus chilensis*; OF_m, male *Otaria flavescens*; OF_f, female *Otaria flavescens*; GG, *Galeorhinus galeus*. Shaded numbers are non-significant.

Considering the importance in diet of prey by habitat, a cluster analysis allowed identification of three supersets of predators (Koen-Alonso 1999). These groups were related to the consumption of demersal–pelagic prey (OF_m, DC_m, SA_{mm}, SA_{mf}, and GG), high importance of pelagic prey (SA_{im} and LO), and high importance of demersal–benthic prey (OF_f and DC_{im}).

When size of prey shared by the set of predators is investigated, the dolphin species CC and LO ate the smallest squid but the pattern was more variable in the case of fish prey (Figure 3.4). Commerson's dolphins and immature skates ate the smallest hakes, while mature skates ate the largest. The other predators, including dusky dolphins, ate hakes of intermediate and similar sizes (Figure 3.4a). All predators, with the exception of CC, ate similar sizes of southern anchovies (Figure 3.4b). CC and LO also ate the smallest Argentine squid (Figure 3.4c) and LO ate the smallest Patagonian squid (Figure 3.4d), also suggesting that both dolphins are feeding on smaller and pelagic squid sizes.

Some generic patterns indicate that diversity in coastal waters tends to be higher than in deeper waters, and demersal communities tend to have higher diversities than pelagic ones (Margalef 1974, Laevastu and Favorite 1988, Woottom 1990). The available data indicate that these patterns are also present in the northern and

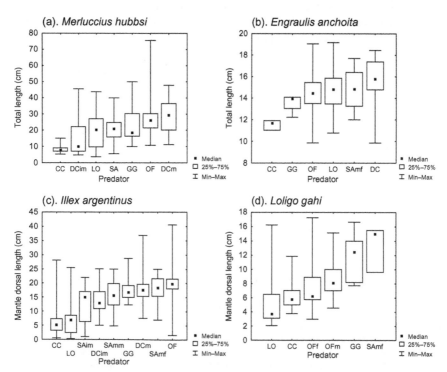

FIGURE 3.4 Sizes consumed by the dusky dolphin and other top predator trophospecies of (a) Argentine hake, (b) Argentine anchovy, (c) Argentine shortfin squid, and (d) Patagonian squid.

central Patagonian marine community (Angelescu and Prenski 1987). With these patterns in mind, the comparative analysis of these top predators suggests that trophic diversity reflects the prey diversity of the environments where the trophospecies feed. Hence, all these predators could be considered as generalist and opportunistic in the context of the environmental domain that they utilize as trophic habitat. The generally medium to low overlap values suggest that these top predators also use the same abundant and frequent trophic resources, but in different ways according to their trophic habitats. This differential use of common food resources could lead to a separation of their trophic niches, perhaps reducing the possibility of ecological competition among them.

MIXED-SPECIES FORAGING ASSOCIATIONS

Observations of foraging activity of dusky dolphins indicate that they increase prey availability for a number of other predators, including seabirds, other dolphins, pinnipeds, sharks, and predatory fishes. Pinnipeds, sharks, and predatory fishes engage in apparent kleptoparasitism, with the associated species taking advantage of dusky dolphin coordinated hunting efforts, which has little or no apparent advantage to the dolphins (Markowitz 2004).

Foraging interactions between dusky dolphins and seabirds have been noted in a number of locations, including Golfo San José and Golfo Nuevo, Argentina, Admiralty Bay and Kaikoura, New Zealand, and off the west coast and Cape coast of South Africa (Würsig and Würsig 1980, McFadden 2003, Markowitz 2004, Markowitz et al. 2004, Degrati et al. 2008, see also Chapters 2 and 6; Figures 3.5 and 3.6). Seabirds commonly associated with diurnal coordinated foraging groups appear to use dolphin foraging activity as an indicator of localized prey abundance. In the Marlborough Sounds, New Zealand seabirds most often accompanied dolphin groups actively feeding (96% of feeding groups were accompanied by birds compared with 57% of non-feeding groups), and were less likely to be feeding themselves when found in the absence of dolphin groups (McFadden 2003).

New Zealand fur seals (*Arctocephalus forsteri*) feed on both the small fishes herded by the dolphins and predatory fishes attracted to the foraging activity (McFadden 2003). In Kaikoura, where <1% of dusky dolphin groups were observed to engage in daytime foraging, 25% of interactions with fur seals occurred when dolphins were feeding; in the Marlborough Sounds, 96% of fur seal interactions with dusky dolphins occurred in a feeding context (Markowitz 2004). Common dolphins (*Delphinus delphis*) occasionally engage in apparently coordinated fish herding with dusky dolphins. South American sea lions took advantage of dusky dolphin herding behavior on southern anchovies in Patagonia (Würsig and Würsig 1980). Sea lions were recorded in 23% of a total of 169 groups of dolphins feeding (M. Degrati, personal communication). Consequently, dusky dolphin foraging efforts influence prey accessibility for a large number of other predators.

FIGURE 3.5 Dusky dolphins herding and preying on southern anchovies (*Engraulis anchoita*) in Golfo Nuevo, Argentina. Common gulls (*Larus dominicanus*) capture fish from the surface of water (photos from Marine Mammals Laboratory, CENPAT).

FIGURE 3.6 Dusky dolphins increase prey accessibility to other predators in mixed-species foraging aggregations. There are two photos, left and right, per figures a, b, and c. (a) Dusky dolphins and fluttering shearwaters, (b) dusky dolphins and Australasian gannets, (c) dusky dolphins and New Zealand fur seals (photos taken in Admiralty Bay, Marlborough Sounds, New Zealand, by T. Markowitz).

DUSKY DOLPHINS AS HOSTS IN PARASITIC RELATIONSHIPS

Parasitism is an ecological concept dealing with one kind of the wide range of relationships between organisms and their environment. The way of looking at parasitism and finding its place among ecological interactions is based on trophic relationships (Bush et al. 2001). Parasites can affect behavior, growth, fecundity, and ultimately mortality of their hosts. In turn, parasites, and particularly those transmitted through predation, have the potential to organize their host communities. These parasites can affect the relative abundance of different animal species in the same way that a top predator could, justifying the inclusion of parasitism as a biotic force capable of determining the biodiversity of communities (Poulin 1999).

Parasites also provide data on the population ecology of their host, such as natural mortality factors, stock identity, distribution and diet, social behavior, and spatial segregation (Perrin and Powers 1980, Dailey and Otto 1982, Dailey and Vogelbein 1991, Balbuena and Raga 1994, Balbuena et al. 1995, Aznar et al. 2001). They have been successfully used as biological stock indicators of host populations, under the assumption that different parasite communities would indicate that the host populations are (more or less) isolated (Dailey and Vogelbein 1991, Balbuena et al. 1995, Berón-Vera et al. 2001).

The parasites of dusky dolphins were studied off Chile and Peru (Van Waerebeek 1992b, Van Waerebeek et al. 1993), and some data from New Zealand are also available (Cipriano 1992). In Argentina, Dans et al. (1999) studied gastrointestinal helminths of dolphins incidentally caught in the trawl fishery off Patagonia. Comparison among populations or stocks of dusky dolphins is difficult because information is scarce and heterogeneous (Table 3.4). In most regions where dusky dolphin parasites were studied (Table 3.4), helminths were removed from the contents of stomachs and intestines, and the ventral zone was also examined for blubber cysts. However, liver, lungs, heart, and skulls were not systematically examined, and parasites were not identified at the species level in all regions. The terms "prevalence" and "mean intensity"conform with Bush et al. (1997).

In Argentina, one study included a sample of dusky dolphins incidentally caught in the trawl fishery within the Golfo San Jorge (between 45°00'S and 46°30'S and between the coastline and the 100m isobath), between 1990 and 1995 (Dans et al. 1999). Two individual strandings occurred in locations to the north (around 42°00'S and 64°00'W), in September and October 2005. The gastrointestinal helminth community seems to be depauperate, with low species richness. The communities are dominated by *Anisakis simplex* (Nematoda) and *Braunina cordiformis* (Trematoda, Brauninidae), although *Synthesium* spp. (Trematoda, Campulidae) also showed a high prevalence (Table 3.4). *Pholeter gastrophilus* (Trematoda, Troglotrematidae), *Corynosoma australe*, and *C. cetaceum* (Acanthocephala) may be considered as rare. *A. simplex* also showed the highest intensity of infection ($\bar{X} = 104.9 \pm 97.2$ SD), while *B. cordiformis*

TABLE 3.4 Species and prevalence (P%) of helminths found in dusky dolphins by region

Helminth species	Peru		Argentina[a]		New Zealand	
	P%	Location	P%	Location	P%	Location
Nematoda						
Anisakis simplex			100	Stomach/intestines		
Anisakis sp.	40	Stomach				
Unidentified					100	Stomach/intestines
Trematoda						
Braunina cordiformis	8.5	Stomach/intestines	84	Stomach/intestines	60	Stomach
Synthesium sp.1			52	Stomach/intestines		
Pholeter gastrophilus	8	Stomach	8	Stomach		
Unidentified					12.5	Stomach/intestines
Acanthocephala						
Corynosoma australe			4	Stomach		
Corynosoma cetaceum			4	Stomach/intestines		
Cestoda						
Phyllobotrium delphini	70	Blubber				
Phyllobotrium sp.					80	Blubber
Unidentified	3.7	Intestines				

[a]Northern and southern Patagonia combined.

and *Synthesium* spp. showed lower and similar intensities ($\bar{X}$ = 50.5 ± 45.5 SD, and $\bar{X}$ = 42.6 ± 51.0 SD respectively). Blubber in the ventral zone was also examined, although cysts of cestodes were not found.

Even though information is not fully comparable, some differences seem to exist with the New Zealand and Peruvian populations (Cipriano 1992, Van Waerebeek 1992b, Van Waerebeek et al. 1993). *Anisakis* spp. and *B. cordiformis* have been recorded in Peru but with lower prevalences of 40% and 8.5%, respectively (Table 3.4), and lower intensities (*Anisakis* spp. $\bar{X}$ = 2.96 ± 3.41 SD, *B. cordiformis* $\bar{X}$ = 2.17 ± 2.26 SD) than those observed in Patagonia. No campulids or acanthocephalans were reported from Peru, whereas cestodes occurred in intestines in 3.7% of the Peruvian dolphins but not in the Argentine ones. New Zealand dusky dolphins, most of which were collected off Kaikoura, were infected with unidentified nematodes in stomachs and intestines (100% of stomachs and 66% of intestines surveyed), *B. cordiformis* in stomachs (60% of stomachs), and unidentified trematodes in liver, intestines, and stomachs. Plerocercoid blubber cysts of cestodes were present in 80% of dolphins in New Zealand (presumably *Phyllobothrium* spp.) and 60% of dolphins in Peru (*P. delphini;* Table 3.4).

Blubber cysts of cestodes, *Phyllobothrium* spp. in New Zealand and *Phyllobothrium delphini* in Peruvian and South African populations, were not found in Argentine dusky dolphins. However, in Argentina, blubber cysts were found in the genital region in other pelagic or deep diving dolphins. Qualitative differences presented here may indicate the segregation between the Peruvian and Argentine dolphins, while presence or occurrence may be indicating the exploitation of different ecological systems. Additional data on the host's biology and parasite faunas are needed to reach conclusions about the stock identity of dusky dolphins in South America, as potentially indicated by parasites.

In Patagonia, Argentina, dusky dolphin parasite communities were also studied along with a set of other top predators in order to describe potential transmission routes through prey (Dans et al. 1999; Berón-Vera et al. 2007, in press). This approach is a first step in incorporating parasites into food webs. Then we can add parasites as nodes in already existing food webs, or we may try to build food webs around systems for which we already have a good understanding of infectious processes (Lafferty et al. 2008). Dusky dolphins, as well as other marine mammals of this marine community, feed upon the most important and abundant prey present within the environment, many of which are commercially important. Consequently, they may play a relevant role in the transmission of parasites that affect those commercial species that are important prey of these predators, such as Argentine hake and Argentine shortfin squid.

Within a regional perspective, the parasites that infect dusky dolphins along Patagonia change with latitude along the continental shelf. From the sample collected within Golfo San Jorge, a total of 3936 helminth individuals belonged to five species: *A. simplex* s.l., *B.cordiformis*, *P. gastrophilus*, *Synthesium* sp., and *C. australe*. The same species were found in the two individuals stranded in northern Patagonia, with the exception that *C. cetaceum* was found instead of *C. australe.* All the individuals of *A. simplex* were adults in the south, whereas third-stage larvae appeared in the north. This fact is likely related to differences in host diet and may be reinforced by latitudinal differences in parasite infections of intermediate/paratenic hosts (Timi and Poulin 2003). Larval *A. simplex* infected dusky dolphins from northern Patagonia, in contrast to the adult stages present in those from Golfo San Jorge. Timi (2003) and Sardella and Timi (2004) observed an increase in frequencies of *A. simplex* in southern anchovy and Argentine hake at higher latitudes. This species rarely parasitizes fish in Buenos Aires province and northern Patagonia (between 35°S and 42°S), while it is common in hosts south of 43°S. According to this, we may consider the southern anchovy as the major paratenic host in its transmission to dusky dolphins. *C. cetaceum* is possibly associated with intermediate hosts (i.e., prey) besides host–parasite specificity in the nearby neritic area of Buenos Aires Province. Among potential prey, the South American striped weak fish, *Cynoscion guatucupa,* harbors juvenile *C. cetaceum* (Timi et al. 2005), so it could be assumed that the cetaceans become infected while consuming this prey.

As a common pattern in dusky and other dolphins, the mean number of species taken as prey is just above one-half the component species richness (Dans et al. 1999, Berón-Vera et al. 2001). Low species richness and diversity are characteristic in these dolphins, and may be directly related to their narrow diets. In general, these marine mammals concentrate only on a few most abundant prey species out of their overall diet, while other species rarely appear. This may directly reflect parasite aggregation in the preys consumed.

DUSKY DOLPHINS AS PREY

The natural predators of dusky dolphins are sharks and killer whales. The effect of this predation on the behavior of dusky dolphins is just receiving attention, and not much is known. From a trophic ecology standpoint, there are not enough quantitative data and predation rates are not available. Therefore, the importance of dusky dolphins as prey on predator populations as well as the importance of mortality due to predation on dusky dolphin populations cannot be assigned. This subject is particularly important since the effect of predation on population dynamics is well known in some pinnipeds, and in some cases predation might represent an important source of mortality and a regulation factor in population growth (see Chapter 7 and Estes et al. 2006 for further detailed discussions).

As part of a study of feeding habits of top predators of the Argentine marine community, dolphin body pieces such as flukes and tails were found in stomach contents of broadnose sevengill sharks (*Notorynchus cepedianus*) collected along the Patagonian coast (42–46°S; Crespi et al. 2004). Sharks were also observed on several occasions when dolphins were feeding in Golfo Nuevo (Mariana Degrati, personal communication) as well as in Golfo San José (Würsig and Würsig 1980). It is also known that killer whales prey on dusky dolphins (Würsig and Würsig 1980). In New Zealand, off Kaikoura, relatively high levels of predation were documented. During an observation period of 11 days, killer whales were observed to chase dolphins on 4 days, resulting in seven captures (Constantine et al. 1998). During this study, it was documented that killer whales employed a strategy of sneaking up behind tour boats in order to capture dolphins.

Information about dusky dolphin responses to predation is summarized in Chapter 7. Generally, it appears that the threat of predation has a great effect on grouping, ranging, and behavioral patterns of dusky dolphins in New Zealand. However, predation *per se* is a relatively rare event in this region, suggesting that dusky dolphin anti-predator strategies are quite effective.

DUSKY DOLPHIN INTERACTIONS WITH HUMANS

Humans are also large predators in marine ecosystems, and their relationship with dusky dolphins can be approached in the same way that other predators are studied. Dolphins have been fished for human consumption or incidentally caught in fishing gear, so we can study those events as a predator–prey relationship. Dolphins also feed upon species that are the target of fisheries, so we can explore the possibility of competitive relationships. Furthermore, in the same way that indirect effects traveling through a complex food web can give unexpected results when analyzing the net effect that one species may have on another, the same applies when considering the reciprocal effects between humans and dusky dolphins.

In terms of predation, for over three decades dusky dolphins have been taken directly in the multi-species small cetacean fisheries off Peru. These animals were one of the only sources of red protein for human consumption for low-income Peruvian families. Protection measures came into place in 1990 but were not well enforced and harvest continued. At present, illegal dolphin hunting at a lower level than before is still going on, together with incidental entanglement in fishing gear. Dusky dolphins are caught in a wide variety of gear, ranging from simple artisanal nets to industrial trawlers (Crespo et al. 1994, 1997, Dans et al. 1997, Van Waerebeek and Würsig 2002). The impact of incidental mortality on dusky dolphin populations can be significant even if the numbers of dolphins caught do not appear to be particularly high. For example, an average of 70 dolphins per year was the estimated rate of incidental mortality in Argentine waters during the 1990s (Crespo et al. 1997, Dans et al. 1997a). Simulations of the dynamics of the Argentine population of dusky dolphins suggest that the estimated mortality rate could be already too high or in the upper limit of what could be considered acceptable from a conservation perspective (Dans et al. 2003a; see also Chapter 11).

In terms of potential competition with fisheries, the simplest exploration is to investigate the overlap of major dolphin prey with those that are also important for fisheries. For example, in Kaikoura, New Zealand dusky dolphins are largely dependent on myctophid fishes, and to a lesser extent on squid and hoki. Although there is no commercial fishery for myctophids, squid (including the arrow squid *Nototodarus* spp. and *Todarodes* spp.) are fished commercially in the waters south of the island. Hoki is fished both recreationally and, to a lesser extent, commercially in Kaikoura. In the Marlborough Sounds, there is some potential for trophic overlap of dusky dolphins with the pilchard fishery, as pilchard (*Sardinops neopilchardus*) has been noted to be a frequent prey item in behavioral studies of foraging dusky dolphins (Markowitz et al. 2004). Inter-annual variation in dusky dolphin abundance and distribution in the Marlborough Sounds appears to be closely linked to prey availability (Markowitz and Würsig 2004). In this region, more complex indirect interactions may also take place between dusky dolphins and aquaculture. Green-lipped mussels (*Perna canaliculus*) are extensively farmed, which can produce a decrease in local phytoplankton concentration, and in turn influence the local abundance and distribution of small fishes. Thus, presence of dusky dolphins could be a plausible indicator for the potential effects of fishing and aquaculture practices on ecosystem health in the region (Markowitz et al. 2004).

Another example of potential competitive interactions between human activities and dusky dolphins can be found in Patagonia (Crespo et al. 1997, Dans et al. 2003b). From the comparative study of the diet of several top predators of this marine community, it was clear that Argentine hake and Argentine shortfin squid were both important prey species as well as the main target of the fisheries. Southern anchovy, the main prey for dusky dolphins, is also a commercial species, but to a much lesser extent. Its importance in the trophic sense is largely as a key prey for commercial species, as well as for many other top predators (Figure 3.7).

The role of southern anchovy in the Argentine marine community clearly exemplifies how convoluted the interactions through food webs can be. From the simpler perspective of competition and linear food chains, if the fishery exploits species that feed upon southern anchovy, one could expect that the removal of these species by the fishery would reduce predation pressure on the southern anchovy stock, and hence, favor a higher availability of anchovy. If this actually occurs, then fisheries could be enhancing the food supply for anchovy eaters such as dusky dolphins. However, the fishery also produces mortality of dolphins through incidental catches. So, what could be the net balance of these two opposing phenomena?

A first approach to the question of gains and costs of fisheries overlap was attempted by using a simple and preliminary multispecies model (Koen-Alonso et al. 2000b, Dans et al. 2003b). The simplified food web considered in this exploration included three top predators (beaked skate, southern sea lion, and dusky dolphin), southern anchovy, Argentine shortfin squid, and hake, and the catches of the fisheries (subset of the food web depicted in Figure 3.7). When only the fisheries catches of hake, squid, and anchovy were considered, this preliminary study suggested that dusky dolphins would show a fluctuating positive trend, where fluctuations were directly associated with the strong link between dolphins and anchovy. Anchovy, like many other forage fishes, has an intrinsically highly fluctuating population dynamic. However, when incidental

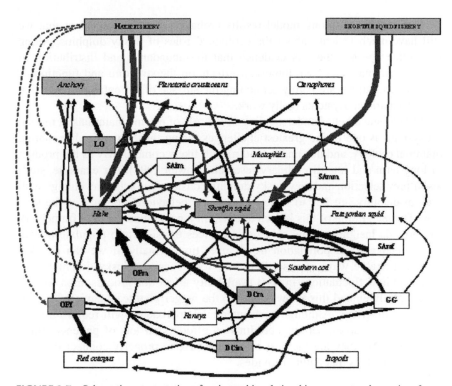

FIGURE 3.7   Schematic representation of main trophic relationships among trophospecies of
the marine community in northern and central Patagonia, southwestern South Atlantic. Red lines
indicate main predation relationships of fisheries (broken arrows represent marine mammal
incidental catch), blue lines indicate predation relationships of main target, hake, and shortfin
squid. The width of the line represents the magnitude of the relationship for the predator. Gray
shadowed boxes represent predators considered in the system modeled by Koen-Alonso et al.
(2000b) (Dans et al. 2003b).

mortality of dolphins was included in model simulations, the results indicated
that the estimated annual catches always produce a significant negative trend,
reducing about 20% of the population size in 50 years of simulation.

Unfortunately, the scarcity of data has prevented the inclusion of dusky dolphins in more sophisticated model development. For example, Koen-Alonso
and Yodzis (2005) implemented another multispecies model for key species
of the Patagonia marine community, but in this case only South American sea
lions were included at the top of the tritrophic system constituted by hake,
anchovy, and squid. Nonetheless, one interesting suggestion from this later
modeling work is that the hake fishery may have a negative effect on the
anchovy stock. If the counterintuitive effect suggested by this later model is
real, then the different impacts of fisheries instead of being opposite as we
speculated above, could actually proceed in the same direction and affect dusky
dolphins in a negative manner.

Contradictions among model results probably arise from the fact that we still have much to learn about the ecological roles of dusky dolphins. Along with other species, there is evidence that the abundance and distribution of marine mammals can have important effects on the structure and function of ecosystems, but we are barely scratching the surface of the entangled question of how natural ecosystems really work.

Developing a better understanding of the role of marine mammals in marine ecosystems is one of the greatest challenges facing those interested in marine mammal ecology, and this understanding will likely come slowly as the product of long-term and interdisciplinary research (Bowen 1997). For example, studies of prey selection and how dusky dolphin consumption rate may change with prey availability are key to understanding the behavioral ecology of the species. The integration of this knowledge within a population ecology framework can provide insight on how behavioral choices translate into population-level energy budgets and population trends over time. Finally, if we can develop some reasonable ways of describing how food becomes dusky dolphins over time, and how fluctuations in food availability affect this process, then we can meaningfully embed this knowledge into the bigger picture as depicted by our food web diagrams. Although such integrated approaches will require a large amount of resources, serious coordination efforts, and lots of patience, time, and hard work, they are the best means we have to properly connect the dots between the clues that nature provides.

ACKNOWLEDGMENTS

The authors are indebted to the many people who made this long-term research on marine fisheries interactions possible. Institutional support was given by the Centro Nacional Patagónico (CONICET) and the University of Patagonia. Work was funded partially by the Research Council of the University of Patagonia, CENPAT, Agencia Nacional de Promoción Científica y Tecnológica. The authors are also indebted to Néstor García, Susana Pedraza, Mariano Coscarella, A. Schiavini, Natalie Goodall, and R. González for their support. The fishing companies Harengus S.A. and Alpesca S.A. allowed observers to work on board fishing vessels, collect samples and retrieve marine mammal carcasses.

NOTE

1. The ecological groups considered were benthic (the prey species lives on the bottom), demersal (the prey species lives near the bottom but not linked to it) and pelagic (the prey species lives in the upper layers of the water column without relationship with the bottom). Demersal prey species were divided into demersal–pelagic (the prey has a diel vertical migration pattern, dispersing in the water column during the night and remaining close to the bottom during daylight hours) and demersal–benthic (the prey species does not migrate vertically).

Chapter 4

Acoustics of Dusky Dolphins (*Lagenorhynchus obscurus*)

Whitlow W.L. Au,[a] Marc O. Lammers[a], and Suzanne Yin[b]

[a]*Institute of Marine Biology, University of Hawaii, Kailua, Hawaii, USA*
[b]*Hawaii Marine Mammal Consortium, Kamuela, Hawaii, USA*

It is still pitch black at five in the morning as I groggily stumble out of my field home "Muritai" (which means "Sea Breeze" in Māori) to check the weather. I know you can't blindly trust the forecast from the night before; weather conditions can change especially quickly where mountains meet the sea. I'm met by no wind, yay, and a clear night sky, full of a wealth of southern constellation stars. It looks and feels like it will be a good data day.

Rochelle Constantine (whom most of us call "Brave," because she is), a long-time colleague from the University of Auckland, rouses the rest of the team, composed of five Earthwatch volunteers who have come to work with us. While the volunteers take turns in the two bathrooms and Brave brews coffee, I check the field equipment, making sure that camera, hydrophone, GPS monitor, and sound recorder are stocked with fresh batteries, and that binocs and data sheets are ready to go.

Once we've packed our bags with lunch, water, and the field gear, we split up. Brave takes the car and two volunteers 20 km south to Omihi to observe the behavior of the dolphins from the shore station. I take a crew of three out on our little inflatable boat, hopefully to take photo-ID shots and get some underwater recordings of their communication and echolocation sounds. I've been coming to Kaikoura on and off since 1994, to learn more about dusky sounds and to investigate potential long-term effects of tourism. This stint is for my Master's degree, so I'm particularly dedicated to the work.

It takes only 15 minutes on the water to find a dusky group of about 30 animals, five or so kilometers from the slipway. The visibility is excellent, and we spot them even 1 km away. I am perpetually struck by the beauty of these dolphins, with their striking coloration of gray, white and black, their blunt black-tipped rostrum, their fast surfacing ways. We turn off the engine, write down the GPS position and observational conditions, lower the hydrophone into the water, and start recording.

We immediately obtain clear and noise-free recordings of social communication burst pulses and (largely echolocation) click trains, that are a cacophony of squeaks and squawks that make up a good portion of dusky dolphin vocal repertoire. They sound somewhat like a dinner party of people where everyone is talking at the same time, but through buzzing kazoos, that little toy-like membranophone. We are particularly interested to discover whether dusky dolphins are frequent whistlers. Our impression is that they are not, but we need more evidence.

The small group moves inshore, and soon their calls are obscured by the whooshing sounds of shore-break and the staccato clicks of snapping shrimp. We pull the hydrophone out of the water, start up the outboard and head south to Black Rock, where Brave has been tracking the main group of a bit over 300 dolphins. Here we try another acoustic pass and again, I hear the familiar sounds of squeaks, squawks and clicks—but no whistles.

Suddenly, I simultaneously hear loud rubbing and crunching noises, and quickly pull the headphones off my ears to see a tautly-stretched cable under the boat. A dusky has grabbed the hydrophone element in its mouth and is slowly swimming away with it, while several other dolphins are rubbing on the hydrophone cable. We often see duskies carrying strands of seaweed on bodies, passing it from their rostrum, to pectoral fins down to fluke. They seem to make a game of it, playing "keep away" from other dolphins. This time they are playing with the wrong toy!

I gently take hold of the cable to put just a little bit more tension on the line, as I know that if I pull back too quickly or too hard, the dolphin will bite down on the element, sending salt water into highly sensitive electrical parts. With persistent but gentle tugging, I slowly retrieve the hydrophone cable, coiling it back into the research vessel. There are clear little indentations in the black acrylic of the hydrophone element where the dolphin held it in its mouth. But it did not bite through!

I wait a few minutes and then deploy the hydrophone again, hoping that the mischievous dolphins have passed and I can continue recording. As I resume monitoring, the clear and unmistakable sound of a whistle, and then many whistles, overlapping over and over again, fill my headset. I jerk my head upwards and scan the group. Sedge, one of the Earthwatch volunteers, calls and points to a group of six dolphins with triangularly shaped dorsal fins and long, prominent rostrums, quite unlike the blunt ones of duskies, slowly traversing our bow. Common dolphins! Yes! When we hear lots of whistles, we always find commons mixed in with the duskies. But if we do not hear any whistles, we don't find commons. Teasing apart what species of dolphin is making what sound is a tricky endeavor but we now have a bit more evidence for our theory that common dolphins may be the major source of whistles.

We spend about 30 minutes more, recording and taking photographs, but quickly stow our gear when we feel the wind start to freshen, shift direction, and small whitecaps appear around the boat. This is the afternoon seaward breeze that often kicks up in summer, and it is time to head back to harbor.

When we get home, there will be cleaning of equipment, labeling and storing of gathered data, hot showers and Brave's famous spaghetti; along with the sharing of the day's events with the shore team, who have experienced the dolphins from a very different perspective. After that, an early night to bed as the weather forecast sounds good for tomorrow as well!

Suzanne Yin, Muritai, Kaikoura, South Island, New Zealand, January 17, 2001

INTRODUCTION

Acoustic energy propagates in water more efficiently than almost any form of energy. Therefore, acoustics offers the most effective method for an animal to perform various functions such as communication, navigation, obstacle avoidance, and prey avoidance and detection. The dolphins' dependency on acoustics is probably the strongest influence in the evolution of the acoustic processing centers within the dolphin brain (see Ridgway 2000). According to Ridgway, the hypertrophy of the auditory system may be the main reason for the dolphin's large brain. The parts of the dolphin brain that are homogolous to areas of the human brain associated with acoustic processing are about seven times larger in dolphins (Hillix and Rumbaugh 2004).

Most odontocetes emit three fundamental types of sounds: whistles, burst pulses, and echolocation clicks (see Au et al. 2000). Dusky dolphins (*Lagenorhynchus obscurus*) are among this group of odontocetes that emit the three basic types of signals. There are some species of odontocetes that do not emit whistles; these are from the family Phocoenidae (harbor porpoise *Phocoena phocoena*, finless porpoise *Neophocaena phocaenoides*, Dall's porpoise *Phocoenoides dalli*), and from the genus *Cephalorhynchus* (Commerson's dolphin *Cephalorhynchus commersonii* and Hector's dolphin *Cephalorhynchus hectori*). Rounding out this list of non-whistling odontocetes is the largest, the sperm whale (*Physeter macrocephalus*).

There is some evidence that beaked whales can emit whistles (Caldwell and Caldwell 1971, Lynn and Reiss 1992, Rankin and Barlow 2007); however, this family of odontocetes has not been studied much. We will return to the topic of whistling and non-whistling odontocetes after we discuss dusky dolphin acoustic emissions.

We are aware of only five peer-reviewed references on dusky dolphin acoustics; three by the current authors (Yin 1999, Yin et al. 2001, Au and Würsig 2004), and two others, by Wang (1993) and Wang et al. (1995). We summarize these and some of our latest information here.

WHISTLE SIGNALS

Wang et al. (1995) analyzed recordings collected in 1988 and 1992 during the austral summer of New Zealand, and recordings from Argentina collected in 1975, 1976, and 1986. Group sizes ranged from fewer than 12 up to 400 dolphins. The recording system had an upper frequency limit of 20 kHz. Measurements of whistle parameters were taken of 491 whistles from New Zealand and one whistle from Argentina, and compared with six other species of dolphins. Wang et al. (1995) suggested that whistle contours resembled those of bottlenose dolphins (*Tursiops truncatus*), with stereotyped loops. Because the overwhelming number of whistles collected were from the apparently dusky-only groups in New Zealand, the authors theorized that whistles were important for dusky social communication in New Zealand, but not for duskies in Argentina, where whistles were rarely heard.

A second study looking at dusky whistles was initiated as part of a long-term study on the behavioral ecology and effects of tourism on dusky dolphins off Kaikoura, South Island, New Zealand (Yin 1999, Yin et al. 2001). We recorded vocalizations of dusky dolphins from small research vessels (<6m in length) and larger swim-with-dolphin tourist vessels (>12m in length) over four seasons (1995–1997, 2001). While recordings of duskies were collected between November and April, the majority of recordings focused on the period from January through June, corresponding to austral summer and fall seasons. Group sizes for duskies off Kaikoura range from small groups of 3–20 animals, to large groups of over 1000. Generally, one "main group" consisting of approximately 200–300 duskies, is found in this area. Small numbers (<30) of common dolphins (*Delphinus delphis*) are frequently found in association with this large group of duskies.

When we recorded sound emissions from dusky-only groups and from mixed groups of dusky and common dolphins, very few whistles were recorded from the dusky-only groups, while the majority of whistles were recorded from mixed groups of dusky and common dolphins (Figure 4.1, Table 4.1). Despite the fact that many more hours were spent recording dusky dolphins (39 hours), only 21 whistles were recorded (Table 4.1).

The time–frequency structure (spectrogram) of a dusky whistle is shown in Figure 4.2. The mean and standard deviation of the whistle parameters for dusky dolphin whistles are shown in Figure 4.3. From this latter figure, we conclude that dusky dolphin whistles generally vary between 7 and 16kHz, a frequency span of 9kHz, with durations between 0.29 and 0.76 seconds. Frequency measurements (start, end, high and low frequencies, and duration) are significantly different from whistles recorded from groups containing only dusky dolphins (Table 4.2).

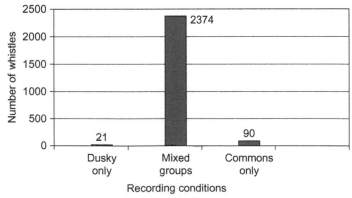

FIGURE 4.1   Number of whistles by type of dolphin groups. Note that we recorded dusky-only groups almost three times longer than mixed groups, but obtained less than 1% of the total number of whistles (from Yin et al. 2001).

TABLE 4.1 Summary of effort in recording whistles in the presence of dusky dolphins (*Lagenorhynchus obscurus*), common dolphins (*Delphinus delphis*), and mixed groups of dusky and common dolphins

	Hours of recording	Number of whistles	Whistles per minute
Dusky only	39.4	21	0.01
Mixed groups	9.0	2374	4.40
Common only	0.3	63	3.15

There was unequal effort, with over six times more dusky-only recordings, but without a subsequent increase in whistles. Rates of whistles per minute were also substantially lower for groups containing only dusky dolphins. Whistle rates are not standardized by group size; however, group sizes for both common and dusky groups were approximately 200–300 animals.

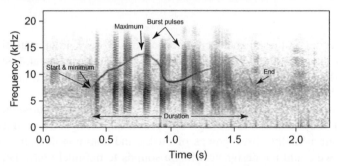

FIGURE 4.2 A spectrogram of a dusky dolphin whistle. Note the presence of burst pulses recording simultaneously with the whistle. The start frequency and minimum frequency are the same in this case, but this is not generally true for all whistles (from Yin, 1999).

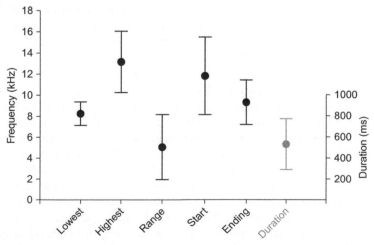

FIGURE 4.3 Mean and standard deviation of the whistle parameters associated with the whistles of dusky dolphins (after Yin 1999).

TABLE 4.2   Mean values of whistles measured in the three recording conditions

	Lowest frequency (kHz)	Highest frequency (kHz)	Frequency range (kHz)	Duration (s)	Start frequency (kHz)	End frequency (kHz)
Dusky only	8.4	13.3	4.9	0.5	12.0	9.6
Mixed groups	8.7	15.6	6.9	0.8	12.4	13.0
Commons only	9.7	15.8	6.0	0.8	12.2	12.8

S. Yin, unpublished data.

Differences in these parameters between dusky-only group with the other groups again suggest that the data of Wang (1993) and Wang et al. (1995) may have overlooked the presence of common dolphins, attributing (many or most or all?) whistles made by common dolphins to duskies.

The raw tape data from the Wang et al. (1995) study were subsequently re-examined and we found that all groups previously designated as dusky-only groups were actually mixed groups of dusky and common dolphins, and that therefore we could not designate whistle sounds to definitely have been made by dusky dolphins.

Recordings from Argentina that were also part of the Wang et al. (1995) study did not have common dolphins present, and only one whistle was recorded. Furthermore, more recent recordings from Argentina from 1998 revealed no whistles from small groups of duskies (K. Dudzinksi, unpublished data). However, whistles were recorded from mixed groups of duskies and commons (K. Dudzinksi, unpublished data).

Furthermore, whistles produced by dusky dolphins have distinct differences from those produced by common dolphins (Figure 4.4). In general, dusky whistles were simpler in contour, showing fewer loops and less repetition (Figures 4.2 and 4.4). Only once did we record more than two whistles from a dusky-only group during a recording period. From the contours of these spectrograms, at least four whistles had distinctly different contour shape (Figures 4.2 and 4.4a,b,c). The contours in Figure 4.4b and c are very similar. Unfortunately, we do not have more data that would allow us to define the whistle repertoire of dusky dolphins.

Based on these results, we believe that whistles do not play a major role in the communication system of New Zealand dusky dolphins. The acoustic behavior of this population appears to be more similar to that of dusky dolphins off Argentina.

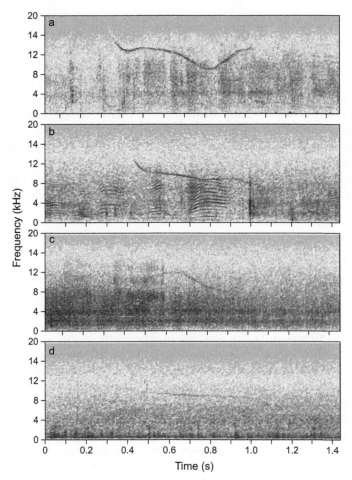

FIGURE 4.4   Four additional spectrograms of dusky dolphin whistles.

BURST PULSE SIGNALS

The presence of burst pulses is obvious in the spectrograms of Figures 4.2 and 4.4. Although burst pulses were not specifically studied for dusky dolphins, our experience in the field indicates that burst pulse signals are emitted regularly by duskies. The primary difference between echolocation clicks and burst pulse clicks is the number of clicks produced per unit time (Lammers et al. 2003a, 2003b) and difference in amplitude (Au et al. 1987). Burst pulse signals are broadband click trains similar to those used in echolocation, but with inter-click intervals of only a few milliseconds (0.5–10 ms) (Lammers et al. 2003a, 2003b). This criterion for defining burst pulses was based on previous efforts to examine patterns of click production in spinner dolphins (*Stenella longirostris*) by Lammers et al. (2003a, 2003b). The interclick intervals had a bimodal distribution separated at around 10 ms. This bimodal distribution was interpreted as being indicative of two distinct patterns of click train production: a burst pulse pattern with interclick intervals consistently less than 10 ms and a sonar click train pattern with intervals greater than approximately 15 ms. Because interclick intervals associated with burst pulses are considerably shorter than the processing period generally associated with echolocation, and because they are often recorded during periods of high social activity, these click trains are believed to play an important role in communication (Popper 1980, Overstrom 1983, Herzing 1988).

Echolocation clicks are used for sensing the surrounding environment, so they are generally emitted only after the echo of the previous click has been received by the dolphin, plus a short delay time.

ECHOLOCATION SIGNALS

Characteristics of echolocation signals

Echolocation no doubt plays an important role in the acoustic ecology of dusky dolphins, especially when foraging at night for relatively small prey over the Kaikoura Canyon. Dusky dolphins in other parts of the world, including other parts of New Zealand, also prey on schooling fish during daylight hours (Würsig and Würsig 1980), no doubt utilizing their echolocation capabilities. However, there is very little information on the structure and form of echolocation signals utilized by dusky dolphins, except for the reference of Au and Würsig (2004).

The echolocation characteristics of delphinid species have been studied primarily in captivity (Au 1993). Measurements from stationary dolphins in captivity have shown that echolocation clicks are emitted in a directional beam and signals measured off-axis are distorted with respect to the signals measured along the major axis of the beam. Therefore, it is difficult to obtain accurate measurements from free-ranging fast-moving dolphins in the wild. Another complicating factor is associated with the broadband nature of echolocation signals, and the possibility that the center frequency of echolocation clicks may vary with the intensity of the emitted clicks. Au et al. (1985) found that the higher intensity clicks emitted by a beluga whale (*Delphinapterus leucas*) in Hawaii had higher frequencies than the lower intensity clicks used by the same animal in San Diego Bay. Au et al. (1995) also found a nearly linear positive relationship between the center frequency and source level echolocation clicks produced by a false killer whale (*Pseudorca crassidens*). Therefore, any measurements of the spectra of echolocation clicks should be accompanied by source level estimates.

Au and Würsig (2004) used a four-hydrophone array with the hydrophones arranged as a symmetrical star to measure the echolocation signals of dusky dolphins near the Kaikoura Peninsula, New Zealand. Such a sensor geometry was used successfully by Aubauer (1995) in tracking echolocating bats in the field and by Au and various colleagues (Rasmussen et al. 2002, Au and Benoit-Bird 2003, Au and Herzing 2003) in measuring echolocation signals of dolphins. The array structure resembled the letter "Y," with each arm 45.7 cm in length and separated by an angle of 120° (Figure 4.5). The arms of the array were constructed out of plastic (PVC) pipe (1.27 cm outer diameter) with a spherical hydrophone connected to the end of each pipe and the cable running through the center of the pipe. Another hydrophone was connected at the geometric center of the "Y."

The PVC pipes fit into holes drilled into a 2.54-cm-thick plastic (Delrin) plate. The range of a sound source can be determined by measuring the time of arrival differences between the signal arriving at the center hydrophone and the three other hydrophones. If the arrival time difference between the center and the other hydrophones is denoted as τ_{0i}, where $i = 1, 2$, and 3, then the range, R, of the source can be expressed as:

$$R = \frac{c^2(\tau_{01}^2 + \tau_{02}^2 + \tau_{03}^2) - 3a^2}{2c(\tau_{01} + \tau_{02} + \tau_{03})} \tag{1}$$

where c is the speed of sound, and a is the distance between the center hydrophone and the three other hydrophones (Au and Herzing 2003).

Echolocation signals were digitized with two Gage-1210, 12-bit dual simultaneous sampling data acquisition boards that were connected to a "lunch box" computer via two EISA slots. The data acquisition system operated at a sample rate of 500 kHz with a pre-trigger capability. When the computer signaled the Gage-1210 to collect data, four channels of acoustic signals were simultaneously

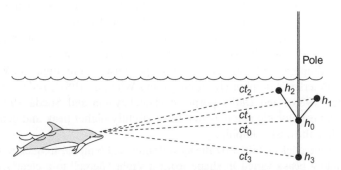

FIGURE 4.5 Schematic of the hydrophone (h) array and the travel time for an acoustic signal propagation from the dolphin to the different hydrophones in the array (from Au and Benoit-Bird 2003). The range of the dolphin to each hydrophone is denoted as ct_i, where c is the speed of sound in water and t_i is the signal propagation time from the dolphin to the ith hydrophone. Range was determined by measuring the difference in the time of arrival of a signal at each hydrophone.

and continuously digitized, with the results going into separate circular memories on each Gage-1210 board. When an echolocation signal was detected by the center hydrophone, it triggered the data acquisition board. In total, 200 pretriggered points and 200 post-trigger points were collected for each channel and downloaded into the computer. A total of 80 clicks could be downloaded for each episode before the data had to be stored on the hard drive.

A total of 23 files containing 1456 echolocation clicks were collected, representing 23 instances in which a dolphin made an echolocation run at the hydrophone array. The average number of clicks collected per file was slightly over 63. Only echolocation events in which the amplitude of the signals received by the center hydrophone was either the highest or within 3 dB of the highest were accepted for analysis. This criterion was chosen to ensure that a dolphin beam was directed at the array. The beam patterns measured for three different odontocete species (Au 1993, Au et al. 1995) indicate that when the major axis of the beam is directed to within ±5° of the center hydrophone, the signal received by the center hydrophone will either have the highest or will be within 3 dB of the highest amplitude for echolocation signals for ranges between approximately 5 and 10 m between the array and dolphins. Each signal passing the 3 dB criterion but outside the 5–10 m range was examined visually for any distortions. Results from bottlenose dolphins, beluga whales (Au 1993), and false killer whales (Au et al. 1995) indicate that off-axis clicks often have sudden reversals in the waveform or large asymmetry between positive and negative excursions, or longer intervals between successive cycles in the waveform when compared with on-axis clicks. These clicks that are off axis by more than about ±5° will appear distorted and can be distinguished easily from on-axis clicks. A total of 618 clicks met the appropriate criteria.

Four spectral series of dusky dolphin echolocation click trains are presented in a waterfall format in Figure 4.6. The spectra suggest that portions of the signals in a click train can be relatively stable in shape but also include portions that are highly variable and complex. Most of the energy in the spectra were found between 30 and 130 kHz, much higher than for spectra of echolocation signals measured for most other dolphins in the field (Rasmussen et al. 2002, Au and Herzing 2003). Smaller non-whistling odontocete species such as Hector's dolphin (Dawson 1988), harbor porpoise (Kamminga and Wiersma 1981), Commerson's dolphin (Kamminga and Wiersma 1981), finless porpoise (Kamminga 1988), and Dall's porpoise (Hatakeyama and Soeda 1990) emit narrow-band echolocation signals that have slightly higher peak and center frequencies than the dusky dolphin.

Most of the clicks emitted by dusky dolphins had bimodal frequency spectra. The secondary peaks varied in shape from a slight "bump" to a clearly defined peak. A secondary peak existed if the slope of the spectrum for two consecutive increasing frequency bins changed from positive to zero or negative values. A bimodal spectrum is defined as one in which the amplitude of the secondary peak is greater than half the amplitude of the primary peak. Au et al. (1995) separated

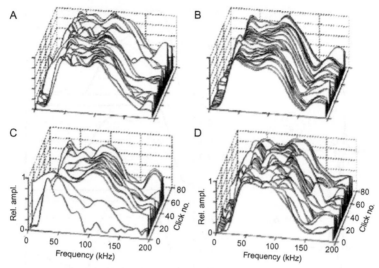

FIGURE 4.6   Four waterfall spectra of dusky dolphin echolocation click trains (from Au and Würsig 2004).

echolocation signals into four types based on the number of peaks in the frequency spectra. Type 1 and 2 signals had low peak frequencies (below 64 kHz); type 1 being unimodal and type 2 being bimodal. Type 3 and 4 signals had high peak frequencies (above 64 kHz); type 3 being bimodal and type 4 being unimodal.

Examples of representative signals of type 1–4 are shown in Figure 4.7, with waveforms on the left and frequency spectra on the right. All the clicks were brief, generally less than 70 µs in duration, with broad frequency spectra. The percentages of the various type signals are also shown next to the waveforms. Clicks with bimodal spectra are obvious in the spectra plots; most of the clicks had bimodal spectra (88.3%). The type 1 signals occurred 8.7% of the time, and the type 4 signals occurred only 2.9% of the time. Some of the bimodal spectra have relatively high peak frequencies (>90 kHz) whereas some have low peak frequencies (<40 kHz). The click waveforms resemble those of other odontocetes such as bottlenose dolphin, beluga whale (Au 1993), false killer whale (Au et al. 1995), and white-beaked dolphin (*Lagenorhynchus albirostris*; Rasmussen et al. 2002).

The peak-to-peak source level as a function of range between an echolocating dolphin and the array is shown in Figure 4.8. As the dolphin's range to the array decreased, the source level also decreased. The solid curve in Figure 4.8 is the one-way spherical spreading curve fitted to the data in a least-square fashion, and is represented by the equation

$$SL = 177.8 + 20\log(R) \tag{2}$$

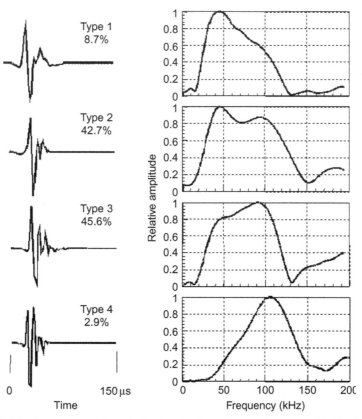

FIGURE 4.7   Some representative echolocation signal waveforms and spectra emitted by the dusky dolphin in Kaikoura. Type 1 and 2 signals have low peak frequencies (<64 kHz) with type 1 being unimodal and type 2 being bimodal. Types 3 and 4 have high peak frequencies (>64 kHz) with type 3 being bimodal and type 4 being unimodal. A bimodal signal is one in which amplitude of the secondary peak is greater than half the amplitude at the peak frequency (from Au and Würsig 2004).

where SL is the source level in decibels re $1\,\mu$Pa at 1 m, and R is the range in meters. The increase in source level as a function of $20 \log R$ compensates for the one-way spherical spreading loss as the signal propagates outward from a transmitter. The highest peak-to-peak amplitude recorded for the echolocation signals was 210 dB re $1\,\mu$Pa emitted by a dusky dolphin at a range of 25 m. The results also suggest that the dolphins were echolocating on the hydrophone array and not on some other objects, since the source level decreased as the range to the array decreased. Also, the interclick intervals were always greater than the two-way travel time from the animals to the array and back, which is consistent with the notion that the dolphins were echolocating on the array. The fitted curve was constrained to vary as $20 \log R$; however, the "best-fit" logarithm's curve would be very similar to the $20 \log R$ fit, with the r^2 values differing only in the third decimal place.

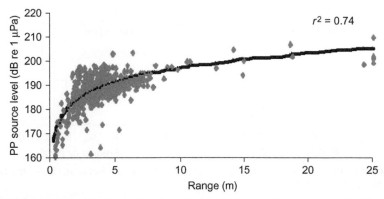

FIGURE 4.8　Scatter plot of source level as a function of the range between an echolocating dolphin and the hydrophone array. The solid curve represents the one-way spherical spreading curve-fitted to the data in a least-square with $r^2 = 0.74$ (from Au and Würsig 2004).

The distributions of peak and center frequencies of the echolocation signals are shown in Figure 4.9. Peak frequency is defined as the frequency at which the frequency spectrum of a signal has its maximum amplitude. Center frequency is defined as that frequency which divides the energy in a frequency spectrum into two equal parts (the centroid of the spectrum). The peak frequency histogram has a low-frequency peak between 50 and 60 kHz and a high-frequency peak one octave higher between 100 and 110 kHz, reflecting the bimodal characteristics of dusky dolphin echolocation signals. The mean and standard deviation of the peak frequency for 618 clicks were 73.8 ± 27.3 kHz.

The center frequency histogram had a well-defined peak between 90 and 100 kHz. Ninety-two per cent of the signals had center frequencies greater than 80 kHz, and 66% had center frequencies greater than 90 kHz. The center frequency extends to much higher frequencies than the peak frequency, and this property is indicative of signals with bimodal spectra; the secondary peak causes the spectrum to be broader, introducing a large asymmetry in the spectrum. Center frequency is a more representative measure of signals with bimodal spectra (Au et al. 1995), since a slight shift in the spectrum could move the peak frequency over one octave away. The mean and standard deviation of the center frequency were 80.5 ± 8.7 kHz.

Histograms of the 3 dB bandwidth and the root mean square (RMS) bandwidth are shown in Figure 4.10. The 3 dB bandwidth is the width of the frequency band between the two points that are 3 dB lower than the maximum amplitude of a spectrum. The 3 dB points are also referred to as the half-power points, since a level 3 dB below the maximum in a continuous signal is exactly one-half of the power or energy in a signal. Its distribution is rather scattered with a peak between 20 and 25 kHz, but with distributions reaching out to 100 kHz. The mean and standard deviation of the 3 dB bandwidth were 67.4 ± 27.4 kHz.

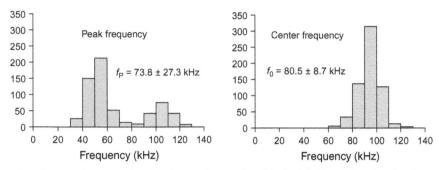

FIGURE 4.9   Histogram of peak and center frequencies of dusky dolphin echolocation signals (from Au and Würsig 2004).

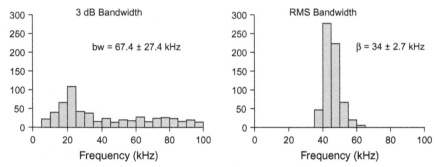

FIGURE 4.10   Histogram of 3 dB and RMS bandwidths of the dusky dolphin echolocation signals (from Au and Würsig 2004).

The RMS bandwidth is the standard deviation of the spectrum about the center frequency, and its distribution is clustered between 40 and 50 kHz with 80% of the signals having bandwidth in this range. The 3 dB bandwidth for bimodal spectra can often provide a misrepresentation of the width of the signal, since this bandwidth might cover only the frequency range about the peak frequency. The RMS bandwidth is probably a better measure of the width of signals, with bimodal spectra since the effects of local maxima and minima are accounted for in the calculation. The mean and standard deviation of the RMS bandwidth were 34.0 ± 8.7 kHz.

An analysis examining the relationship between center frequency and peak-to-peak source levels indicated that there is not a direct relationship between center frequency and source level. However, when examining the average source levels for the different signal types (Figure 4.11), the higher frequency signals have higher source levels. By the manner in which the different types of signals were defined, the type 1 signal will have the lowest center frequency and the center frequency will progressively increase, with type 4 having the highest frequency. The mean source level of the type 1 signals was significantly different

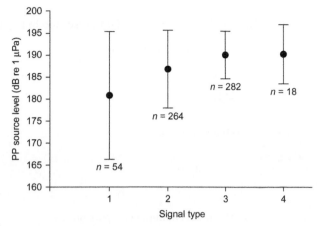

FIGURE 4.11 The mean and standard deviation of source levels for the different signal types (from Au and Würsig 2004).

($P < 0.05$, one-way ANOVA) from that of all the other signal types (Table 4.3a,b). The mean source level of the type 2 signals was significantly different ($P < 0.05$, one-way ANOVA) from that of the type 1 and type 3 signals, and the type 4 signals were significantly different ($P < 0.05$) from the type 1 signals.

ECHOLOCATION BEHAVIOR

Dusky dolphins in New Zealand exhibit two seemingly distinct foraging patterns. One pattern is represented by the population of animals occurring in waters near Kaikoura (Benoit-Bird et al. 2004, see also Chapter 5). These dolphins usually occur in large groups of tens to hundreds of animals and forage at night in deep waters on a vertically migrating community of squid, fish, and crustaceans, also known as the deep scattering layer (DSL). This is a foraging pattern shared with a number of pelagic species around the world, including spinner dolphins and striped dolphins (*Stenella coeruleoalba*). The other foraging strategy is one adapted to a more coastal shallow-water environment (see also Chapter 3). Dusky dolphins in the Marlborough Sounds area of New Zealand occupy bays and coves and feed diurnally on a variety of schooling fishes, including pilchard (*Sardinops neopilchardus*), sprats (*Sprattus antipodum*), anchovy (*Engraulis australis*), and yelloweye mullet (*Aldrichetta forsteri*). This foraging pattern is distinct not only because of the prey involved, but also because it occurs in relatively shallow waters (<50 m) where the dolphins often herd schools of fish into so-called "prey balls" at the water's surface, a behavior thought to represent a form of cooperative foraging (see also Chapter 6).

A pilot study was conducted in the Marlborough Sounds to determine spatial distribution and movement patterns of animals engaged in this form of feeding. A portable video/acoustic array system was constructed for this study (Figure 4.12).

TABLE 4.3   Results of the statistical test of the data presented in Figure 4.10

(a) One-way ANOVA

Source	DF	SS	MS	F	P-value
Corrected model	3	4484.93	1495.00	21.77	<0.001
Error	614	42165.80	68.70		
Corrected total	46650.77				

(b) Dunnett C multiple comparison

Variable 1	Variable 1		Mean difference	Standard error	95% confidence interval	
					L. Bound	U. Bound
Type 1	Type 2	$P < 0.05$	−5.84	1.24	−11.28	−0.39
	Type 3	$P < 0.05$	−9.21	1.23	−14.52	−3.90
	Type 4	$P < 0.05$	−9.40	2.26	−16.31	−2.49
Type 2	Type 1	$P < 0.05$	5.84	1.24	0.39	11.28
	Type 3	$P < 0.05$	−3.38	0.71	−5.04	−1.71
	Type 4	ns	−3.56	2.02	−8.29	1.16
Type 3	Type 1	$P < 0.05$	9.21	1.23	3.90	14.52
	Type 2	$P < 0.05$	3.38	0.71	1.71	5.04
	Type 4	ns	−0.19	2.01	−4.76	4.38
Type 4	Type 1	$P < 0.05$	9.40	2.26	−2.49	16.31
	Type 2	ns	3.56	2.02	−1.16	8.29
	Type 3	ns	0.19	2.01	−4.38	4.76

Three hydrophones with broadband frequency sensitivity between 100 Hz and 200 kHz were spaced 3.1 m apart in a horizontal line array using a PVC frame. A video camera was fixed at the center of the array and the signals from the hydrophones and video were cabled to a data acquisition computer and a video camcorder on the research vessel.

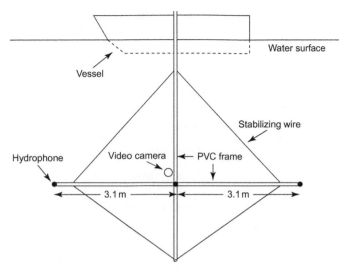

FIGURE 4.12   Schematic of the video/hydrophone array system that was used to examine foraging behavior of dusky dolphins. The vessel and the array are not drawn to scale.

Groups of foraging dusky dolphins were studied in Admiralty Bay between July 28 and August 3, 2006. A 7.2-m vessel was used to locate and approach animals occurring in the bay. When groups of animals were encountered, they were followed at a distance until their behavioral pattern could be established. Groups that were determined to be feeding were approached to within a few tens of meters and the array was deployed overboard to a depth approximately 2 m below the surface. The computer and video operator began recording as soon as the array had stabilized and continued until the animals left the area. Acoustic recordings were made in a continuous series of 10-second segments at a sampling rate of 400 kHz.

Acoustic data were analyzed by locating instances of two or more animals echolocating simultaneously. This criterion was used to establish the spatial relationship in the horizontal plane between animals at that specific moment in time. Echolocation pulses from each animal were identified on each of the three recording channels using the program CoolEdit and the signal's exact time of arrival at each hydrophone was manually measured. These measurements were then entered into a custom-written Matlab program that uses the difference in time of arrival at each hydrophone to calculate a two-dimensional position for a sound source.

Groups of dusky dolphins were encountered each day of the study; however, none were observed foraging on prey balls close to the surface. On several occasions, groups of dolphins were found traveling and also foraging on prey close to the bottom or in the middle of the water column. Although prey-balling behavior could not be visually observed or confirmed, the acoustic recordings yielded several instances of multiple animals echolocating simultaneously. Figure 4.13 shows the localization results from three such instances. In total,

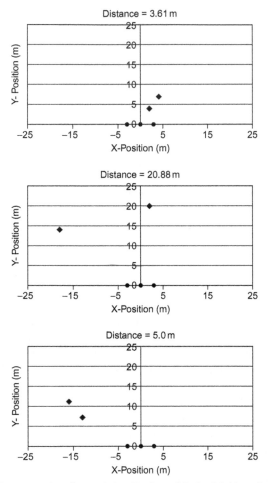

FIGURE 4.13   Three examples of acoustic localizations of dusky dolphins echolocating simultaneously. Each coordinate plot shows the three hydrophones of the array (circles) and the position
of each dolphin (diamonds) relative to the middle hydrophone. "Distance" refers to the measured
distance between the animals.

35 recordings were identified with pulse trains from two or more animals echolocating simultaneously. Of these, 13 had signals that could be reliably localized; the rest produced ambiguous results. The reliably obtained pulse trains
were used to measure the horizontal distance between animals as they foraged
and traveled by the array. The distribution of the measured distances between
animals is presented in Figure 4.14.

   Although the pilot study did not obtain echolocation information while
duskies were prey ball foraging, some interesting results were obtained. First,
the effort demonstrated that the technique of localizing dolphins using a portable
stationary array works. This is a valuable finding because it means the technique
can be applied to a variety of behavioral situations for which the instantaneous

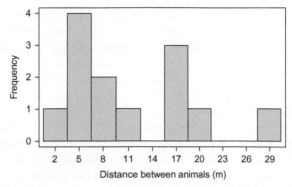

FIGURE 4.14   Distribution of distances between simultaneously echolocating dusky dolphins.

position of animals relative to one another is desired. Second, although the sample size was too small to draw firm conclusions, it is interesting that the distribution of distances between animals shown in Figure 4.14 had a bimodal pattern, with most animals being spaced 5–8 m or 17–20 m apart from one another. A very similar result was found in a previous study examining the spatial distribution of echolocating spinner dolphins (Lammers et al. 2006). This result could be indicative of a difference in spacing between foraging and traveling animals, since they were recorded in both contexts, or it could perhaps represent a more general pattern related to how delphinids space themselves relative to one another when echolocating. Additional data from the Admiralty Bay animals, as well as comparative data from the Kaikoura population, would help resolve these questions and advance the effort begun here.

DISCUSSION AND CONCLUSIONS

Whistles do not play an important part in the social systems of New Zealand dusky dolphins, and the question is why? In recent years, a number of hypotheses have been developed to explain differences in sound emission behavior among the different species and families of odontocetes. Collado-May et al. (2007) discuss the loss of tonal production within the delphinid group from a phylogenetic and social basis. They conclude that increased whistle production is linked to large group size and social structure. Given the large group sizes of dusky dolphins seen off New Zealand, it would seem that duskies would frequently whistle. But our studies found little evidence for whistling in dusky-only groups, even within the large (200–300 dolphin) main group or even the "super" groups of over 1000 animals. Morisaka and Connor (2007) theorized that the loss of whistles from the phonation repertoire of different odontocete species and the rise of narrow-band frequency clicks is linked to predation pressure by killer whales (*Orcinus orca*). Killer whales are occasionally seen off Kaikoura and have been observed to feed on duskies (Constantine et al. 1998, see also Chapter 7).

From the perspective of whistle production, dusky dolphins can almost be labeled as a non-whistling odontocete; however, they can indeed whistle. Whether the non-whistling dolphins and porpoises can actually produce whistles is an open question. If they could, why are they not, even infrequently, observed to whistle? Could there be some anatomical factors that prevent the production of whistle signals? The theory of Morisaka and Connor (2007) seems attractive when we consider the fact that the narrow band echolocation signals produced by non-whistling dolphins and porpoises are generally out of the hearing range of killer whales (Szymanski et al. 1999). However, if dusky dolphins rarely whistle because of predation avoidance, why do they still use broadband echolocation signals that would have components that fall within the hearing range of killer whales? Furthermore, common dolphins share overlapping habitats with dusky dolphins as well as killer whales, yet they often whistle, raising the question as to why have not common dolphins selected out whistle signals according to the theory of Morisaka and Conner (2007)?

Clearly, more information is needed to address the reason for low or no whistle production among small odontocetes. The one characteristic of dusky dolphins that is shared among the non-whistling odontocetes is the relatively small size of these animals. The porpoises in the family Phocoenidae and dolphins in the genus *Cephalorhynchus* are among the smallest odontocetes. Dusky dolphins are almost as small as the non-whistling odontocetes; and perhaps to further advance the size-related argument relative to whistling, the much larger white-beaked dolphins in Icelandic waters (presently listed as the same genus as dusky dolphins) often emit whistles (Rasmussen et al. 2006).

The variation of source level as a function of 20 log R range for echolocation signals produced by dusky dolphins is consistent with echolocation signals of other dolphins that have been measured in the field with the symmetrical star array (Au and Benoit-Bird 2003). Au and Benoit-Bird (2003) attributed variation of the source level as a function of range to a method of obtaining a time-varying gain function in the dolphin sonar system. The strength of echoes from a fish school containing many individual targets will decrease with range as a function of 20 log R (Urick 1983). Therefore, the dolphin sonar system compensates for energy lost in the backscattered signal by increasing its source level. This time-varying gain phenomenon is not specific to a particular dolphin population or any geographic region, but has been confirmed by measurements in Iceland with white-beaked dolphins (Rasmussen et al. 2002), in the Bahamas with Atlantic spotted dolphins, *Stenella frontalis* (Au and Herzing 2003), and in British Columbia with killer whales (Au et al. 2003).

Source levels utilized by dusky dolphins are much lower than the signals used by white-beaked dolphins of the same genus. The difference in source levels can be determined by the value of the constant in Equation 2, which is 177.8 dB for the dusky dolphin and 189.6 dB for the white-beaked dolphin (Rasmussen et al. 2002). These values represent a difference of 11.8 dB, measured with the same array and electronic system for both species. Also, the maximum source level recorded for the dusky dolphin was 210 dB re 1 µPa compared with 219 dB for white-beaked dolphin (Rasmussen et al. 2002). The difference in source levels may not be surprising when comparing the sizes of the two species. The white-beaked dolphin has a robust body with a short thick rostrum. Adult white-beaked dolphins weigh between 220 and 350 kg and have lengths between 2.2 and 2.8 m (Reeves et al. 1999) compared with approximately 100 kg and 1.75 m for the dusky dolphin. Therefore, the white-beaked dolphin is over twice as heavy as the dusky dolphin, yet only about 1 m longer, but with a much wider girth. Unfortunately, information on the relationship of animal size and source levels of echolocation signals does not exist. However, the Hector's dolphin, one of the smallest dolphin, has been reported to emit echolocation signals with typical source levels of about 150 and 175 dB (Dawson 1988). The harbor porpoise, also one of the smallest odontocetes, emits average source levels of 191 dB (Villadsgaard et al. 2006). These levels are lower than those used by the dusky dolphin.

Dusky dolphins project broadband short-duration echolocation signals similar to those of other odontocetes that also produce whistles. Prior to this study, the symmetrical star array had been used to measure the echolocation signals of five different species of dolphins in the field. The majority of the echolocation signals had bimodal spectra. The echolocation signals of dusky dolphins in the waters of Kaikoura also exhibited a strong bimodal tendency, with 88.3% of the signals being bimodal. It is becoming apparent that bimodality is an inherent feature in the sound generation mechanism of dolphins. Type 3 signals were emitted most frequently (45.6% of the time); however, the number of type 2 signals emitted was close behind, occurring 42.7% of the time.

The frequency characteristics of bimodal echolocation signals are best described by their center frequency and RMS bandwidth, rather than peak frequency and 3 dB or half-power bandwidth. The standard deviation of the peak frequency values of 27.3 kHz is three times as high as the standard deviation of 8.7 kHz for center frequency. The difference in standard deviation is even greater when comparing 3 dB bandwidth with the RMS bandwidth. The variance in the 3 dB bandwidth is relatively high, with a standard deviation of 27.4 kHz compared with a standard deviation of only 2.7 kHz for the RMS bandwidth. These results suggest that more stable measures of the spectra of echolocation signals are center frequency and RMS bandwidth.

Most of the signals had center frequencies between 90 and 110 kHz, which represent some of the highest center frequencies for broadband signals emitted by free-ranging delphinids. This characteristic of high center frequency may be related to the size of dusky dolphins. A general rule in sound production is that the smaller the animal, the higher the frequency of sounds produced. Wang et al. (1995) found a strong correlation between maximum frequency of whistles and body lengths of seven species of dolphins.

The broad bandwidth of the echolocation signal provides a good range resolution capability (Au 1993) that should enable dusky dolphins to forage for small schooling fish to perform fine target discrimination in a shallow water environment where bottom reverberation can be troublesome. Good range or time resolution would also be beneficial for dusky dolphins when foraging on mesopelagic prey such as myctophid fishes and small squid (Cipriano 1992). Although there does not seem to be a relationship between source level and center frequency as for the false killer whale (Au et al. 1995), Au and Herzing (2003) found that only when the source level increased beyond 210 dB re 1 μPa, did a relationship between center frequency and source level emerge for the Atlantic spotted dolphin. The highest level signal for the dusky dolphin was only 210 dB. Perhaps if higher source levels were used, a relationship between center frequency and source level may emerge. However, there is a definite trend in which the source level increased as the signal type increased from 1 to 3, which seems to indicate that there should be some kind of relationship between center frequency and source level. Grouping the signals into types definitely helped in bringing out the relationship between source levels and center frequency.

The use of two different types of arrays of hydrophones with very different geometry and hydrophone spacing have provided interesting data on characteristics and use of echolocation by duskies in the field. The results from the array data clearly demonstrate the utility of using a multi-hydrophone array to measure echolocation signals of dolphins in the wild. The symmetrical star array is relatively compact and easy to handle, and can provide information on whether a specific received echolocation signal originated in the vicinity of the major axis of the animal's transmission beam. Time of arrival differences between hydrophones were easily ascertained because of the rapid onset of the echolocation signals. Our results have demonstrated the value of using an array to obtain reliable data on echolocation signals in the field. The linear array was longer than the symmetrical star array and a bit more difficult to handle in the field, yet it provided different types of data that could be used to infer spacing between simultaneously echolocating dolphins.

We conclude this chapter by stating that much more work on the acoustics of dusky dolphin should be done. There is so much that can be learned of this species by using acoustic techniques. A better understanding on when and in what context whistles are emitted is needed. The amount of time burst pulse signals are emitted and in what context is another area that should be addressed. How echolocation signals are used in foraging and in communication and how much of the total signal emission effort consists of echolocation signals is another vital field of study. Understanding the acoustics of dusky dolphins will increase our understanding of dusky dolphin behavior and natural history, and may even help to elucidate questions of use of acoustics for different habitats, prey and predator relationships, and related delphinid species.

ACKNOWLEDGMENTS

The content of this chapter came from several projects performed over several years, so there are many colleagues and helpers to acknowledge along with multiple sponsors. Many thanks to Bernd and Mel Würsig, Kirsty Russell, Keith Algie, L.J. Smith, Glenn Gailey, Wendy Markowitz, the entire Dolphin Encounter staff and all of the Earthwatch volunteers who helped us collect data during our five seasons in Kaikoura. We thank Ian Bradshaw, Jackie Wadsworth, and Lynnette and Dennis Buurman of "Dolphin Encounter." We thank Kelly Benoit-Bird and Kimberly Andrews for their help with the ANOVA statistics. Thanks also to author Dr. Roland Aubauer for his suggestion of using a symmetrical star geometry, and for various discussions associated with his research on detection of flying bats. The assistance of Michiel Schotten in testing and calibrating the array is also appreciated. For ongoing field and laboratory logistics, we thank the Edward Percival Field Station of the University of Canterbury, Jack van Berkel, station manager. Finally, we acknowledge the assistance of Mridula Srinivasan and Robin Vaughn in obtaining the data in Admiralty Bay, Marlborough Sounds. Sponsors for the various projects included the Center for Field Studies of Earthwatch, the Marlborough District Council, and the New Zealand Department of Conservation. Funds from a Fulbright research/teaching award to Bernd Würsig were also used. Whitlow Au was partially funded by the Office of Naval Research, Dr. Robert Gisiner, program manager.

Chapter 5

Dusky Dolphins Foraging at Night

Adrian D. Dahood[a] and Kelly J. Benoit-Bird[b]

[a]*Marine Mammal Research Program, Texas A&M University, Galveston, Texas, USA*

[b]*Biological Oceanography, Oregon State University, Corvallis, Oregon, USA*

When we tell non-biologists that we study dolphins, they often conjure up images from enchanting educational or adventure documentaries showing tanned researchers and leaping dolphins. In reality, the study of dolphin foraging is a challenging endeavor that is not always glamorous (Figure 5.1), especially when it happens in the dark, in deep waters with oft-turbulent waves. The following timeline gives a vignette of what it is really like to study night-foraging dolphins.

FIGURE 5.1 The authors, with help from labmates, practice deploying the towfish in the stable, relatively dry, boat ramp parking lot. The boat shown was used for day surveys. At night, work was conducted on a much larger fishing vessel. Even with all of the helping hands, the towfish is heavy and awkward to move.

17:30	While the team members who conduct observations only during the day enjoy a leisurely glass of wine and savor every last mouthful of curry, we dine on bland non-seasickness-inducing plain white rice and water.
18:15	We two young women, weighing less than 60 kg each, attempt to move 150 kg of gear into the minivan.
18:45	At the boat slip, we carry this gear over narrow paths and down too many slippery stairs to our waiting boat and skipper. Pete greets us with a cigarette in his mouth and bare feet. Everyone thinks we're nuts. The tour guides, who are normally interested in our activities, laugh at us. We can't even pay graduate students from the local marine lab to help. No one goes out at night in Kaikoura.
19:30	All gear is in place and secured for the trip. All electronics have been tested and seem to be working. We head out from port. We will be the only boat on the water until the fishing boats launch at sunrise tomorrow.
20:00	It is approaching sunset and the weather is beautiful. At the start of our first line, we make sure that the towfish, which carries three transducers, is "flying" level and recheck all electronics. Something isn't working. We check the connections, battery terminals, and settings. We fiddle with the pins in the serial port, do a little dance, spin in a circle three times and . . . voilá, the computer now reads the GPS.
20:25	The first "data alarm" goes off. We look for dolphins for 5 minutes (a handful approach the boat), note environmental conditions, and check the electronics. We will repeat this every half hour.
22:25	One of us is seasick and lying face down on the engine cover, which has little x's embossed on its metal. It is cold and raining. Pete feels sorry for "sickie" and drags a cushion out on deck so she no longer gets little "x" prints all over her face. When the "data alarm" goes off, we pop up, take environmentals, check the towfish and electronics, then "sickie" goes right back to lying face down.
00:05	It is time for Pete's smoke break and our turn at the wheel. Perhaps, if we had grown up playing video games we would have been better prepared to steer a boat at night with nothing but the GPS for guidance. The swells roll and the point of the bow veers widely. Thankfully, Pete is just a shout away.
02:25	How many more hours? We have been up and working since 5:00 am yesterday with only a brief nap. Must stay awake . . . Oh look, dolphins leaping in the moonshine!
03:25	One of the echosounders is not operating well; the computer output is a mess. We ask Pete to slow down, and pull in the towfish. Ah-Ha! we caught seaweed, and this is messing up the echo reception.
04:15	We see whitecaps coming in from the south; the southerly storm due in this evening is coming early. Come on wind. Hold off a few hours more. We're almost done.

05:00   Pete decides to take us home before the weather gets ugly out here. It has been a good night. We sampled for 9 hours and nearly completed all planned transects.

05:45   We are back at the dock, battling the rapidly freshening southerly blow as we unload those 150 kg of gear. It seems heavier this morning. We still need to download data before bed.

The exciting part of the work will happen weeks or months later while sitting at a lab computer thousands of miles from Kaikoura. This is when the first "picture" of dusky dolphin feeding will begin to appear as calibrated, colored dots on the monitor. These dots will tell us where along the coast dolphins were and what the food was like; they will tell us how dolphins were spaced underwater and how deep they were. We will learn about prey density and nightly migration patterns. We will "see" foraging in action and learn more about how duskies relate to their prey. We doubt the Discovery Channel will be coming to film our work. If they change their minds, we will save them a spot on the engine cover and a rickety computer chair.

Adrian Dahood, Galveston Island, October 2008

INTRODUCTION

Foraging at night requires that predators locate and capture prey in low light or near total darkness. For many animals, vision is fundamental both in the detection and capture of prey and in the detection and avoidance of predators (Beauchamp 2007). Social foragers use vision to identify group members at a distance and maintain group cohesion (Beauchamp 2007). These abilities are severely limited at night. Despite this, animals choose to forage in the dark either to reduce predation risk or gain access to a food resource that is not available during the day (Benoit-Bird et al. 2004, Beauchamp 2007). Night-foraging animals must adapt their foraging tactics to compensate for decreased visibility.

To forage successfully at night, animals develop specialized morphological adaptations. Night foragers across several taxa have large eyes and large pupils to let in maximum amounts of light; additionally, their eyes have significantly more rods, the photoreceptors that allow vision in dim light, than color-sensitive cones (Warrant 2004). The eyes of night-foraging monkeys are significantly larger than those of day-foraging ones, and lack the ability to detect color (Bicca-Marques and Garber 2004). Many night-active mammals and fishes also have a tapetum lucidum, a reflective surface behind the retina to enhance the collection of scarce light (Warrant 2004). Some night foragers such as bats (Griffin et al. 1960), oilbirds (Konishi and Knudsen 1979), and odontocetes (Au 2004) have evolved high-frequency echolocation systems and appear to rely on sound, rather than vision, to sense prey in the dark.

In addition to morphological adaptations, animals also adjust their behavioral regimes to allow for night foraging. Some animals feed both day and night

(Donati et al. 2007, Stimpert et al. 2007), while others switch feeding from diurnal to nocturnal based on perceived predation risk (Shimek and Monk 1977) or prey availability patterns (Mougeot and Bretagnolle 2000). Species that primarily use senses other than vision to forage might more readily switch to night foraging. Greater flamingos (*Phoenicopterus ruber ruber*; Beauchamp and McNeil 2003) and sea otters (*Enhydra lutris;* Shimek and Monk 1977, Ribic 1982) are both tactile foragers and often forage at night to supplement low caloric intake during the day. Dusky dolphins (*Lagenorhynchus obscurus*) rely on echolocation and switch between day and night foraging depending on habitat and available prey (Benoit-Bird et al. 2004, Markowitz et al. 2004).

In aquatic environments, low light conditions regularly occur in the middle of the day at depth or in turbid, shallow waters (Lalli and Parsons 1995). Animals foraging in these conditions often have adaptations more commonly associated with night foragers (Warrant 2004). Elephant seals (*Mirounga* spp.; Heithaus and Dill 2002b) and king penguins (*Aptenodytes patagonicus*; Wilson et al. 2002) feed during the day hundreds of meters below the surface and have large eyes, suggesting that at these foraging depths there is sufficient light for visual predation, often by bioluminescence of prey. King penguins choose to forage at depth during the day rather than wait for their preferred prey, lanternfishes (family Myctophidae), to rise to the surface at night, perhaps to take advantage of the small amount of daylight visible at foraging depths and the decreased daytime speed of their prey (Wilson et al. 2002). In contrast, Indian river dolphins (*Platanista gangetica*), which live in shallow and turbid waters of the Indus and Ganges Rivers, are especially active during the day but have small, poorly developed eyes without a lens (Mass and Supin 2002, Smith 2002). They appear to forage largely by using well-developed echolocation.

When there is insufficient light, animals may rely on sound to sense their environment using active echolocation or passive listening. Echolocating animals can detect and identify prey, predators, and obstacles at a distance based on the characteristics of returning sound waves (Au 2002, see also Chapter 4). Because of their ability to echolocate, odontocetes are well suited to foraging at night, and several species commonly do so. Night-foraging dolphins include Hawaiian spinner (*Stenella longirostris*; Norris and Dohl 1980), pan-tropical spotted (*Stenella attenuata*; Baird et al. 2001), dusky (Benoit-Bird et al. 2004), striped (*Stenella coeruleoalba*; Ringelstein et al. 2006), bottlenose (*Tursiops truncatus*; Klatsky et al. 2007), and common (*Delphinus delphis*; Pusineri et al. 2007) dolphins. There are surely others.

Resting in protected nearshore waters during the day and foraging in deeper waters at night may reduce predation risk to small dolphins, including duskies (Norris and Dohl 1980, Würsig et al. 1997). Major predators of dolphins tend to be killer whales (*Orcinus orca*; Constantine et al. 1998) and large sharks of several species (Norris and Dohl 1980, Heithaus and Dill 2006, see also Chapter 7).

Mammal-eating killer whales seem to hunt using combinations of vision, echolocation, and passive listening (Baird 2000); low night-time light levels would impede their ability to see their prey. Little is known about mammal-eating killer whale diel dive cycles. Until recently, it had been assumed that like fish-eating killer whales, they are most active during the day (Baird 2000, Baird et al. 2005). However, new evidence suggests that mammal-eating killer whales in southeast Alaska hunt as much during crepuscular periods or at night as they do during the day (Volker Deecke, 2008, personal communication).

To detect prey at a distance, sharks rely on scent and (close up) pressure changes sensed by their lateral line; the final strike is directed by changes in electromagnetic fields sensed by head-based ampullae of Lorenzini (Gardiner and Atema 2007). Because sharks do not depend highly on vision for prey detection or capture, they would be almost equally able to detect prey in the dark as in daylight. Many shark species, including some that may prey on marine mammals, are active during the night (Hulbert et al. 2006). Recent tagging studies have shown that some large sharks migrate to the surface at night and are presumably foraging (Hulbert et al. 2006, Chapman et al. 2007, Weng et al. 2007). Therefore, night-time foraging may be no less dangerous than daytime foraging, and may be even more so in areas where large shark abundance is high.

For marine mammals and birds, an advantage to dangerous night foraging is the predictable nightly ascent of food associated with the diely migrating deep scattering layer (DSL), a community of fish and invertebrate mesozooplankton and micronekton which serves as a rich prey source for many marine predators (Johnson 1948, Cushing 1973, Pearcy et al. 1977, Pieper and Bargo 1980). The animals in the DSL occur in sufficiently dense concentrations to scatter sound waves, giving the illusion of a bottom to depth-sounding devices (Johnson 1948, Cushing 1973). The DSL is too deep for most air-breathing marine predators to easily access in daytime. However, the diel vertical migration (DVM) of the DSL typically brings this abundant prey base from depths of hundreds of meters to within tens of meters of the surface (Enright 1979).

Duskies living near Kaikoura, New Zealand are able to track the DSL at night as it approaches the surface (Benoit-Bird et al. 2004) and preferentially capture lanternfishes and squids within it (Cipriano 1992). Duskies foraging at night likely rely on echolocation and passive listening to detect and localize individual prey (Au and Würsig 2004), but at very close range, duskies may be able to see their prey, aided perhaps by moonlight or bioluminescence of the lanternfishes (Gago and Ricord 2005). This differs from typical daytime feeding tactics, when duskies rely on vision and light to capture schooling fishes (Würsig and Würsig 1980, Würsig et al. 1990, Vaughn et al. 2007, 2008).

As a species and as individuals, dusky dolphins are able to drastically change their feeding tactics in relation to bathymetry and prey availability (Box 5.1). In shallow water environments, typically areas shallower than 60 m, such as Admiralty Bay, New Zealand and Golfo San José, Argentina, duskies forage during daylight and coordinate behavior to herd schooling fishes into tightly packed prey balls near the surface (Würsig and Würsig 1980, Vaughn et al. 2007, see also Chapter 6). When capturing fish from a prey ball, duskies typically face their ventral surfaces

BOX 5.1

Daily and Seasonal Habitat Use Patterns

Nicholas M.T. Duprey and Griselda Garaffo

Animals choose a habitat that meets their food needs, reduces predation, reduces competition with others, and allows for successful procreation. Variables such as depth, sea surface temperature, salinity, distance from shore, and primary productivity are often used to describe habitat use patterns in cetaceans (Ballance et al. 2006, Redfern et al. 2006, Garaffo et al. 2007), but these variables may or may not be directly responsible for the patterns.

Small-scale differences in bathymetry can have significant effects on delphinid distribution. For example, Indo-Pacific humpback dolphins (*Sousa chinensis*) in Algoa Bay, South Africa occur near shallow-water rocky reefs where they have higher feeding success (Karczmarski et al. 2000a). Indian Ocean bottlenose dolphins (*Tursiops aduncus*) of Shark Bay, western Australia prefer deeper channels where they have a better chance of escaping tiger shark (*Galeocerdo cuvier*) attacks (Heithaus and Dill 2006).

Daily habitat use patterns of dusky dolphins are at least partly linked to their prey, but probably also to the threat of predation (see Chapter 7). Duskies off Kaikoura that feed on the deep scattering layer (DSL) of lanternfishes and squid can only do so at night, when the DSL is close enough to the surface for dolphins to meet it (Benoit-Bird et al. 2004). But, they do not stay "out there" in the Kaikoura Canyon during daytime, preferring to come closer to the edge of the Canyon and over the shallows, probably to avoid or minimize shark and killer whale (*Orcinus orca*) predation (although sharks are much reduced in the area, and were probably a threat in the past, not now).

In areas where duskies feed on schooling fishes in daytime, there seems to be a progression of being close to shore at night, when prey balling does not take place— probably again for predator avoidance near the shore—and movement into somewhat deeper prey-rich waters during the day. This trend was quite clear in Golfo San José, Argentina in the 1970s (Würsig and Würsig 1980), but may not be as strong in recent years, nor as strong in Admiralty Bay, New Zealand (see Chapter 6).

Seasonal differences in prey distribution, predation risk, or the reproductive cycles of dolphins may explain aggregations in particular areas during certain times of year (Ballance et al. 2006, Redfern et al. 2006). Mothers with newborn calves may seek refuge in very shallow waters, even though there may not be large-scale prey changes in that season (Mann et al. 2000, Degrati et al. 2008, Weir et al. 2008). Other seasonal patterns may well be prey-related, and duskies of the Argentine coastline occur near the shore more in summer than winter (Garaffo et al. 2007), possibly due to the greater nearshore occurrence of anchovy in summer (Whitehead et al. 1988, Dans, personal communication).

In Admiralty Bay, seasonal differences in the depth range of prey species may result in different foraging tactics by dusky dolphins (Vaughn et al. 2007). In winter, pilchard (*Sardinops neopilchardus*) appear to occur deeper in the water column and duskies prey more individually on them. In spring, pilchard occur near the surface and dolphins forage by coordinating their efforts to herd prey into fish balls (Vaughn et al. 2007). Thus, duskies change their foraging tactics relative to how shallow or deep prey are, as opposed to changing geographic location during the winter to spring progression.

Kaikoura's dusky dolphins exhibit strong seasonality in their occurrence patterns. In general, dolphins prefer shallow nearshore waters during summer and move off-shore in winter (Cipriano 1992, Markowitz 2004, Dahood et al. 2008). South of Kaikoura, arrow squid (*Nototodarus sloanii*), an important winter prey of duskies (Cipriano 1992), move offshore into deeper waters (200–400 m) in winter (Beentjes et al. 2002). Overall, however, it is unclear how much shifting of prey, predator pressure (for example, Dill 1987, see also Chapter 7), and spring and summer calving seasonality (see Chapter 9) may be responsible for seasonal movements.

While prey access and predator avoidance may be largely responsible for dusky dolphin movements on daily and seasonal scales, "social reasons" are probably also important. For example, Garaffo et al. (2007) found that in Golfo Nuevo, Argentina, group size and composition significantly influenced the habitat selection of dusky dolphins. Smaller groups and those consisting of mothers with calves occurred in the shallowest areas closest to shore, while mixed groups (groups of calves, adults, and juveniles) and larger groups occurred in deeper areas further from shore. Degrati et al. (2008) suggest that the segregation of mothers with calves from larger groups of adults and juveniles may be explained by differences in activity requirements. They found that mothers and calves rested and milled more than other group types, activities they could not engage in as well in the generally faster moving and more socially active larger groups. Deutsch (2008) mentioned a similar possibility (see also Chapter 9), and factored in the possibility of nursery groups attempting to avoid sexually active males.

towards the fish. They may use the light reflecting off their white bellies to help control prey (Würsig et al. 1990), a behavior that has been recorded in fish-eating killer whales (Similä and Ugarte 1993). The same individuals that have been documented prey balling in Admiralty Bay in the winter migrate south and feed at night in deep waters of the Kaikoura Canyon during summer (Benoit-Bird et al. 2004, Markowitz 2004). When foraging at night on the DSL near Kaikoura, they can no longer rely on vision to coordinate activities, and prey detection, containment, and capture techniques must change radically from daytime tactics.

Foraging strategies and choices about when to use specific tactics affect individual and population level fitness (Krebs et al. 1974). To understand these choices, it is important to study both the temporal and spatial patterns of predators and prey. Dolphins feed on highly mobile prey out of sight of researchers, making it difficult to quantify both the available prey base and dolphin foraging behaviors. These difficulties are compounded when dolphins forage at night in deep waters. However, even in the dark, tagging can provide information on the diving behavior of individual dolphins, and active acoustic techniques make it possible to simultaneously track dolphins and their prey base (Benoit-Bird et al. 2009). Using these tools, we are just beginning to understand the night-time component of dusky dolphin flexible foraging strategy.

METHODS

Three studies have been conducted on night foraging in duskies. The studies were conducted near Kaikoura, New Zealand, between 1984 and 2006, and employed a variety of methods (Figure 5.2 depicts the study area). We briefly summarize the methods used in each study.

Würsig et al. (1991) and Cipriano (1992)

As part of the initial studies of duskies in New Zealand, researchers radiotagged duskies (Würsig et al. 1991, Cipriano 1992) and studied the gut contents of incidentally caught or beachcast dead ones (Cipriano 1992) between January 1984 and February 1988. These approaches allowed them to quantify both individual night foraging behavior and diet composition. Ten duskies near Kaikoura were

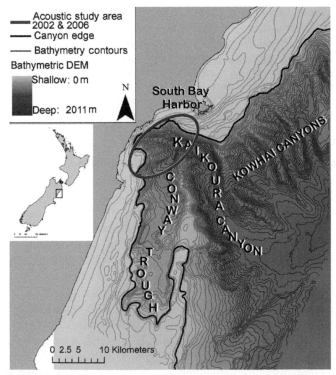

FIGURE 5.2   Kaikoura is located on the northeast coast of the South Island of New Zealand. The Kaikoura Canyon head comes within 500 m of shore, in the center of the acoustics study area. The bathymetric digital elevation model (DEM) displays depth as a continuous color gradient. Radio tags were monitored from the cliff tops near South Bay Harbor. Bathymetric data were provided by New Zealand's Institute of Water and Atmospheric Sciences (NIWA) and are discussed in detail in Lewis and Barnes (1999).

outfitted with radiotags. The tags were monitored day and night from elevated shore stations, providing the first glimpse of night-time behavior for this New Zealand deep water area, and with comparison possibilities to radiotagging carried out a decade earlier in Argentina (Würsig and Würsig 1980). Dive duration was quantified by timing signal silences caused by dolphins diving below the surface. Dives less than 30 seconds were assumed to be part of the respiration pattern, and those greater than 59 seconds were considered feeding dives (Würsig et al. 1991, Cipriano 1992). The stomachs and guts of 26 individuals were removed and opened. All hard parts, otoliths, squid beaks, eye lenses, intact crustaceans, and mandibles/jaws were sorted, counted, and preserved. Prey were identified to species when possible, and the minimum number of individuals consumed was estimated for each species or group (Cipriano 1992).

Benoit-Bird et al. (2004)

Over the course of four nights during the austral winter of 2002, researchers towed an echosounder along transect lines that crossed the Kaikoura Canyon (Figure 5.3). The echosounder used a pulse 130 μs long and 200 kHz, and transmitted its signal in a 10° cone directed downwards. The echosounder could detect dolphins and prey as deep as 156 m. It was connected directly to an onboard laptop computer, and returning echoes were displayed in real time as a series of calibrated colored dots. Additionally, onboard observers scanned for dolphins and recorded dolphin surface behavior.

Benoit-Bird et al. (2009)

In the late austral summer of 2006, a team set out to examine the effects of night-time light patterns on DVM and dusky foraging behavior. On seven nights, spread unevenly across two lunar cycles, the team towed a newer generation three-echosounder array along transects that crossed the Kaikoura Canyon. The transects were located in the same general area, but were not the same as those used in the previous study. A 38 kHz echosounder used a pulse length of 512 μs, projected its signal in a 12° conical beam, and could detect targets down to depths of 1000 m. The 120 and 200 kHz echosounders each used a pulse length of 256 μs, projected signals in a 7° conical beam, and could detect targets down to 300 and 200 m, respectively. At the conclusion of this study, the team was able to compare both dusky and DSL behavior between summer 2006 and winter 2002, and, in limited fashion, across lunar periods.

RESULTS

Each study allowed for descriptions of a particular aspect of dusky night foraging, and generated plenty of questions for future research. The first study quantified dive times of individuals and, paired with gut content analyses, indicated

FIGURE 5.3   This cartoon depicts the typical arrangement of the boat, the towfish, and the targets during acoustic surveys. The towfish was lowered to approximately 1 m below the surface, and towed slightly behind the boat at 3–5 knots. The echosounders use conical sound beams pointed downwards, towards the bottom. The sound bounces off objects in its path, such as dolphins or fish, and returns to the echosounders.

that duskies were carrying out deep foraging dives at night (Cipriano 1992). The second study confirmed that duskies were diving at night on the rising DSL and explored how duskies interacted with the DSL during dives (Benoit-Bird et al. 2004). The third study examined how night-time light regimes, which influence DVM, might affect dusky dolphin interactions with the DSL (Benoit-Bird et al. 2009). We briefly discuss the major findings here.

Würsig et al. (1991) and Cipriano (1992)

Tracking confirmed that dolphins moved offshore and into deeper waters at night. The frequency of both short and long dives increased at night and was interpreted as an increase in deep foraging dives coupled with more "recovery" respirations. Based on dive times alone, Cipriano hypothesized that at night duskies dove to 75–200 m, depths consistent with foraging on the rising DSL (Würsig et al. 1991, Cipriano 1992). Gut content analyses of stranded dolphins confirmed that duskies were primarily eating lanternfishes and squids, species commonly associated with the DSL, and illustrated that diet composition changed seasonally. The diversity of lanternfish species was greatest in summer and the proportion of squid eaten increased in winter (Cipriano 1992).

This work supported the hypothesis that duskies feed at night on the DSL, but provided no direct observations of foraging behavior. It was still unclear when and at what depths duskies interacted with the DSL, and if they did so individually or in coordinated groups.

Benoit-Bird et al. (2004)

This study provided initial observations of duskies overlapping spatially and temporally with the DSL. Dusky dolphins and the DSL were readily distinguished from each other and the background by unique characteristics of the returning echoes (Figure 5.4). The simultaneous acoustic and surface visual observations showed that dusky dolphin surface behavior was not a good indicator of night-time foraging. Surface behaviors commonly associated with foraging began 2 hours before duskies overlapped with the DSL or other possible prey concentrations in the upper 130 m of the water column. Once the DSL crossed this depth threshold, duskies tracked it, diving about 1 m deeper than the shallowest area of high prey concentration, until the high concentration region sunk below 130 m. The DSL was accessible to dusky dolphins for about 12–13 hours and attained a minimum depth of 29–49 m during the winter study period.

As the DSL changed depth, foraging tactics changed. Dolphins dove individually when the layer was deep and formed coordinated subgroups of up to five animals when the layer was near the top 40 m of the water column. Group size increased as the DSL approached the surface, and declined when the DSL started its descent after midnight. When dolphins were interacting with the DSL, they were most often detected in small subgroups; when dolphins were detected outside of the DSL, they most often swam alone. Of approximately 960 dolphin detections, including non-foraging animals, approximately 300 were individual dolphins, 240 were pairs, 200 were groups of three, 160 were groups of four, and 60 detections were of groups of five (from Figure 4 in Benoit-Bird et al. 2004). The remote echosounding technique could not determine whether prey herding was taking place (Benoit-Bird et al. 2004).

This brief winter study illustrated that dolphin dive behavior was closely tied to the movements of the DSL, and that foraging tactics are flexible on the scale of one night. However, the diel vertical migration of the DSL changes with seasons and lunar phase (Clarke 1970, Blaxter 1974, Koslow 1979, Haney et al. 1990, Balino and Aksnes 1993), and it was still unknown how these changes would affect dusky diving behavior.

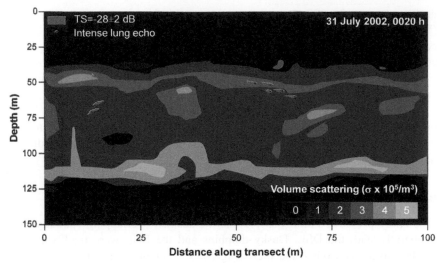

FIGURE 5.4   Echoes returning from the deep scattering layer (DSL), dolphins, and other objects are received by the echosounders and plotted in real time by a shipboard computer. These data were collected during the winter of 2002. The DSL is shown in blue-green as sound scattering isosurfaces. Echoes from dusky dolphins are indicated by the yellow points consistent with a narrow range in target strength (TS) values. These values surround a single extremely strong echo that is likely from the lungs of each dolphin. The coordinated dolphin subgroups that characterized the 2002 study are clearly visible (redrawn from Benoit-Bird et al. 2004).

Benoit-Bird et al. (2009)

Length of night-time appeared to affect DSL behavior and the amount of time duskies forage. The general characteristics of the summer DSL, including relative abundance and minimum depth of approximately 35 m, were similar to the winter DSL, but in summer the layer was shallower than 150 m for only 7–9 hours. As in winter, duskies tracked the DSL, diving within the layer, though they were not consistently 1 m deeper than the shallowest area of high concentration. No other significant seasonal differences were discernable in the DSL (Benoit-Bird et al. 2009), although it is possible that the layer composition or distribution of preferred prey changed seasonally and went undetected. Such changes could influence seasonal changes in dusky movement patterns and behavior.

Unlike winter 2002, no coordinated subgroups were detected during the summer 2006 surveys. However, duskies appeared to be part of larger and loosely organized groups (Benoit-Bird et al. 2009). It is possible that subgroup formation is influenced by how long a layer is accessible and how deep it is (and how much time and energy the dolphins need to expend to reach food between needing to breathe every several minutes). Distribution of prey items within the DSL may influence the effectiveness of subgroups, but it has not been possible to assess the distribution of specific prey types within the Kaikoura DSL.

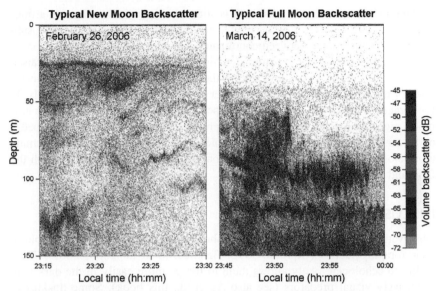

FIGURE 5.5 Presented here are the raw echo returns from the 120 kHz echosounder gathered during the 2006 surveys. Depth is on the y-axis; the top 150 m of the water column are displayed. The x-axis shows local time. Fifteen minutes of data are shown for each night. The color of the dots represents the intensity (dB) of the echoes and is a rough indication of the density of scatterers. These "snap shots" illustrate the trend of more scattering, and therefore more organisms, with increasing lunar illumination. There is also a clear trend of decreased scattering in the upper 50 m of the water column with increasing lunar illumination. At 11:30 at night during the new moon, there is a concentration of organisms in the upper 50 m. During the full moon, the area of concentration is deeper than 50 m at midnight.

Despite limited sampling, lunar cycle trends were evident in both DSL behavior and dusky dolphin relative abundance. Maximum and mean backscatter, measures of the density of animals detected, both decreased with decreasing lunar illumination (i.e. approaching the new moon) (Figure 5.5). Additionally, the number of larger potential prey, on the scale of 10 cm, present in the DSL increased with decreasing lunar illumination. Finally, the amount of time the DSL was accessible to duskies increased with decreasing lunar illumination; the layer was in the top 150 m of the water column for approximately 9 hours during new (no) moon and 7 hours during full moon. There was a general trend of dusky dolphin relative abundance decreasing with decreasing lunar illumination, and no dolphins were observed during the one new moon night sampled.

Over the past 20 years of dusky research, it has been speculated (Cipriano 1992) and confirmed (Benoit-Bird et al. 2004) that duskies interact with the DSL as it rises at night. It was also speculated based on seasonal diet changes (Cipriano 1992), and confirmed by active acoustic techniques (Benoit-Bird et al. 2009), that duskies employ seasonally specific foraging tactics. Seasonal and monthly changes in night-time light seem to affect the DVM of the DSL in

Kaikoura (Benoit-Bird et al. 2009). It is still unknown whether there is a minimum time the DSL must be accessible for duskies to form coordinated subgroups, or if the choice between coordinated and solo feeding is more strongly influenced by changing prey patterns within the DSL. More questions are raised by the lack of dolphin sightings during the new moon when the DSL is available for the longest period during summer. It is unlikely that duskies simply do not forage during new moon, and we suspect that our transects on only one new moon night simply have missed the dolphin feeding area that night. To better understand dusky night-time foraging choices, future work will need to examine more closely changes in DSL composition and the effects of light on both the DSL and dusky behavior.

DISCUSSION

Observations of feeding dolphins have shown duskies to be highly flexible in their foraging tactics (Cipriano 1992, Benoit-Bird et al. 2004, 2009). Their ability to echolocate likely allows them to forage more easily in the dark than exclusively visual predators (see also Au et al., this book). Some duskies in New Zealand switch between shallow-water prey herding and individual foraging in Admiralty Bay in the winter to deep-water night foraging when in Kaikoura (Markowitz 2004).

When night foraging, duskies can forage in a coordinated manner or in loosely organized groups (Benoit-Bird et al. 2004, 2009). The possible decreasing abundance of foraging dolphins approaching the new moon (but with scarce data) hints that duskies may use lunar phase-specific foraging tactics. Additionally, the seasonal change in diet suggests that duskies may use seasonally specific foraging tactics to accommodate a changing prey base. The decision to forage at night in Kaikoura is influenced by the predictability and abundance of prey in the DSL. The mechanisms for choosing the particular night foraging tactics are unknown, although the choice may be influenced at least in part by night-time light levels (Benoit-Bird et al. 2009).

According to optimality models, animals choose foraging tactics to maximize net caloric gain and thereby improve fitness (Pyke et al. 1977). Duskies rarely forage during the day in Kaikoura (Cipriano 1992, Markowitz 2004), but commonly forage at night on the shallowest area of high prey concentration (Benoit-Bird et al. 2004, 2009). Day foraging by herding prey is likely unprofitable in this area; the Kaikoura Canyon dominates the bathymetry (Lewis and Barnes 1999), making shallow water habitats favored by schooling prey scarce. Deep-water fishes and invertebrates associated with the canyon form an abundant prey base for duskies, but they are out of the dolphins' diving range until the DSL enters the top 150 m of the water column at night (Benoit-Bird et al. 2004; for night-time foraging data off South Africa, see Chapter 14). Deep dives are energetically expensive (Williams et al. 1999); duskies experience a higher net gain per prey item when the DSL is closest to the surface. Therefore, shallow foraging dives at night are likely the most energetically profitable foraging tactic in Kaikoura.

In the patchy open ocean environment, duskies may forage in groups to improve foraging efficiency (Eklov 1992) or prey detection capabilities (Baird et al. 2001), or to reduce their individual risk of predation (Hamilton 1971). Off Kaikoura, duskies were detected foraging in groups both during winter and summer. In winter, formation of coordinated subgroups increased with patchiness within the DSL; there was no evidence of active concentration of prey by dolphins (Benoit-Bird et al. 2004), although it is possible that such efforts or resulting prey concentrations went undetected. The size of dolphin subgroups increased as the DSL approached the surface, implying that groups become more effective as prey density decreases and as dolphins spend less time traveling to and from the surface to the prey (Benoit-Bird et al. 2004).

Predation pressure may influence the type of groups formed. Larger coordinated groups may be detected more easily by predators (Ioannou and Krause 2008). In summer, when killer whale sightings (Dahood et al. 2008) and large black shark and seal shark (family Dalatiidae) catches are highest (Peter Bradshaw, 2006, personal communication), duskies forage in larger loosely organized groups (Benoit-Bird et al. 2009). These dispersed groups may allow duskies to detect their predators more readily while not drawing the predators' attention as much as a closely coordinated subgroup (see also Chapter 7). Because there is little evidence of prey herding in Kaikoura at night, it is likely that the advantage of group foraging lies in increased prey detection capabilities, increased safety from predators, or both.

Night-time light regimes appear to influence foraging tactics employed by duskies, likely through changes in DSL composition and the timing and extent of diel vertical migration (Clarke 1970, Blaxter 1974, Balino and Aksnes 1993). Most studies of duskies in New Zealand have focused on seasonal changes in behavior or diet (see Cipriano 1992, Markowitz et al. 2004, Dahood et al. 2008), and acoustic studies confirm that there are significant seasonal differences in the timing of DVM (Benoit-Bird et al. 2009). Generally, as lunar illumination decreases, the time the DSL is in the upper 150 m of the water column increases, the proportion of larger animals in the DSL increases, and dusky relative abundance decreases (Benoit-Bird et al. 2009). Together, these findings suggest, but do not prove, that duskies use lunar phase-specific foraging tactics.

While data on dolphin abundance over lunar phase were limited, there was a trend of decreasing dolphin relative abundance with decreasing lunar illumination. There are several potential explanations for the scarcity of dolphin sightings approaching the new moon. While night-time light levels are low, at times there may be sufficient illumination to allow effective vision. In low lunar illumination conditions, foraging efficiency may be improved by limiting the prey's ability to detect dolphins, by the decreased depth of the prey, or by the greater availability of larger prey items with higher number of calories per item, or all of these. Under such conditions, dolphins may satiate quickly and move out of the area to pursue other activities. Conversely, foraging in low light conditions may be inefficient if the dolphins rely on moonlight to detect prey. Under such conditions, dolphins may choose to give up foraging in dangerous canyon waters to remain in safer or more profitable areas. The relationship between night-time light levels, DVM, and dusky behavior is complicated. We do not know if the dolphins' foraging abilities are affected directly by changing light levels, or if they are responding to light-induced changes in the DSL. This uncertainty raises tantalizing questions about the flexibility of dusky dolphin foraging tactics, particularly at relatively short time-scales.

At every temporal scale studied (nightly, seasonal, and now lunar), duskies show flexibility in foraging behavior. Duskies are able to switch between day and night foraging, an ability that has been noted in other visual and non-visual foragers (Shimek and Monk 1977, Ribic 1982, Beauchamp and McNeil 2003). Light appears to be important to night-foraging dolphins and influences the tactics they use (Benoit-Bird et al. 2009). Unlike day foraging, where it is well established that light is essential for prey detection and capture (Würsig et al. 1990, Vaughn et al. 2007) and the maintenance of social groups of land animals (Beauchamp 2007), the complicated relationship between duskies, their DSL prey, and night-time light is still little understood.

ACKNOWLEDGMENTS

We thank C. Waluk, I. Annsman, M. Kunde, P. Bradshaw, and B. Würsig for their invaluable assistance with fieldwork. We thank R. Vaughn, K. McHugh, B. Würsig, D. Biggs, and K. Winemiller for reviewing earlier versions of this chapter. We thank A. Fritz and Jake for their support while we were writing this chapter.

Chapter 6

Dusky Dolphins Foraging in Daylight

Robin L. Vaughn,[a] Mariana Degrati,[b] and Cynthia J. McFadden[c]

[a]*Marine Mammal Research Program, Texas A&M University, Galveston, Texas, USA*
[b]*Centro Nacional Patagónico, CONICET, Chubut, Argentina*
[c]*Leander, Texas, USA*

We follow a group of 5–8 resting and socializing dusky dolphins in Outer Admiralty Bay, New Zealand. They meander around awhile, and then join lazily with other dolphins. Suddenly, they begin rapid travel towards the middle of the bay. Fast travel is interspersed with occasional porpoising leaps, allowing the dolphins to efficiently breathe while burst swimming. Several minutes later, we see a distant group of dolphins doing high clean ("noiseless") leaps, seemingly diving deep to their prey. After 8 minutes of travel, our dolphins finally reach the distant group, and they join in diving deeply, possibly to herd prey to the surface. Clean leaps are interspersed with horizontal burst-swims, and dolphin activity is spread over an area about 100 m wide, as they seem to be trying to contain prey. Slowly the feeding/herding activity becomes less mobile, and it appears that the dolphins have herded a prey ball into a stationary aggregation.

We slip into the water, swim close to the feeding activity, and see about 15 dolphins diving down. In the emerald clearness of the water, a tiny prey ball ascends into sight. The top surface of the tightly-packed prey ball is only 4 m^2, which appears to be too small to feed the dolphins, gannets (about six), and shearwaters (about 17) that are consuming fish or scraps. The dolphins are concurrently herding and capturing pilchards as they glide by the prey ball, tilting their bellies towards the ball just before taking a fish. Their movements are graceful, almost like an underwater circular dance, all clockwise around the ball as seen from above. Occasionally, a bubble train streams behind a gliding dolphin, or a vertical column of bubbles is emitted. Their efforts appear to be uniformly focused on containing and eating as many fish as possible, and they feed continually as the prey ball slowly ascends. After about 12 minutes, the prey ball has reached the surface. The dolphins continue to feed for several more minutes, then, perhaps satiated, they split into subgroups and slowly travel away. Shortly after the dolphins finish feeding, the prey ball, apparently no smaller after the attack, swims away horizontally. In the water below us, a group of spiny dogfish continues to feed on scraps of fish.

Robin Vaughn, Admiralty Bay, New Zealand, October 12, 2005

INTRODUCTION

Foraging by social carnivores consists of locating, containing/herding, and capturing prey (Heithaus and Dill 2002b). Ecological factors influence when it is beneficial for individuals to coordinate behaviors to forage, and comparisons within and between species increase our understanding of how ecology influences foraging behavior (Krebs and Davies 1993). Groups may locate prey more rapidly than individual foragers due to coordination of activities or mutual parasitism. Groups may also better contain and capture prey than individuals, although competition increases with group size. For marine predators, foraging in coordinated groups can facilitate successful feeding on patchily distributed schooling fishes, which are often difficult to locate, contain, and capture.

Prey herding by dusky dolphins (*Lagenorhynchus obscurus*) has been studied in the Marlborough Sounds (mainly Admiralty Bay, 40°57'S, 173°55'E), New Zealand, and in Golfo San José (42°20'S, 64°20'W) and Golfo Nuevo (42°40'S, 64°40'W), Argentina. An examination of dusky dolphin foraging tactics and ecological conditions in both locations provides an opportunity for increasing our understanding of the relationships between behavior and the environment.

In this chapter we review what is known about daytime foraging, and form hypotheses on how ecology may influence daytime foraging behavior. First, we discuss variation in marine foraging behaviors, focusing on prey herding. Second, we describe how dusky dolphins coordinate behaviors to forage on schooling fishes in Admiralty Bay and in Argentina. Third, we describe differences in dusky dolphin foraging behaviors between these study sites. Fourth, we examine how ecological differences between study sites may explain foraging behaviors. Throughout the chapter, we include interspecies comparisons to place dusky dolphin daytime foraging behavior within a broader context, and to aid in the formation of hypotheses for future studies.

HERDING AS A MARINE FORAGING SPECIALIZATION

Predators herd schooling fishes by swimming around and under them to frighten them into tighter prey balls, or to move them. Delphinids exhibit diverse foraging tactics, and at least 14 of 33 species herd prey (see Wells et al. 1999, Connor 2000, Heithaus and Dill 2002b for summaries). Herding has also been observed in predatory fishes (Hiatt and Brock 1948, Schmitt and Strand 1982, Misund 1993, Robinson and Tetley 2007), penguins (Harrison et al. 1991), humpback whales (*Megaptera novaeangliae*; Clapham 2000), and minke whales (*Balaenoptera acutorostrata*; Kuker et al. 2005).

Predators may herd prey to cut off prey escape options by trapping prey against permanent (e.g., the surface of the water, the shore, or a dock) or temporary (e.g., a curtain of bubbles or other predators) barriers (Wells et al. 1999). Killer whales (*Orcinus orca*) coordinate behaviors to trap harbor seals (*Phoca vitulina*) underwater in rock crevices or caves, by taking turns spending time at depth (Baird and Dill 1995). Driver bottlenose dolphins (*Tursiops truncatus*) herd schools of fish against barrier dolphins in Florida (Gazda et al. 2005). It may also be easier for predators to capture fish from tightly herded prey balls due to fish disorientation or depolarization that occurs due to fright, decreased oxygen within the prey ball, or the use of sounds by predators (Würsig 1986, Norris and Schilt 1988). The most effective herding tactic depends on prey escape tactics and predator foraging specializations. Yellowtail amberjack (*Seriola lalandei*) in California herd schooling jack mackerel (*Trachurus symmetricus*) against the shore, to counter the latter species' escape tactic of schooling in open water (Schmitt and Strand 1982). Conversely, in the Gulf of California, yellowtail amberjack herd grunts (*Lythrulon flaviguttatum*) away from shore, to decrease the probability that this latter prey will escape into reef crevices.

Cetaceans use diverse behaviors to herd prey horizontally or vertically in the water column. Horizontal herding includes tail slaps or breaches at the periphery of prey balls (Atlantic spotted dolphins, *Stenella frontalis,* Fertl and Würsig 1995), and swimming in circles around the prey ball (common, *Delphinus delphis*, and Atlantic spotted dolphins, Clua and Grosvalet 2001). Vertical (usually towards the surface) herding includes emitting curtains or streams of bubbles (humpback whales, Clapham 2000; killer whales, Similä and Ugarte 1993), and swimming under prey balls with white bellies facing prey (killer whales, Similä and Ugarte 1993; Atlantic spotted dolphins, Fertl and Würsig 1995). Sound and pressure waves may also be useful for herding prey. Killer whales in Norway and Iceland emit a low-frequency call just before they use tail slaps to stun fish, and both the sound and slap appear to cause schooling herring (*Clupea harengus*) to cluster more tightly (Simon et al. 2006).

HOW DUSKY DOLPHINS COORDINATE BEHAVIORS DURING FORAGING

During the day, Argentine and bay-living New Zealand dusky dolphins primarily feed on schooling fishes, in contrast to their night-time feeding behavior (for example in the deep waters of Kaikoura; see also Chapter 5) that is focused on a deep scattering layer (DSL). Daytime foraging behavior studies have been conducted in Golfo San José, Argentina from 1973 to 1976 (Würsig and Würsig 1980, Würsig 1982, 1986, Würsig et al. 1991), in Golfo Nuevo and Golfo San José, Argentina from 2001 to 2005 (Degrati 2002, Degrati et al. 2003, 2008, Garaffo et al. 2007), and in Admiralty Bay and adjacent Current Basin, New Zealand from 1998 to 2006 (McFadden 2003, Benoit-Bird et al. 2004, Markowitz 2004, Markowitz et al. 2004, Vaughn et al. 2007, 2008, 2009). Most of these studies have focused on above-water data. A comparatively small amount of underwater data were recorded in Golfo San José from 1973 to 1976, and more in Admiralty Bay from 2005 to 2006 (Vaughn et al. 2007, 2008, 2009).

Bay-living dusky dolphins in Argentina and New Zealand have broad similarities in foraging behaviors and ecology. The above study locations occur at the same approximate latitude (41°–42°S), and feature relatively shallow, enclosed bays, with roughly similar prey species aggregated in patches. Dolphins in both study areas at times coordinate behaviors to locate, herd to the surface, and capture prey; and they forage in conjunction with various bird species, pinnipeds, and sharks.

Prey species

Dusky dolphins in both locations feed primarily on fishes from the order Clupeiformes. In Argentina, dusky dolphins consume primarily anchovies (*Engraulis anchoita*; Koen Alonso et al. 1998), but also feed on young, schooling, pelagic hake (*Merluccius hubbsi*) and pelagic Argentine shortfin squid (*Illex argentinus*) over Continental Shelf waters. In the Marlborough Sounds, dusky dolphins consume pilchard (*Sardinops neopilchardus*), anchovies (*E. australis*), garfish (*Hyporhamphus ihi*), and yellow-eyed mullet (*Aldrichetta forsteri*; Duffy and Brown 1994, Vaughn et al. 2007), but pilchards appear to be the most important prey (Vaughn et al. 2007). Admiralty Bay contains pilchard, sprat (*Sprattus* spp.), and anchovies (McFadden 2003), but pilchard may be more abundant than anchovies (Robertson 1978, Paul et al. 2001).

Seasonality

In Argentina and New Zealand, dolphins seasonally herd fish schools into stationary balls at the surface. In Argentina, dolphins herd prey mainly from spring to fall (Würsig and Würsig 1980); in New Zealand, they herd prey from late winter to mid spring (Vaughn et al. 2007). In both areas, dolphins are present for months before prey herding occurs (Würsig and Würsig 1980, Vaughn et al. 2007), which suggests that prey are present during this time, but not distributed such that herding is beneficial.

Locating prey

Dusky dolphins locate prey directly using echolocation or vision, and indirectly, by cueing into other dolphins that have located prey. In both locations, dolphins search for prey in small groups that are in close proximity to each other, although there are differences in coordination between groups when locating prey (see below). In both Argentina and New Zealand, foraging groups also arrange themselves in a linear fashion while searching for prey.

Dusky dolphin herding techniques

After locating prey, dusky dolphins at times appear to chase prey in a coordinated manner, as evidenced by synchronous diving and bursts of rapid movement

(Figure 6.1; Markowitz 2004). This behavior increases the effectiveness of chasing prey, since dolphins can then encircle fishes and herd them into stationary balls at the surface (Figure 6.2). In turn, the herding of prey balls appears to facilitate the capture of individual fish. Prey in a stationary, tight ball appear to be easier to capture than mobile prey, and prey that are closer to the surface are more accessible than prey found deeper in the water column. Dolphins in New Zealand herd fish not just to the surface, but also horizontally against the shore, a point of land, or the hull of a boat (Duffy and Brown 1994, McFadden 2003).

It is often not clear whether the dolphins are intentionally herding prey, or if the prey are herded as a by-product of prey capture behaviors, even when prey balls can be observed underwater. In New Zealand, dolphins do appear to intentionally herd prey to the surface (at least at times). Prey balls typically ascend during feeding bouts, and dolphin behaviors appear to cause them to ascend since dolphins swim by the bottom half of prey balls more frequently than the top half (Vaughn et al. 2008). Behaviors observed in New Zealand that may function in herding and subsequent capture include turning white bellies towards a prey ball as a fish is captured, blowing columns of bubbles (although this may also function in communication), and attacking fish concurrently with other dolphin/s from opposite sides of the prey ball (Box 6.1).

Prey capture techniques

Dusky dolphins in both locations coordinate behaviors to capture prey. In Argentina, one or a few dolphins at a time attack the prey ball by ascending from depth, capturing 1–5 fish per dolphin during any one dive (Würsig and Würsig 1980). Fish are captured by swimming through the center of the prey ball or by the edge. Dolphins that are not capturing prey appear to continue herding, at times via behaviors such as noisy leaps or tail slaps at the outside of the prey ball.

FIGURE 6.1   After locating prey, dusky dolphins at times coordinated behaviors to chase prey. Dolphins may spread out to chase fish (top, Golfo Nuevo, Argentina), when large schools are present, whereas dolphins may cluster together to chase fish (bottom, Golfo San José, Argentina, photo courtesy of Paula Faiferman) when small schools are present.

FIGURE 6.2   After chasing a fish school, dusky dolphins at times were able to encircle them, then to herd the fish school into a stationary ball at the surface. Dolphins appeared to coordinate behaviors during prey herding, based on behaviors observed above-water (top, Golfo Nuevo, Argentina), and underwater (bottom, Admiralty Bay, New Zealand).

In New Zealand, dolphins in the vicinity of the prey ball appear to mainly capture prey, although prey capture behaviors may also function in herding. During prey captures, more than one individual often attacks a prey ball at one time, but in coordinated fashion. Dolphins typically circle clockwise (as seen from above) around the prey ball in a pinwheel manner as they capture individual fish along the edge of the prey ball (Figure 6.3, top).

Clockwise circling during prey capture may be related to a mammalian right-sided preference (e.g., the right-handed preference of humans). Bottlenose dolphins that strand fish schools on banks capture them by sliding up onto banks on their right sides (Petricig 1995). Similarly, gray whales (*Eschrichtius robustus*) typically roll onto their right side to feed on bottom-dwelling invertebrates (Woodward and Winn 2006). At any rate, the clockwise circling facilitates dolphins staying in a loose school and not getting into each other's way.

When feeding on large prey balls, dusky dolphins at times coordinate behaviors to attack a small focal region of concentrated prey, and occasionally may swim directly through the middle or through the periphery of the ball to capture prey. New Zealand dusky dolphins appear to swim directly through

BOX 6.1

Prey Ball Herding

Bernd Würsig and art by Emily Gillespie

It is difficult to illustrate the stages of prey ball herding using photography because of low water clarity, the rapidity of action, and the air–water interface. At times, an artist's rendering better represents the situation. The illustration here, taken from many video and still photo sequences, shows a small dusky dolphin subgroup beginning to herd a small group of anchovies. The dolphin on the right is nosing towards the fish school, possibly producing an intensive series of acoustic click trains, or buzzes, that may serve to frighten and tighten the school. The next dolphin down is flashing her belly towards the school, thereby visually frightening the fish. The larger middle one has just left the bottom of the school, and still has a fish in his mouth, sticking out of the left side of his gape. The two in the distance farther below are descending below the fish to bring the ball to the surface, and possibly to aggregate more fish from depth (bottom dolphin).

Almost all dolphin rotation around a prey ball tends to be clockwise as seen from above, possibly due to dolphins generally being "right-oriented," and perhaps also as a strategy for them to not get in each other's way.

There is much leaping around the periphery of prey balls, and this seems to serve as a way for dolphins to surface from depth, overshoot the surface to breathe, and use the weight of their bodies to go down to depth again. The leap shown here could be one of these, but it also has a side-sliding emphasis that may have it end in a right-sided slap onto the water near the small fish school. This may serve to further tighten the school, and may also communicate to other dolphins farther away. By now there should be birds beginning to aggregate above the activity, but these have been left out in favor of concentrating on the dolphins and fish school . . .

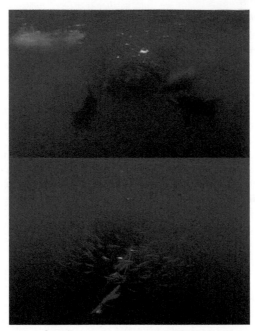

FIGURE 6.3   Dusky dolphins in New Zealand typically capture fish from the edge of prey balls in a pinwheel formation (top), but at times capture fish by swimming directly through prey balls (bottom).

prey balls more often when only one individual is capturing fish at a time (Figure 6.3, bottom).

Dusky dolphins tilt their bellies towards prey balls during prey captures, which appears to allow them to effectively use vision to capture individual fish (similar to Herzing 2004). The visual fields of the left and right eyes overlap just at and below the jaw (Norris et al. 1994). It is also possible that dusky dolphins tilt their bellies towards prey balls to startle the fish into tighter balls. Tilting their sides towards prey balls is probably also effective for herding fish, as an experiment on *Spheniscus* penguins reveals (Wilson et al. 1987). These penguins have distinct black and white patterns on their sides that are similar to those of dusky dolphins, but which differ from the sides of a typical counter-shaded penguin. An anchovy (*E. capensis*) school exposed to a penguin model with this unique side coloration became depolarized more frequently and for longer than when a less disruptively colored penguin model was used (Wilson et al. 1987), possibly because the coloration interfered with the ability of fish to determine the locations of their neighbors.

Multi-species associates

Seabirds associate with dusky dolphins in both locations, although surface-feeding seabirds are most numerous in Argentina, while diving seabirds predominate in

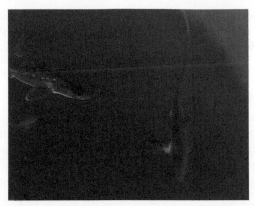

FIGURE 6.4   Based on three observations in New Zealand, dusky dolphins appeared to leave prey balls when thresher sharks (*Alopias vulpinus*, shown at right) arrived. Smaller dogfish sharks (*Squalus acanthias*, shown at left) typically fed below and behind prey balls, so usually did not interfere with dolphin feeding.

New Zealand. In Golfo San José, Argentina in the 1970s, seabirds that associated with feeding dusky dolphins were, in order of decreasing numbers: black-headed gulls (*Larus dominicanus*), cormorants (*Phalacrocorax* spp.), terns (*Sterna* spp.), and procellariiformes species (i.e., shearwaters, petrels, and albatrosses; Würsig and Würsig 1980). In Golfo Nuevo, Argentina, associated seabirds were shearwaters and black-headed gulls, terns, Magellanic penguins (*Spheniscus magellanicus*), and other procellariiformes species (M. Degrati, unpublished data). In New Zealand, associated seabirds were shearwaters, gannets (*Morus serrator*), gulls (*Larus* spp.), spotted shags (*Phalacrocorax punctatus*), and terns (Vaughn et al. 2007).

Pinnipeds and sharks also associate with feeding dusky dolphins in both locations. In Argentina, South American sea lions (*Otaria flavescens*) are often present at feeding bouts (Würsig and Würsig 1980); in New Zealand, fur seals (*Arctocephalus forsteri*) are often present (Markowitz 2004). In Argentina, large sharks (3–5 m) were infrequently observed in Golfo San José (Würsig and Würsig 1980); in New Zealand, spiny dogfish (*Squalus acanthias*) are frequently present, and barracuda (*Sphyraena* spp.) and large thresher sharks (*Alopias vulpinus*) were infrequently observed (Figure 6.4; Markowitz 2004, Vaughn et al. 2007).

DIFFERENCES IN DUSKY DOLPHIN FORAGING BETWEEN STUDY SITES

Despite the above broad similarities in dolphin foraging behaviors between study sites, finer scale scrutiny demonstrates that there are behavioral differences between locations. In Argentina, dolphins travel greater distances when searching for prey, spend more time searching for prey, converge into larger feeding groups after locating prey, have longer feeding bouts, feed in larger groups, and may even exhibit a greater degree of coordination than in

New Zealand. Several of these characteristics may simply be related to the sizes of bays, as these are much larger in Argentina (and many more dolphins "fit" into them) than for the Admiralty Bay study area of New Zealand.

Locating prey

In Golfo San José, dolphins sometimes move up to 50 km in one day (Würsig et al. 1991, see also Würsig and Bastida 1986), and movement patterns in Golfo San José and Golfo Nuevo suggest that dolphins search over longer distances and for longer durations to locate prey (Würsig et al. 1991, Degrati et al. 2003). In contrast, in New Zealand, dolphins travel over shorter distances to locate prey (Figure 6.5), although movement patterns have not been well-quantified. Behavioral budgets of dolphins in Argentina indicate that they

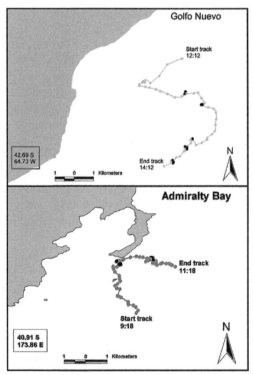

FIGURE 6.5   Typical movements of dusky dolphin groups during focal follows in Argentina (top), April 13, 2001, and in New Zealand (bottom), October 9, 2005. Similarities between locations in search patterns suggest that prey is distributed patchily. Differences between locations in distance traveled while searching for prey may be indicative of differences in prey patchiness. Black points indicate feeding behavior, while gray points indicate non-feeding behavior. Each figure represents 2 hours, and each point represents dolphin group location at 2-minute intervals. Maps were created using ArcView GIS version 3.2 (a) and 3.3 (b) (Environmental Systems Research Institute, Inc., Redlands, CA).

spend a large percentage of their day traveling (Table 6.1), which suggests that they spend more time searching for prey than in New Zealand.

Dolphins in Argentina coordinate behaviors when locating prey. In Golfo San José during the 1970s, dolphins foraged for prey by spreading out in small groups that were generally within acoustic range of each other (≤1 km; Würsig and Würsig 1980, Würsig 1986), then joined together into much larger groups once prey were detected. Dolphins appeared able to cue into a feeding aggregation of other dolphins and seabirds from a distance of ≤0.5 km via noisy leaping sounds (e.g., acrobatic leaps, slaps), ≤1 km via whistle-like sounds, and ≤8 km via in-air vision (e.g., cueing off diving seabirds and leaping dolphins). Dolphins performed noisy leaps, possibly but certainly not proved to "call" other dolphins to help them herd prey, and this was the most frequent type of leap during feeding bouts (Würsig and Würsig 1980). Other delphinids also appear to use noisy leaps for communication over long distances, which provides further support for this possibility. For example, white-beaked dolphins (*Lagenorhynchus albirostris*) trap fish schools between two groups by rapidly swimming towards each other (Evans 1982). This foraging tactic is at times preceded by tail slaps or noisy leaps, which may serve a communicative function. In all of these cases, noisy activities may also be a part of the rapid movements of the dolphins as

---

TABLE 6.1 Dusky dolphin (*Lagenorhynchus obscurus*) group sizes, daily activity budgets, and lengths of feeding bouts in Admiralty Bay, New Zealand and Golfo Nuevo and Golfo San José, Argentina

Location	% groups ≤20 dolphins	Max. no. dolphins/ group	% day traveling	% day feeding	Typical length feeding bout	Max. length feeding bout
Admiralty Bay	96%[a]	50[a]	About 25%[b]	About 25%[b]	Mean = 8 min[c]	42 min[d]
			29%[a]	18%[a]	Mean = 5 min[d]	
Golfo Nuevo	54%[e]	300	40%[e]	16%[e]	Mean = 8 min[f]	28 min[g]
Golfo San José	—	300[h]	—	—	—	3 hours[h]

[a]Pearson (2008)
[b]Markowitz et al. (2004)
[c]McFadden (2003)
[d]Vaughn et al. (2007)
[e]Degrati et al. (2008)
[f]Degrati et al. (2003)
[g]Degrati (2002)
[h]Würsig and Würsig (1980).

they "work" a prey ball, or may be primarily structured to frighten the prey and cause them to tighten more effectively (Würsig and Würsig 1980).

More recent research in Golfo Nuevo (Degrati 2002) and Admiralty Bay (McFadden 2003, Markowitz 2004) suggests that dolphin groups coordinate behaviors to a lesser degree when locating prey than the very large groups of prey and dolphins studied in the 1970s. Recent studies have found that clean leaps are the most frequent type of leap during feeding, and noisy leaps that would be expected to function in communication seldom occur (Degrati 2002, McFadden 2003, Markowitz 2004). Additionally, in Admiralty Bay, groups have only a slight tendency to increase in size during feeding bouts (McFadden 2003), in contrast to the large increases in feeding group sizes found in Argentina during the 1970s (Würsig and Würsig 1980). Since there is less tendency to aggregate concomitant with fewer noisy leaps, this may be further support for noisy leaps at least at times serving a broader communication function.

Degree of coordination during prey herding and capturing

There may be differences between locations in degree of coordination during herding and prey capturing. In Golfo San José, Argentina in the 1970s, dolphins appeared to first herd prey to the surface, not capturing prey until it was contained (Würsig 1986). In contrast, in New Zealand, dolphins concurrently herd and capture prey. However, it is also possible that prey were taken contained and eaten simultaneously in Argentina, and that the observers were simply not able to assess this adequately in real time (no video records were taken for those early studies). In the latter potential situation, no real difference between Argentina and New Zealand may exist.

Length of feeding bouts and group sizes as they relate to capturing techniques

Feeding durations were longer in Golfo San José in the 1970s than in Argentina or New Zealand in the 2000s (Table 6.1). Feeding groups of up to 300 dolphins were observed off Argentina (Würsig and Würsig 1980), while the maximum size of feeding groups in New Zealand in 2005 and 2006 was just 30 dolphins (Vaughn et al. 2007).

Coordinated behaviors may make it easier for dusky dolphins to capture fish when they feed in large groups. Dusky dolphins in Argentina at times herd and feed on anchovy until the fish become lethargic (Würsig 1986), while this fish lethargy has not been observed in New Zealand. The fish may become lethargic due to stress from prolonged feeding bouts, which may decrease the amount of oxygen around the fish school. Large groups (200–500) of Atlantic spotted dolphins in the North Atlantic also herd schooling fishes until they are lethargic (Martin 1986), which suggests that fish lethargy may occur when large numbers of dolphins feed for long durations.

HOW DOES ECOLOGY INFLUENCE DUSKY DOLPHIN FORAGING?

Possible similarities between locations in how ecology influences dusky dolphin foraging

Depth of prey

Depth of prey appears to influence costs and benefits of herding prey to the surface. In Admiralty Bay during winter, there is some evidence that fishes occur in deeper subsurface schools (Baker 1972a), coinciding with times when prey are rarely brought to the surface (Vaughn et al. 2007). In contrast, during spring, fishes appear to school closer to the surface (Baker 1972a), and dolphins more frequently herd prey to the surface (Vaughn et al. 2007).

Multi-species associates

Seabirds differ in their foraging behaviors (Figure 6.6), which probably influences how difficult it is for dusky dolphins to herd prey into stationary balls at the surface. However, most seabirds do not generally influence dolphin foraging

FIGURE 6.6 Differences in seabird foraging behaviors likely influence how difficult it is for dolphins to herd prey into stationary balls at the surface. As an example, in Argentina, large numbers of gulls and penguins feed with dusky dolphins. Gulls capture fish by grabbing them from the surface (bottom left), whereas penguins appear to herd prey (bottom right). Additionally, gulls feed alone (top left) and in large aggregations, but do not appear to coordinate behaviors. In contrast, penguins exhibit much sociality (top right), which may facilitate prey herding. Prey balls are probably more likely to descend when gulls are feeding, and less likely to descend when penguins are feeding. Above-water photos courtesy of Paula Faiferman.

behaviors to a measurable degree, with two exceptions. One exception is large plunge-diving gannets in New Zealand, which may scatter fish schools or cause them to descend (see below), thus making it more difficult for dolphins to herd prey. A second exception is Magellanic penguins in Argentina, which likely herd prey balls to some degree (see above), and thus may feed with dolphins in synergistic fashion.

Pinnipeds appear to interfere with the ability of dusky dolphins to effectively contain prey. Dolphins typically capture prey from the periphery of prey balls, although they may swim directly through prey balls in some (relatively rare) situations. In contrast, New Zealand fur seals typically dive directly into the center of prey balls, scattering fish by movements of their large pectoral flippers. Additionally New Zealand dusky dolphins seem to avoid fur seals to some degree, at times veering away from the pinnipeds during prey capture attempts. However, fur seals do not generally interfere with prey mobility in overall highly disruptive fashion (and not to measurable degree, Vaughn et al. 2009), perhaps due to the small number of fur seals typically present during feeding bouts (mean = 2 fur seals, Vaughn et al. 2007).

Large sharks do not seem to interfere with dolphin feeding in Argentina (Würsig and Würsig 1980), although broadnose sevengill sharks (*Notorynchus cepedianus*) represent a risk of predation for dusky dolphins inhabiting these bays (Crespi-Abril et al. 2003). In New Zealand, there is some evidence for avoidance of thresher sharks (Figure 6.4) by dusky dolphins (Vaughn et al. 2009). Based on three observations, dolphins appear to leave prey balls when thresher sharks arrive at the prey balls.

Possible reasons for differences in dusky dolphin foraging between locations

Differences in dolphin foraging behaviors between locations may be due to ecological differences between the convoluted small bays that comprise the Marlborough Sounds region of New Zealand, and the large bays found along the open coast of Argentina. In Argentina, prey appear to be distributed more patchily, and in larger schools. Bays are much larger, and greater numbers of dolphins are present in the bays. Additionally, the large, disruptive, plunge-diving gannets that feed with dolphins in New Zealand are not present off Argentina. Magellanic penguins are present in Argentina, and they may herd prey in a manner that is generally similar to dolphins, and are not as disruptive as plunge-diving birds.

How prey patches are distributed

Prey patches appear to be more widely dispersed in Argentina than in New Zealand, since movement patterns of dolphins off Argentina indicate that they travel greater distances when searching for prey than do New Zealand dusky dolphins (see above and Figure 6.5). Dolphins in Argentina also spend a greater percentage of their day traveling (Table 6.1), which suggests that prey patches are distributed farther apart in Argentina.

Differences in prey patchiness between locations may account for differences in dolphin foraging behaviors between locations, since social carnivores tend to exhibit a greater degree of coordination during foraging when prey are more difficult to locate (e.g., distributed patchily or unpredictably) (Gowans et al. 2008, see also Chapter 16). Chimpanzees (*Pan troglodytes*) that hunt meat (mainly *Colobus* spp.) in forest environments use a greater degree of coordination than those that hunt in savanna–woodland environments (Boesch and Boesch 1989). Chimpanzees in forest environments coordinate movements to capture prey by converging on prey from below, working together to corner prey in a tree. In contrast, chimpanzees in savanna–woodland environments do not strategically surround prey, but instead pursue prey from a single direction. Boesch and Boesch (1989) hypothesized that this difference in degree of coordination occurs because low visibility makes prey more difficult to find in forest habitats. There may also be more escape routes available to prey in forest habitats due to the three-dimensional nature of the forest canopy, necessitating increased coordination to result in a successful hunt.

Sizes of patches of prey

Prey patches in Argentina appear to be larger than those in New Zealand. In Golfo San José in the 1970s, a single school of anchovy could feed up to several hundred dolphins and thousands of seabirds (Würsig and Würsig 1980, Würsig 1986). Large aggregations of dolphins in Argentina (Table 6.1) at times fed in several nearby locations in a single area (e.g., 300 m across), possibly on multiple prey balls (Würsig 1986), or on different regions of a single prey ball (Degrati et al. 2008). Sonar data indicate that anchovy schools in Golfo Nuevo are up to 100 m in length (M. Degrati, unpublished data). In contrast, small groups of dolphins (Table 6.1) off New Zealand typically feed on a single small patch of prey (mean two-dimensional area = $8 m^2$; Vaughn et al. 2007).

Larger groups of dolphins such as those found off Argentina may be needed to herd large prey balls. In New Zealand, scarce large prey balls tend to descend during feeding while common small prey balls tend to ascend (Vaughn et al. 2009). Thus, it appears easier for the small groups of dolphins that are found off New Zealand to herd smaller prey balls.

Habitat area in relation to dolphin abundances and grouping patterns

The above differences in foraging behavior between locations are probably partly related to habitat area. Golfo Nuevo and Golfo San José, Argentina are deeper and much larger than Admiralty Bay and Current Basin, New Zealand (Table 6.2). Dolphin abundances in the Argentine Bays are also larger. In the 1970s, dolphin abundance in Golfo San José was at least 300 dolphins (Würsig 1986),

TABLE 6.2   Physical characteristics of Admiralty Bay, New Zealand, and Golfo Nuevo and Golfo San José, Argentina

Location	Area (km²)	Typical depth (m)	Maximum depth (m)
Admiralty Bay	160	30–50	105[a]
Current Basin	30	20–30	105[a]
Golfo Nuevo	2500	40–100	184[b]
Golfo San José	750	30–60	about 80

[a]Depth at French Pass, a deep channel that connects Admiralty Bay to Current Basin (McFadden 2003).
[b]Mouzo et al. (1978).

and is now about 200 dolphins, while abundance in Golfo Nuevo is about 300 dolphins (Loizaga et al. 2006). Admiralty Bay contains a mean of 220 dolphins during any given week during winter (Markowitz et al. 2004), but these are never together at one time. Differences in dolphin abundances between Argentina and New Zealand may also be related to differences in group sizes (Table 6.1). Group sizes in New Zealand are smaller than those found in Argentina, and smaller dolphin groups appear less able to successfully manage large prey balls than larger dolphin groups.

Longer feeding bout durations are associated with larger dolphin groups in both locations (Würsig and Würsig 1980, Degrati 2002, McFadden 2003), but it is unclear to what extent this is due to increased ability of dolphins to contain a prey ball versus a larger number of dolphins feeding until they are satiated. In New Zealand, larger numbers of dolphins appear better able to contain prey balls (Vaughn et al. 2009), and this may be due to coordinated foraging behaviors.

Individuals form groups for diverse reasons, including to forage more effectively, reduce risk of predation (see also Chapter 7), and for social or reproductive reasons (see also Chapter 8; Connor 2000, Acevedo-Gutiérrez 2002, Gowans et al. 2008). Optimality models predict that individuals form groups to forage when benefits outweigh costs (Krebs and Davies 1993). Some social carnivores forage in optimally sized groups (e.g., wild dogs, *Lycaon pictus*, Creel and Creel 1995). For other carnivores, group sizes are larger than optimal due to factors such as increased prey abundance (e.g., lions, *Panthera leo*, Stander 1992), and group sizes during foraging can also be influenced by risk of predation (Gowans et al. 2008). Smaller foraging groups can experience an increased risk of predation if they are detected, while larger groups may attract predator attention by being more conspicuous.

Predation risk and sociality may influence dusky dolphin group size to a lesser extent than does foraging in both locations. Killer whales and large sharks are the main predators of dusky dolphins in some situations (Constantine et al. 1998,

Heithaus 2001a), but they may minimally influence dolphin group size in the shallower bay waters discussed in this chapter (unlike the situation in open waters off Kaikoura, see Chapter 7). Killer whales were seldom observed in Golfo Nuevo (M. Degrati, unpublished data), Golfo San José (Würsig and Würsig 1980), or Admiralty Bay (R. Vaughn, unpublished data). Likewise, the relative scarcity in winter and early spring of sharks large enough to prey on dusky dolphins in Admiralty Bay reduces dolphin risk of predation. However, dusky dolphins in Argentina may at times be eaten by sevengill sharks (Crespi-Abril et al. 2003). Although predation does not need to occur frequently to influence grouping patterns (see Chapter 7), inter-group joining and splitting behaviors reveal that group sizes change most often during prey herding in semi-enclosed bay environments (Würsig and Würsig 1980, Pearson 2008).

In both locations, group sizes are probably constrained by abundance of available prey (similar to Lusseau et al. 2004), but this relationship is complex. The large groups found in Argentina appear either necessary to effectively prey, or larger than optimal due to abundant prey. Small groups of dolphins do not appear to effectively herd large patches of prey (Würsig 1986). However, the small prey balls available in Admiralty Bay appear adequate to feed all dolphins present, since prey balls that were observed underwater were not noticeably smaller post-feeding compared to pre-feeding. To optimize foraging, dolphin groups may stay close enough such that they can join together if a "good" prey ball is located, but far enough apart such that they can productively feed on scattered prey when necessary.

Multi-species associates—plunge-diving gannets versus herding penguins

Differences in numbers and types of multi-species associates between locations may also account for differences in dolphin foraging behaviors. One notable difference between locations is the presence of large plunge-diving gannets in New Zealand that are absent in Argentina. In both locations, longer feeding bouts are associated with greater numbers of seabirds (Würsig and Würsig 1980, Markowitz 2004). Since seabirds influence mobility of prey balls in New Zealand (Vaughn et al. 2009), the costs of herding prey probably increase as duration of feeding bouts increases. Relative costs may be greater in New Zealand than in Argentina, since plunge-diving gannets may scatter the prey ball or cause it to descend. Costs of herding prey may also be less in Argentina due to the presence of Magellanic penguins. There is evidence that similar penguins coordinate behaviors to herd prey to the surface (Wilson and Wilson 1990, Harrison et al. 1991), and Magellanic penguins also may do so (Figure 6.6).

CONCLUSIONS

In summary, we hypothesize that differences in dusky dolphin foraging behaviors between Argentina and New Zealand are due to differences in how prey are distributed, sizes of prey patches, dolphin abundances in each bay, and presence of plunge-diving gannets in New Zealand versus herding penguins in Argentina. In Argentina, prey patches appear to be farther apart and larger. This would increase the benefits of herding prey and decrease costs of feeding in larger groups. Since the larger Argentine bays hold more dolphins, dolphins are able to join together into larger groups to effectively herd large fish schools into stationary balls at the surface for longer periods of time. Large plunge-diving gannets that are present in New Zealand but not in Argentina also make it more difficult for dolphins to maintain prey at the surface. Magellanic penguins that are present in Argentina but not in New Zealand may feed with dolphins in synergistic fashion by herding prey, but this is presently a suggestion only. Over the long term, these balances of costs and benefits may have led to the apparently greater degree of cooperation that occurs during foraging in Argentina. Differences between Argentina and New Zealand in other multi-species associations, and depth of prey, appear to influence dusky dolphin foraging behaviors to a lesser degree.

Hypothesis-based and comparative studies among species are an effective way to examine how ecology influences behavior across a broad range of habitats and behavioral adaptations. Research on the relative influences of these factors increases our understanding of the behavior not only of dusky dolphins, but also of other carnivores that forage in coordinated manner. In marine environments, better knowledge of the relationships between ecology and behavior will better enable us to ascertain how human impacts such as dolphin watching (Coscarella et al. 2003, Dans et al. 2008), fisheries (DeMaster et al. 2001, Dans et al. 2003a), noise (McDonald et al. 2006, Weilgart 2007), and climate change (Learmonth et al. 2006, Simmonds and Isaac 2007), potentially affect the ability of delphinids to forage effectively.

ACKNOWLEDGMENTS

Our research on dusky dolphins in Admiralty Bay would not have been possible without funding from the Texas A&M University Departments of Wildlife and Fisheries Sciences and Marine Biology (Animal Use Protocol #2005-48), New Zealand Department of Conservation, Marlborough District Council, National Geographic Society, Earthwatch, and Erma Lee and Luke Mooney grants. The research in Argentina would not have been possible without funding from Agencia Nacional de Promoción Científica y Tecnológica (FONCYT PICT 01-4030A and PICT 11679), Fundación BBVA (BIOCON 04), Fundación Vida Silvestre Argentina, and institutional support from Centro Nacional Patagónico (CONICET) and the University of Patagonia. We thank Heidi Pearson and an anonymous reviewer for feedback that much improved this chapter. Background maps of New Zealand are courtesy of Eagle Technologies, Wellington, New Zealand.

Chapter 7

Predator Threats and Dusky Dolphin Survival Strategies

Mridula Srinivasan[a] and Tim M. Markowitz[b]

[a]Wildlife and Fisheries Sciences, Texas A&M University, College Station, Texas, USA
[b]School of Biological Sciences, University of Canterbury, Kaikoura, New Zealand

It's a day like any other. . . . The ocean off Kaikoura is surprisingly calm. For once, the forecast that is often not quite right matches the scene before us. The waters are a dark blue, glistening in the morning sun, providing ideal conditions to record dolphin behavior. We are eager to get out.

Just as we launch our inflatable Punua Aihe—*"Dolphin Young" in Māori—the dolphin tour boat crew casually remark to us—"There are orcas in the bay!" We hurry up to get the boat launched and out on the sea . . . for an "orca" day, a day with mammal-eating killer whales, is an exciting event in Kaikoura.*

We navigate across the waters towards the last known location of the orcas, using our hand-held GPS to guide us to the coordinates provided by our tour boat friends. Finally, we get to see the fins that tell it all. . . . Black and whites, with shades of gray smoothly break the glassy water.

As we move closer, the boat crew gets gear and datasheets ready. It's time to follow the orcas as they glide effortlessly along the coast at a speed similar to the top swimming speed of dusky dolphins, a bit above 10 km/h. Where will they go? How many are there? Are there some young ones? Where are the dusky dolphins? Will we see an attack?

Our captain charts the course. Keeping what we hope is a respectful distance of greater than 100 m, we drive our boat parallel to the killer whales. And then the clicking starts . . . the digital SLR camera fires rapidly at high shutter speed. Do the orcas know that they are being recorded for posterity?

Closer inspection reveals nine orcas in the group. The two males dominate the seascape with their huge, majestic dorsal fins . . . much like the male lion's mane, following or alongside the females and juveniles that make up the rest of the pod. Far off in the distance there's a third male, alone. All the orcas seem purposefully cruising in a set direction. For the moment, hunting doesn't appear to be on their minds. There are no dusky dolphins anywhere in sight, just a lone fur seal now and then. The dolphins are probably miles away or safely tucked in the shallows, cleverly hiding to avoid their major predator in these waters. The orca group separates into two, and we stick with the one closest to us. The other group has quietly slipped away in unseen waters. Luckily, we get some identifying photos before they disappear.

The orcas' routine seems set. They take deep dives and reappear ahead or just behind us, roughly every 5–6 minutes. Nevertheless, after a long dive and much unseen travel, it is difficult for us to predict where they will come up next. We slow the boat and scan ahead.

Without warning, a big blob of black and white fills the lens, way too close to focus! The six foot high dorsal fin of an adult male emerges from the water beside our little boat. In those few moments, you realize that the boat doesn't afford much protection should there be a misguided tail slap or one of those occasional half-breaches! We now have a curious orca pod. . . . A moment or two later, a juvenile orca leaves its mother's keep and swims alongside our boat with fish in mouth, perhaps showing off its catch. We don't know what fish, but are lucky enough to get a photo, and our fishermen friends later identify it as a grouper. Soon, fish and mammal are gone into waters below. It appears that dusky dolphins may not be on the menu today.

The hours pass quickly and we have left the shore way behind. The sea is getting rough, and the land is a thin line in the distance. It's time to head back. The orcas go on their way, and we on ours. We have learned a few things, but have to patiently wait for more of such days to start learning when killer whales are truly on the dusky dolphin prowl. Meanwhile, the dusky dolphins have also returned to their daytime routine of zigzagging back and forth near the Kaikoura Canyon. We spot a group as we approach closer to shore, and follow them to systematically record focal group behavior.

It's been a good day, with some filled-in sheets of information. Orca dives and surfacings, speeds, distances apart; dusky meanderings, speeds of the dolphins as well, depths of water. Quite a few identifications of both, useful for who is traveling with whom on this particular day. The data from these few observations may not reveal much, but with continued effort and time, we can gather clues to patterns and processes that govern predator–prey relationships in the submarine canyon ecosystem. It is good to be so close to nature . . .

Mridula Srinivasan, Kaikoura, New Zealand, March 2007

INTRODUCTION

A dusky dolphin (*Lagenorhynchus obscurus*) porpoises away at great speed from a stealthy and powerful killer whale (*Orcinus orca*)—the outcome of such an encounter is either survival, a flesh wound if lucky, or mortality. In a straight ahead race, the killer whale is the faster of the two, its powerful flukes propelling it through the sea at a pace no dusky dolphin can match. But the smaller dolphin is the more maneuverable, and has a few evasive tricks of its own, if it has enough warning time to put them into action. At times, the predator with stamina and strategy overcomes a scared and vulnerable prey, at other times the dolphin outwits its pursuer. On rare occasions, we witness a dolphin caught in the jaws of a killer whale, five times its length and much more than that its weight, flaunting its power over the smaller morsel. More often than not, the dolphins' escape tactics are successful, strategies that involve not just one but hundreds of dolphins, coordinating a "high alert" response to danger.

Dramatic predation events are not a daily occurrence, nor do we researchers regularly witness a successful attack on dusky dolphins by killer whales. When predation events are rare, how can we describe the effect of predator on prey?

For decades, predator–prey interactions have been extensively debated, and described as the principal force driving group living in most taxa (Alexander 1974, Edmunds 1974, Wilson 1975, Bertram 1978, Sih 1980, 1987, Van Schaik and Van Hooff 1983, Norris 1994, Connor 2000). After all, if one misses a meal, one can eat another day. If one misses evading a predator—well, then life is over. Over time, theoretical and empirical contributions have provided credible evidence that predators can affect prey behavior with serious consequences for prey reproduction, habitat selection, mate and food choices (Lima and Dill 1990, Lima 1998), even if the prey are not directly taken; such can be the results of mere predator threat.

The effects of predator intimidation of prey have garnered considerable interest (Brown et al. 1999). Predator presence or behavior can create an environment of fear affecting prey behavior, movement and grouping patterns (Lima and Dill 1990, Lima 1998, Laundre et al. 2001, Wirsing et al. 2008). Several studies also indicate that these predator-driven effects can move down three or more trophic levels, resulting in a so-called trophic cascade (indirect effects of predators on plant biomass via herbivores; Estes and Duggins 1995, Pace et al. 1999, Ripple and Beschta 2004, 2006, Schmitz 2006, Otto et al. 2008). Changes in prey behavior arising from indirect predator effects can have impacts on lower trophic levels either through "trait-mediated" or "density-mediated" pathways (Abrams 1995, Dill et al. 2003, terminology reviewed in Abrams 2007).

There is strong evidence that indirect predator effects can trigger resource declines or alter ecosystems with equal or greater consequence than changes in population dynamics from direct lethal predation, especially in aquatic systems (Kotler and Holt 1989, Werner and Peacor 2003, Preisser et al. 2005, Myers et al. 2007, Peckarsky et al. 2008). The increasing focus on non-lethal or indirect predation risk effects on prey behavior promotes exploration of evolutionary and ecological changes in prey behavior related to ecosystem functions and properties (Luttbeg and Kerby 2005, Creel and Christianson 2008, Schmitz et al. 2008). It also promotes the development of new approaches to better understand the influences of upper trophic level predators in marine ecosystems.

Predators can affect their prey's choice of habitat, movement patterns, social structure, and foraging behavior. Most dolphin studies attribute some of these lifestyle decisions to predation risk effects. For example, dolphins move to shallow waters to avoid deep-water sharks or increase group size to avoid predation (Saayman and Taylor 1979, Würsig and Würsig 1979, Norris and Dohl 1980, Shane et al. 1986, Norris and Schilt 1988, Wells et al. 1987, Connor 2000).

Studies of predator–prey interactions involving dolphins, including long-term investigations, generally have focused on consumptive or lethal predation (see Jefferson et al. 1991 for killer whale interactions with marine mammals, Norris 1994, Mann and Barnett 1999, shark effects reviewed in Heithaus 2001a). The indirect effects of tiger shark (*Galeocerdo cuvier*) predation on multiple prey species, including bottlenose dolphins (*Tursiops aduncus*) in the Shark Bay, Australia ecosystem, are well documented, but this research has not been replicated elsewhere (Heithaus and Dill 2002a, 2006, Wirsing et al. 2007). When such field studies are not viable, alternative approaches and a theoretical shift towards considering indirect predation risk effects will revise our understanding of predator influences on dolphin lifestyle and the ecosystems they inhabit (Heithaus et al. 2008, Wirsing et al. 2008).

In this chapter, we synthesize information on possible indirect effects of predation risk on dusky dolphins. We review available data on their chief predators and the various anti-predator strategies that dolphins employ in their daily lives. While our research focuses on dusky dolphin interactions with predators in Kaikoura, New Zealand, we incorporate relevant details from other locations.

DUSKY DOLPHINS OFF KAIKOURA, NEW ZEALAND

Near Kaikoura (42°S, 173°E), dusky dolphins exhibit a semi-pelagic lifestyle. They feed almost exclusively at night on mesopelagic organisms associated with a deep submarine canyon (Würsig et al. 1989, 1997, Cipriano 1992, Benoit-Bird et al. 2004, see also Chapter 5). Every afternoon, dolphins travel offshore to feed and return to nearshore locations to rest and socialize by day—a lifestyle comparable to that of Hawaiian spinner dolphins (*Stenella longirostris*; Norris and Dohl 1980).

These nocturnal foraging excursions are somewhat stable throughout the year. Daytime distance from shore varies by season, with the dolphins moving close to shore by late austral spring (November), and maintaining closest distance to shore in summer and fall (Cipriano 1992, Markowitz 2004, Dahood et al. 2008; 1995–2007 *Dolphin Encounter* tour boat data courtesy of Ian Bradshaw, Lynette and Dennis Buurman). On winter days, dusky dolphins occur farther from shore, generally range farther alongshore (Figure 7.1), and are typically found in deeper waters (>200 m; Dahood et al. 2008) than in other seasons.

Patch density, composition, and locations where the visually sensitive organisms that constitute dolphin prey come towards the surface at night likely change between nights and seasons. Foraging time also differs between seasons (7–9 hours in summer/fall and 12–13 hours in winter), a consequence of early sunset and late sunrise (Benoit-Bird et al. 2004, 2008). With a shorter summer feeding time, the dolphins would be expected to stay offshore throughout the day. This could potentially reduce food searching, increase foraging efficiency, and provide

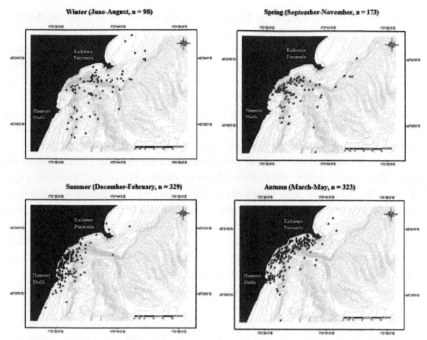

FIGURE 7.1 Seasonal dusky dolphin sightings from research surveys (*n* is sample size) off Kaikoura, New Zealand (1995–2003) (Markowitz 2004) (bathymetry data courtesy National Institute of Water and Atmospheric Research (NIWA), New Zealand).

more time to engage in other activities. But dusky dolphins exhibit a strong diel nearshore–offshore movement, which suggests that other factors may apply.

A common case made for observed movement patterns is predation risk from deep water sharks and killer whales (Würsig et al. 1989, 1997, Cipriano 1992, Markowitz 2004, Weir 2007, Weir et al. 2008). However, little is known regarding the effects of these predators on dusky dolphin behavior. In the following sections, we discuss the principal predators of dusky dolphins off Kaikoura, New Zealand and the potential influence of these predators on dusky dolphin behavior and social organization.

KILLER WHALES

The New Zealand killer whale population is estimated to be 65–167 animals (Visser 2000). It is believed that there are three subpopulations with geographic affiliations to the South and North Island, or both. Stomach content data to gauge New Zealand killer whales' diet diversity are scarce, with most dietary information coming from direct observation of predation events (Visser 2000). Available data indicate that killer whales off New Zealand have a varied diet ranging from rays (frequently observed) and sharks (Fertl et al. 1996, Visser 1999a, Visser et al. 2000) to cetaceans (Visser 2000). It is possible that some pods specialize in dolphin hunting and frequent the Kaikoura region (Visser 2000).

There are no reported observations of killer whale attacks on New Zealand fur seals (*Arctocephalus forsteri*), despite known consumption of fur seals elsewhere (Matkin et al. 2007). On one occasion, when a small pod of killer whales was observed near the Kaikoura Peninsula, fur seals exhibited a threatened response by hauling out on a rock as the killer whales passed by. However, on a different day, a male killer whale swam past a fur seal with no apparent reaction from either animal (M. Srinivasan, personal observation, March 2007).

After years of decline, fur seal populations are on the rise (Bradshaw et al. 2000). Over time, killer whales could more often choose fur seals as prey. For example, Mamaev and Burkanov (2006) reported increasing killer whale attacks on northern fur seals (*Callorhinus ursinus*) off the Commander Islands since 2000, even though there was no evidence of attacks before that.

Killer whales may also target calf humpback whales (*Megaptera novaeangliae*) during their migration past New Zealand. A high killer whale scarring rate for humpback whales off New Zealand has been reported relative to other regions, although only 3 of 8 individuals showed killer whale rake marks (Mehta et al. 2007).

Near Kaikoura, killer whales are rarely observed to prey on dusky dolphins (Constantine et al. 1998, Markowitz 2004). This does not necessarily mean that predation does not happen more often, since it is possible that such events go unobserved; for example, occurring in inclement weather or at night (when no observation-based boat work is possible). However, dusky dolphin behavioral responses suggest that killer whale presence strongly influences the dolphins' distribution, movement patterns, and daily behavioral rhythm, and we follow this thesis here.

Visser (2000) reported a peak in austral summer for killer whale sightings near Kaikoura (1992–1997). Markowitz (2004) also observed killer whale–dusky dolphin interactions to vary by season (1997–2003)—at proportions of spring 21%, summer 55%, autumn 21%, and winter 3%. Long-term killer whale sighting records (1995–2006) from a dusky dolphin tour operator, *Dolphin Encounter*, confirm this seasonal trend (Dahood et al. 2008), with peaks in November (late spring), austral summer and fall, and few sightings in winter (Figure 7.2, *Dolphin Encounter* tour boat data). This seasonal peak in killer whale occurrence off Kaikoura coincides with the time of year when young dusky dolphins are present in greatest numbers.

Markowitz (2004) reported 82% of observed dusky dolphin interactions with killer whales in >50 m water depth, often where the Kaikoura Canyon approaches closest to shore. Killer whales often travel parallel to the shore, generally displaying a directional movement. They are sighted in both shallow (about 20 m) and deep waters (Figure 7.3), and likely follow depth contours. Some killer whale pods are composed of 3–5 individuals, while some are larger (5–30). However, we do not know if group size and depth preference dictate killer whale prey choice and hunting strategy.

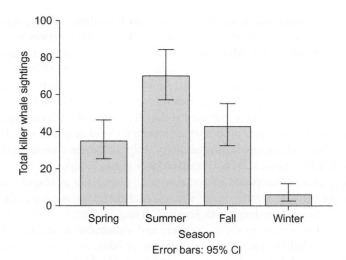

FIGURE 7.2   Total number of killer whale sightings near Kaikoura recorded during dolphin tours 1995–2007 (data from *Dolphin Encounter*, Kaikoura, New Zealand, courtesy Ian Bradshaw, Lynette and Dennis Buurman).

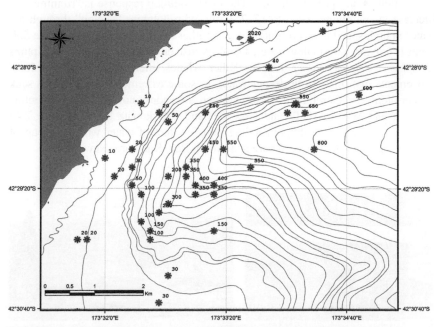

FIGURE 7.3   Selected killer whale sightings to highlight variation in depth choice (depth in meters) near Kaikoura, New Zealand (1995–2007) (data from *Dolphin Encounter*, Kaikoura, New Zealand, courtesy Ian Bradshaw, Lynette and Dennis Buurman; bathymetry data courtesy National Institute of Water and Atmospheric Research (NIWA), New Zealand).

Long-term systematic research suggests that regardless of age, sex, social status, or group affiliation, dusky dolphins seek shallower waters in summer and fall (Cipriano 1992, Markowitz 2004). This generally correlates with opportunistically observed killer whale sightings off Kaikoura (Figures 7.1 and 7.2). Thus, prey may respond to habitat differences as a measure of predation risk, rather than actual predator presence or abundance (Boinski et al. 2003, Verdolin 2006). So prey may choose habitats that allow them to escape easily, rather than simply avoiding all possible predation (Ydenberg and Dill 1986). It is possible that both factors influence dusky dolphin habitat choices and tactics during peak killer whale season, discussed later in this chapter.

Nursery groups composed of females with young calves respond dramatically to predation risk from killer whales, and therefore are arguably a best indicator of varying predation risk levels. Calving in dusky dolphins typically occurs from late spring to early summer, and vulnerable mothers and calves tend to stay in shallow water (≤20 m; Markowitz 2004, Markowitz et al. 2004). Weir (2007) noted that nursery groups were harder to find on days when killer whales were reported in the area, a trend also seen in subsequent field seasons. She also described an incident of a lone calf, stressed and in shallow water, possibly due to an earlier sighting of killer whales in the area (see also Chapter 9).

Dusky dolphins exhibit a number of responses to killer whales that appear to function as anti-predator tactics. The most common response is "running," an increase in swimming speed to 15–25 km/h, with dolphins "slicing" through the water in high speed bursts (Figure 7.4a). Another common tactic used by the dolphins is moving into shallow water, a strategy also employed by common dolphins (*Delphinus delphis*; Visser 1999b). Killer whales often parallel the dolphins' movements just offshore in deeper water. In some extreme cases, dolphins seek shelter along the shoreline in <1 m of water, a response noted in both Argentina (Würsig and Würsig 1980) and New Zealand (Markowitz 2004). Figure 7.4b shows an example of a dolphin hiding in a few feet of water while a group of killer whales patrolled just offshore near South Bay, Kaikoura, New Zealand.

(a)

(b)

(c)

FIGURE 7.4   These photographs show three examples of dusky dolphin responses to killer whales off Kaikoura: (a) "running" or "slicing" in a high-speed burst, (b) seeking shelter in shallow water, and (c) "mobbing," engaging in apparent predator assessment/harassment.

Small groups of adult dusky dolphins have been observed to approach killer whales, swimming rapidly around and in front of them, before departing at high speed (Figure 7.4c). The particularly rapid nature of these interactions suggest that the dolphins may employ a form of "mobbing" or "scouting" in defense against predation by killer whales. In this scenario, the dolphins may assume different social roles in response to potential predators, with most individuals (large group) fleeing along the shoreline while a small "scouting party" or "defense squadron" engage in predator assessment, distraction, and/or harassment. Potential "scouting" and "mobbing" behavior have been suggested for other marine mammals in response to killer whales (Baird 2000), but we are not confident of actual intent pending further work.

Killer whales attacking dolphins rely on stealth and surprise. Predator inspection may benefit dolphins by removing the element of surprise, or by sensing the motivation of the predator (Fitzgibbon 1994). However, such encounters could easily turn costly for the inspecting dolphin. Future studies may investigate the costs and benefits of such behaviors, and whether these apparently inspecting individuals belong to certain age/sex groups and are affiliated with larger or smaller groups.

SHARKS

Ascribing dolphin habitat selection to deep-water shark predation risk requires knowledge of shark habitat use and behavior (Lima 2002). Some sharks can easily move between shallow and deep waters; for example white sharks (*Carcharadon carcharias*; Weng et al. 2007), tiger sharks (Heithaus 2001b), and sevengill sharks (*Notorhynchus cepedianus*; Ebert 1991). The degree of risk posed by these sharks can differ by species, size, season, temperature, time period, abundance, and habitat, which, in turn, can affect dolphin behavior (Long and Jones 1996, Heithaus 2001a, Ebert 2002, Heithaus and Dill 2002a, Klimley et al. 2002, Heithaus 2004, Lucifora et al. 2005, Wirsing et al. 2006).

There is strong evidence that great white and sevengill sharks exhibit an ontogenic shift towards marine mammal prey once they mature into adults (Long and Jones 1996, Ebert 2002, 2003), making them among the most lethal sharks for marine mammals.

There are five species of sharks that may pose a risk to dusky dolphins near Kaikoura, New Zealand—the great white shark, shortfin mako (*Isurus oxyrinchus*, suspected cetacean predator; Heithaus 2001a), sevengill shark, sleeper shark (*Somniosus pacificus*), and less likely, the blue shark (*Prionace gluca*). There are limited data on these shark species in New Zealand, but their movement and distribution patterns may be similar to those in other locations.

Off Kaikoura, local fishermen and tour operators report that recent shark abundance is low, and apparently decreasing. Blue sharks are the most commonly observed shark, followed by mako sharks. White sharks are rare and there are some observations of sevengill sharks. Shark sightings peak in austral summer and fall. Shark diving operations in Kaikoura closed in the late 1990s due to lack of regular shark sightings. However, no systematic studies have been initiated to monitor shark populations in the area.

Sleeper sharks

Two species of *Somniosus* or sleeper sharks have been reported in New Zealand waters (Francis et al. 1988). Although small sleeper sharks (*S. rostrata*) have been identified near Oaro, just south of Kaikoura, these pose no risk to dolphins. The other species, the Pacific sleeper shark, is a known marine mammal predator and scavenger (Crovetto et al. 1992, Taggart et al. 2005, Sigler et al. 2006). Pacific sleeper sharks typically rise to the surface at night and stay at depths during the day, exhibiting continuous vertical movement (Taggart et al. 2005, Hulbert et al. 2006). They tend to stay in deep waters at low latitudes where they are believed to never come to the surface, whereas in high latitudes, they often come close to the surface (Compagno 1984). Their presence in Kaikoura waters is unconfirmed, with little information from elsewhere off New Zealand.

Sevengill sharks

These sharks are nocturnal, venturing into deeper waters during the day and typically preferring turbid waters (Ebert 1991). They also engage in cooperative hunting, which facilitates attacking fast-moving and larger marine mammals, including dolphins. There is evidence of habitat shift in La Plata river dolphins, *Pontoporia blainvillei*, in response to peak abundance in sevengill sharks (Lucifora et al. 2005). In Argentina, cetacean remains of possibly dusky dolphins in sevengill shark stomachs have been recorded (Crespi-Abril et al. 2003). We know little about habitat-use or behavior of this shark off Kaikoura.

Great white, blue, and mako sharks

White sharks use shallow and deep waters (Weng et al. 2007), while blue and mako sharks are pelagic species. All three species exhibit vertically oscillating movements or "yo-yo" type swimming (Carey and Scharold 1990, Holts and Bedford 1993, Klimley et al. 2002). Blue sharks feed on the vertical migrating layer off California (Tricas 1979). Since blue sharks feed on cephalopods, they may feed at night off Kaikoura as well, which may increase their potential interaction with dusky dolphins.

Blue sharks pose no risk to adult dolphins, unless the sharks are over 2 m and perhaps encounter vulnerable or wounded dolphins. Markowitz (2004) reported dusky dolphins in a nursery group attacking a blue shark, ramming it repeatedly with their rostrums. The small size of the shark (<2 m) was such that it likely posed no threat to adult dolphins, but could be perceived a threat to the calves. Bottlenose dolphins (*Tursiops truncatus*) are also known to exhibit such group-based anti-predator behavior (Connor et al. 2000b). Such behavior could serve as a predator deterrent, help to sense predator motivation, or be a learning tool against more dangerous predators. For example, Graw and Manser (2007) observed that meerkats (*Suricata suricatta*) mob both threatening and non-threatening animals, but adjust their mobbing behavior based on the degree of threat posed by the encountered animal. Also, mobbing behavior varied with age, with young meerkats, more often than adult meerkats, mobbing squirrels (*Xerus inauri*), but exhibiting caution against more dangerous snakes.

Blue and mako sharks have been recorded making the deepest dives during the day (Carey et al. 1981, Carey and Scharold 1990): likely older and larger individuals (Graham 2004). However, blue, mako, and white sharks spend a substantial amount of time at the surface or close to the surface (Klimley et al. 2002, Graham 2004). Preliminary results of a white shark satellite-tagging project being conducted in New Zealand waters show no evidence of white shark tracks near Kaikoura, although this species occurs throughout the region (Clinton Duffy, Marine Conservation Unit, Department of Conservation, New Zealand, personal communication).

Sharks in Kaikoura

In March 2007, the senior author conducted a brief shark presence/absence study in Kaikoura, from a fishing boat. The owner had previous experience with shark diving operations and was familiar with favored shark spots in the area, particularly a zone about 15 kms offshore, just northeast of the Kaikoura Peninsula. Frozen bait (composed of ground up fish) released an oily residue, providing a chum line to attract sharks. Burley bags with cod and perch fish scraps were floated on the sides of the boat as additional bait. Over 3 different days, covering a 10–15 sq km area for about 13 hours, using 5 kg of frozen chum/hour, some small sharks, 2–3 porbeagle sharks (*Lamna nasus*, 1.5 m), and one 2.5 m mako shark took the bait. A similar study conducted in 2003 (February 5–March 13), for about 70 hours, off the same fishing boat, resulted in a single sighting of a 3 m mako shark (Victor Foster and Jodie West, unpublished data available from M. Srinivasan). Further, from 1999 to 2007, a seabird tour operator in Kaikoura, Albatross Encounter, recorded 51 blue and 7 mako with 31 sightings in fall, 17 in summer, 3 in spring, and none in winter.

It is possible that sharks are wary of fishing boats and avoid the bait, or shark abundance in Kaikoura waters is low. Based on current evidence, it appears that the latter scenario is true. So, we conclude that killer whales are likely to be a more potent threat than sharks in Kaikoura waters. Additional factors contributing to the reduced shark presence off Kaikoura may include shark finning and increasing longline tuna fishing by-catch (Ayers et al. 2004, Ministry of Fisheries 2007a,b).

DUSKY DOLPHINS—GROUP LIVING AND VOCAL REPERTOIRE

Group living

Predation risk leads to group living by influencing prey behavior, social relationships, and habitat selection, as well described in several excellent papers (for example, Hamilton 1971, Pulliam 1973, Alexander 1974, Jarman 1974, Bertram 1978, Van Schaik and Van Hooff 1983, Trivers 1985, Elgar 1989, Dehn 1990). Over evolutionary time, prey have developed strategies to find food, water, and potential mates, knowing that danger lurked close by.

Recent studies have shown that the relationship between predation risk and group size (and the consequences in terms of vigilance, detection, dilution, and encounter effects) can vary between species, habitat, and predation pressures (Inman and Krebs 1987, Krause and Godin 1995, Bednekoff and Lima 1998, Creel and Winnie 2005, Liley and Creel 2007). Thus, anti-predator strategies need to be evaluated in context. For example, Wolff and Van Horn (2003) in their study on wolf (*Canis lupus*) effects on elk (*Cervus elaphus*) behavior in Yellowstone National Park (predators present) and Rocky Mountain National Park and Mammoth Hot Springs (predator-free), reported no correlation between percentage of cows vigilant and group size, failing to find support for the "safety in numbers" hypothesis of Pulliam (1973). In Yellowstone, cows with calves spent more time vigilant than foraging relative to cows without calves. In the other two parks, there was no difference in levels of vigilance for cows with and without calves. Their findings were consistent with those of Laundre et al. (2001), who also found no correlation between predator vigilance levels and group size for elk in Yellowstone. Thus, levels of predation pressure can determine the degree of anti-predator behavior, and may or may not influence group size.

Dolphins have few refuges from predators in the marine environment (Connor et al. 2000b). Among dolphins, safety in numbers due to the "selfish herd" effect (Hamilton 1971), and a tight social organization with heightened awareness through sensory integration (reviewed in Norris and Schilt 1988), are essential to predatory detection and avoidance.

Group size

Off Kaikoura, dusky dolphins exhibit larger group sizes in winter (often >1000) when they are found further offshore, than in summer (generally <1000) when they are found inshore (Markowitz 2004). Group size also varies with social type/class. A mix of age and sex classes form the largest groups, comprised of >50 individuals (most often hundreds). Mother–calf nursery groups are typically made up of fewer dolphins (median group size 14; Weir 2007). Nursery groups are rarely observed in winter (Markowitz 2004). Other social groups also have smaller group sizes. For example, mating groups have a median group size of 6 (see Chapter 9), and non-sexually active adult groups have a median group size of 8–10 (Markowitz 2004).

For most dolphin species, correlating group size with predation risk is not straightforward, especially in open habitats (Gygax 2002, Gowans et al. 2008). In general, individuals in large groups may derive anti-predator benefits (Krause and Ruxton 2002). However, larger dusky group sizes in winter may not necessarily correspond to higher predation risk, as killer whale presence is minimal and shark numbers are also apparently reduced. Larger winter group sizes may instead reflect a large influx of new individuals (Markowitz 2004, Würsig et al. 2007) or they may have foraging or social benefits. Creel and Winnie (2005) in their elk studies found that grouping is not always an anti-predator tactic, as elk herd size increased when wolves were absent. Instead, elk aggregation in areas away from cover in the absence of wolves was attributed primarily to foraging needs.

Inman and Krebs (1987) postulated that even though the probability of detection increases with group size, it generally does not increase linearly. After a certain stage, detection becomes independent of group size. Krause and Godin (1995) found that cichlids (*Aequidens pulcher*) preferentially attacked large shoals of guppies (*Poecilia reticulata*), even though percentage of success was relatively low. Other studies on birds and mammals also show preferential attack on larger groups (Brown and Brown 1996).

Leaping behavior

Dusky dolphins are well known for their aerial acrobatics, engaging in perhaps a higher frequency and a wider variety of leaping than any other cetacean species (Würsig et al. 1997).

Noisy leaps are loud and powerful acrobatic events, prevalent in large groups, and appear to act as a long-range signal, directing group movement (Markowitz 2004). When observing dusky dolphins resting in large groups, it is common to see active individuals on the outer edges of the groups frequently engaging in noisy leaping activity. These individuals may act as sentinels while others rest.

Zahavi and Zahavi (1997) suggest that stotting in impala (*Aepyceros melampus*) represents a "handicap," a reliable signal to the predator of the fitness and alertness of their prey. It is possible that "noisy" leaping serves a similar purpose in dusky dolphins, but this seems unlikely because (1) it appears to occur frequently in the absence of potential predators; and (2) it does not appear to occur at an especially high frequency (in fact the frequency declines) when the dolphins are in close proximity of potential predators (at least killer whales, Markowitz 2004). Such showy aerial acrobatics are energetically expensive and predator-attracting, so are reduced in order to redirect energies to fleeing from killer whales.

Protecting calves from predators: nursery groups

Like African elephants (*Loxodonta africana*; Lee 1987, Moss 1988), matriarchal groups of sperm whales (*Physeter macrocephalus*) act together to care for calves and defend them from predators (reviewed in Whitehead and Weilgart 2000). The formation of mother–calf nursery groups, common among dolphins, can offer calves protection from predators by choosing low-risk areas (see also Chapter 9). Nursery groups may also form to increase opportunities for social learning (Wells 1991). Dusky dolphins in nurseries (see Chapter 8) rest more, engage in relatively few bursts of speed, and swim slower on average than dolphins in other groups (Kruskal–Wallis $H = 10\,633$, $P = 0.01$, $n = 41$ nursery groups, 169 large groups of 50+ dolphins, 37 non-mating small adult groups, and 42 mating groups). Slower swimming speeds could partially explain why nurseries form (Figure 7.5).

Isolation of nursery groups from large adult groups may serve a dual purpose: (1) to reduce harassment from larger groups (Weir et al. 2008), and (2) to reduce vulnerability by choosing low-risk areas, and thereby increase ability to escape undetected prior to an imminent threat. A female dolphin with a calf is both energy and speed handicapped (Noren 2008), and an easier target than a fast and independent adult dusky (Markowitz 2004). Look at Figure 7.5: Imagine you were a predator, such as a killer whale, and you were swimming onto the page from the left margin: With whom would you catch up first? Speed and energy considerations for mothers with calves may alter their decision to flee from a lethal shark or furtively move along the shallows to be undetected by killer whales.

Nursery groups maintain tight inter-individual proximity, likely facilitating both defense and care of calves (Markowitz 2004; included within small groups in Figure 7.6). Parallel swimming formations are most common in nursery groups

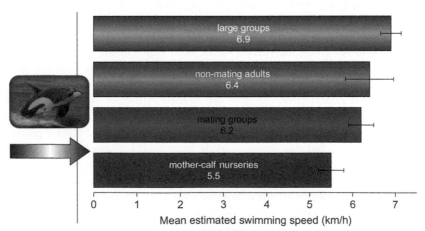

FIGURE 7.5 Mean estimated speeds (km/h, mean values with 1 standard error of the mean) are compared between social groups (large groups 50+ dolphins $n = 169$, non-mating small adult groups $n = 37$, mating groups $n = 42$, nursery groups $n = 41$) of dusky dolphins. Speeds were estimated by GPS tracks while the vessel was positioned alongside the focal group (mating group speed likely underestimated due to erratic, non-linear high-speed behavior).

(Figure 7.7), perhaps functioning to increase search efficiency for potential threats (Norris and Dohl 1980), and to keep calves protectively positioned with older animals on either side (Mann and Smuts 1998). Boisterous social interaction

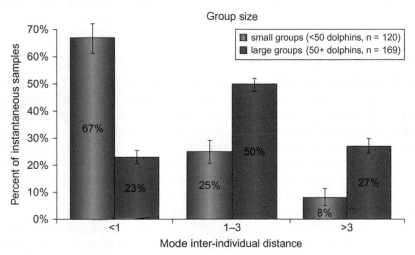

FIGURE 7.6 Mode inter-individual distance (in dolphin body lengths, 1 body length = 1.8 m) is compared for small groups (<50 dolphins) versus large groups (>50 dolphins) of dusky dolphins off Kaikoura. Bars show mean values for the percent of instantaneous samples collected at 2-minute intervals per group encounter with one standard error of the mean.

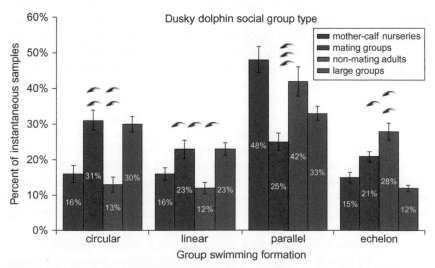

FIGURE 7.7 Swimming formations are compared between different types of dusky dolphin social groups off Kaikoura (large groups 50+ dolphins $n = 169$, non-mating small adult groups $n = 37$, mating groups $n = 42$, nursery groups $n = 41$). Bars show mean values for the percent of instantaneous samples collected at 2-minute intervals per group encounter with one standard error of the mean.

and leaping are minimal in mother–calf nurseries (see Chapter 8), which makes these groups "least obvious" to the casual observer, and possibly also to predators. More subtle interactions between mothers and calves, as documented for bottlenose dolphins (Connor et al. 2000a), occur at a relatively high rate.

Group swimming formation and inter-individual distance

Dusky dolphin groups differ in group inter-individual distance (spread) and swimming formation (shape) depending on season, time of day, activity, and social grouping. Dusky dolphins in large groups maintain a greater inter-individual distance than in smaller groups (Figure 7.6). In general, dolphins spread out further in winter than in summer or spring. Also, spacing increases from daytime to afternoon as the dolphins move offshore to forage (Markowitz 2004). Spreading out in this manner presumably increases the dolphins' encounter rate with predators, as the dolphins potentially cover an estimated area of nearly $20\,km^2/h$ (Markowitz 2004). Thus, dolphins may trade-off predation risk for increased food searching efficiency. But foraging dusky dolphins probably maintain group cohesion and predator detection through acoustic contact, as postulated for dusky dolphins in Golfo San José, Argentina (Würsig et al. 1989).

Markowitz (2004) classified group swimming formations as linear, circular, parallel, or echelon (v-shaped) swimming (based on Weaver 1987; Figure 7.7). Circular formations were a surprisingly regular feature in large groups (Figure 7.7). Although predators such as killer whales and sharks may attack from below, such formations likely provide protection from danger, especially for dolphins toward the middle of the group, and circular formations are predicted according to the "geometry of the selfish herd" (Hamilton 1971). An interesting topic for future study would be to examine whether females with calves in large groups are found nearest the middle of the group.

During the day, resting dolphins adopt circular or echelon positions, but tend to have a linear structure when traveling. During the middle of the day, when the dolphins rest most heavily (Barr and Slooten 1998), they maintain the tightest inter-individual proximity, often in a circular formation (Markowitz 2004). The linear formation, predicted for groups traveling with some leaders and other stragglers, was observed most frequently in large groups, but was not as prevalent as might be expected, occurring less than one fourth of the time (Figure 7.7).

A linear formation may reduce encounters with potential predators compared to being flanked out with a scattered and undefined arrangement. Schaller (1972), for example, observed that hoofed animals walk in a single file, possibly to prevent an unexpected meeting with a wandering predator. Linear structures in dusky dolphins are not generally "single file," but may serve a similar purpose.

Forming mixed-species groups

A common explanation for the formation of heterospecific groups is increased foraging efficiency and greater protection from potential predator attacks (Morse 1977). Predator defense may be improved in heterospecific assemblages by increased predator detection, the dilution effect, and cooperative anti-predator behavior. Predator defense may be an important factor in dusky dolphin formation of interspecific assemblages, as noted for other species, including primates (Gautier-Hion et al. 1983, Dunbar 1997) and gazelles (*Gazella thomsoni*; Fitzgibbon 1990).

Sightings of killer whales in Kaikoura coincide with the time of year when multi-species groups most commonly form between dusky and both common and Hector's dolphins (*Cephalorhynchus hectori*). Common dolphins may use large dusky dolphin groups for enhanced predator detection or defense, and dusky dolphins may tolerate them because they do not compete for resources. On the west coast of New Zealand's South Island, dusky dolphins have been observed joining larger groups of common dolphins, perhaps for similar reasons. Whether these mixed groups form randomly or have social, foraging, or protective benefits needs further study.

Hector's dolphins associate infrequently with dusky dolphins (see also Chapter 15). This may be partially due to variable predation pressures and habitat requirements (Bräger et al. 2003). There is no record of killer whales attacking Hector's dolphins (Jefferson et al. 1991), although they may be prey for sevengill sharks (Cawthorn 1988).

Vocalizations

Dusky dolphins rarely whistle, more often producing clicks and burst pulses (Yin 1999; see also Chapter 4). When milling/resting and traveling, whistles were rarely recorded (Yin 1999), and therefore may not be a significant component of their daytime communication. Could the lack of whistles be a killer whale-avoidance mechanism, perhaps similar to what Morisaka and Connor (2007) hypothesized for non-whistling odontocetes? Further studies on dusky vocalizations during day and night, and in the presence/absence of killer whales, may reveal the anti-predator functions of whistles and other dusky calls.

SUMMARY AND CONCLUSIONS

This chapter is a preliminary review of potential indirect predation risk effects on dusky dolphin behavior, with a focus on waters off Kaikoura, New Zealand. Dusky dolphins are apex predators, but occasionally they fall prey to larger predators such as killer whales. Dusky dolphins, like other prey, have adopted stable (long-term) and flexible (short-term) strategies that appear effective against potential killer whale attacks. For example, they choose shallow water habitats during peak killer whale season, flee or make drastic maneuvers, change group formation, and form nursery groups, which have their own set of tactics. Much remains to be learned about the benefits versus costs of making such choices, as dusky dolphins seek to minimize predation risk while meeting food needs. We look forward to further studies of predator–prey system dynamics, particularly indirect predation risk effects. Such research explorations can strengthen our understanding of predator roles in the behavioral ecology of dusky dolphin societies, and similar social delphinids.

ACKNOWLEDGMENTS

We thank Bernd and Melany Würsig, Robin Vaughn, and Bill Grant for comments and suggestions that helped improve the manuscript. The senior author thanks Sierra Deutsch and Jennifer Bennett for research assistance and field support in 2007. Research (1997–2003) was accomplished with the help of many hard-working full-time field assistants, including Cathryn Owens, Heidi Petersen, Nicholette Brown, April Cognato, Lindsay McOmber, Kim Andrews, Sherri Stanley, Kristen Mazzarella, and Anita Murray. Additional field assistance was provided by many capable and enthusiastic Earthwatch volunteers. We thank NIWA New Zealand for providing valuable bathymetry data. Special thanks to I. Bradshaw, D. Buurman, L. Buurman and staff at *Dolphin Encounter*, Kaikoura, New Zealand for field support and providing valuable dusky dolphin and killer whale tour boat data. Jack Van Berkel at the University of Canterbury Edward Percival Field Station supplied lab space and logistical support. Funding for this project was provided by Earthwatch Research Institute, Marlborough District Council, New Zealand Department of Conservation, National Geographic Society, Mooney Foundation, Texas A&M Galveston Office of Research and Graduate Studies, and through various travel and research scholarship grants from the Department of Wildlife and Fisheries Sciences and Department of Marine Biology, Texas A&M University. Boat maintenance was supplied by John Hocking, Murray Green, and Steve Howie. Bernd and Melany Würsig provided guidance and support throughout the work.

Chapter 8

Mating Habits of New Zealand Dusky Dolphins

Tim M. Markowitz,[a] Wendy J. Markowitz,[b] and Lindsay M. Morton[c]

[a]School of Biological Sciences, University of Canterbury, Kaikoura, New Zealand
[b]Ecology Department, Massey University, Palmerston North, New Zealand
[c]Biological Sciences, Dartmouth College, Hanover, New Hampshire, USA

The ocean erupts with a tightly packed group of dolphins, taking to the air and then plunging below. They streak across the water, back and forth at a blinding pace, throwing spray up to sparkle in the sunlight. At a glance, it's hard to make sense of it, dolphins hurling themselves through the sea in an amazing chase. It's pandemonium, utter mayhem. It seems that everybody is having sex randomly with everyone else. But is there a secret rhythm to this dance? Is there a courtship within the chaos? Is there a ritual hidden in the apparent anarchy? Over the past decade, our vessel-based focal group research has yielded tantalizing hints of such a pattern, glimpses into the sex lives of dusky dolphins that reveal competition between males, cooperation in acquiring mates, and the potential roles of female choice and female social strategies.

Tim and Wendy Markowitz, Edward Percival Field Station Kaikoura, 2008

PROMISCUOUS MATING SYSTEM AND REPRODUCTIVE BIOLOGY

Animal mating systems are generally divided into four broad categories (Brown 1975): monogamous (single mates), polygynous (one male mates with several females), polyandrous (a female mates with several males), or promiscuous (both males and females mate with several partners). Some behavioral ecologists (e.g., Alcock 1998) further distinguish between a structured system of multi-mate polygynandry (involving mate choice and competition) versus promiscuity with relatively random breeding. All of these mating systems occur in birds and fish, with ritualized courtship often a key component in mating, and males investing at least as much in parental care as females (Tinbergen 1953). Polygyny is the most common mating system in mammals, often accompanied by various forms of male–male competition and limited or no male parental investment (reviewed by Krebs and Davies 1993). The evolution of mating systems is often closely tied to the ability of individuals to defend multiple mates or breeding territories, and is consequently linked to animal ecology (Emlen and Oring 1977).

Based largely on post-mortem analyses, past studies of dusky dolphin (*Lagenorhynchus obscurus*) reproductive biology have characterized the mating system as seasonal and promiscuous, citing factors such as the restricted calving period, limited sex size dimorphism, seasonal variation in testis mass, and a high testis mass to body mass ratio indicating sperm competition (Würsig and Würsig 1980, Cipriano 1992, Van Waerebeek 1992b, Van Waerebeek and Read 1994). Cipriano (1992) estimated that female dusky dolphins in New Zealand reach sexual maturity at an age of 7–8 years and length of 160–165 cm ($n = 21$ female specimens). The reproductive cycle of female dusky dolphins lasts roughly 2 years, approximately evenly divided between the gestation and lactation periods (Cipriano 1992, Van Waerebeek and Read 1994). Mean estimated gestation length is 11.4 months (Cipriano 1992). Mating, which occurs for social as well as reproductive reasons, is observed throughout the year, but reproduction is seasonal. Dusky dolphins off Peru likely have an extremely short resting period before becoming pregnant again (Van Waerebeek and Read 1994). This estimated reproductive cycle is shorter than estimates for other dolphins, although a variety of stresses on the Peru population including fisheries by-catch and direct harvest for food could explain this difference (Van Waerebeek and Read 1994, see also Chapter 11). It is unknown whether dusky dolphins in New Zealand have a brief resting period similar to those in Peru, or if their resting period is longer, like that of most other dolphins (for more details of life history parameters, see also Chapters 2 and 14).

Like females, males reach sexual maturity at age 7–8 years (Cipriano 1992). Testes mass accounts for as much as 5% of adult male body weight (4 kg) in the breeding season (Cipriano 1992), with negligible sex size dimorphism (Leatherwood and Reeves 1983). Given these observations, theoretically, aggressive male–male competition should be unlikely, and sperm competition should be likely (Kenagy and Trombulak 1986). However, spending time with dusky dolphins in the wild, witnessing their sexual antics, one cannot help but wonder if there isn't something more going on. Might not all that energy expended by pursuing males and evading females indicate a more elaborate social ritual than random mating? In this chapter, we investigate this possibility based on our field research with dusky dolphins in New Zealand.

STUDY OBJECTIVES

The goal of this research is to learn more about the mating system and sex lives of dusky dolphins. Specifically, we set out to examine:

1. the effect of breeding seasonality on sexual behavior
2. variation in sexual activity by time of day

3. the influence of social group size, sex ratio, and age class composition on mating
4. whether and how males compete behaviorally for access to females
5. whether males cooperate to acquire access to females
6. mechanisms for female choice and female mating strategies
7. the role of aggression and/or mate coercion in dusky dolphin sex
8. differences between reproductive sex (breeding) and social sex (non-reproductive).

KEEPING UP WITH THE ACTION: METHODS AND ADVANCES

We report here results from our vessel-based studies of dusky dolphins off Kaikoura, New Zealand, conducted from 1997 to 2008. In this research we utilized small 4.3–5.5 m vessels with 25–85 hp outboard motors. Focal group approaches and follows were conducted following the protocols recommended by Würsig and Jefferson (1990), with the vessel positioned alongside the dolphins, matching dolphin group heading and speed in parallel so as to minimize disturbance. Dolphin groups were located with the assistance of shore-based monitoring teams engaged in theodolite tracking (Würsig et al. 1991). Over the years, our methods were adjusted to make use of technological advances (see below), but many of the basic methods to describe groups were consistent throughout, including how groups were designated, how age classes were estimated, and how season and time of day were categorized.

Dolphin groups were defined by spatial proximity according to the "10-m chain rule" (Smolker et al. 1992). This "distance measures" definition of group was chosen over a "coordinated activity" definition for its simplicity and because it does not rely on explicit or implicit assumptions about the behavior of a group's members (Mann 1999a). Group age class composition was estimated by size of individuals present based on lengths of aged post-mortem specimens (Cipriano 1992), with age class categorized as: adult/subadult (>2 years, 1.6–1.8 m), juvenile (1–2 years, 1.3–1.5 m), and calf (<1 year, 1–1.2 m). In the field, it was impractical to discriminate between subadults (2–6 years, 1.6 m) and adults (>6 years, 1.7–1.8 m), but we suspect based on their behavior that many of the larger and sexually active individuals were sexually mature. Groups were classified as either large (>50 individuals) or small (<50 individuals). Small groups were further divided into sexually active mating groups (sexual activity was defined as sexual approach with ventral presentation and/or confirmed intromission, see below), non-mating adult groups, and nursery groups (calves present).

Group locations and tracks of group movements were estimated using longitude, latitude, and time data recorded by Garmin global positioning system (GPS) receivers from the vessel as it was positioned alongside the group. Time and location data were recorded at 1-minute intervals and later downloaded to computers for analysis. GPS tracks were plotted and analyzed using Mapsource and GPS Utility software.

For examination of seasonal patterns, seasons were defined as: winter = June–August, spring = September–November, summer = December–February, and autumn = March–May (Markowitz 2004, after Cipriano 1992). For examination of diurnal patterns, the day was divided into "early" <9:00, "morning" = 9:00–10:59, "midday" = 11:00–12:59, "afternoon" = 13:00–15:00, "late" >15:00.

Datasheets and stopwatches

Initially, we relied on well-established methods, collecting focal group behavioral data using interval sampling with paper datasheets and stopwatches (Martin and Bateson 1993, Lehner 1996). Behavioral state, heading, mode nearest neighbor distance, and group swimming formation were recorded by instantaneous sampling at 2-minute intervals and all occurrences of leaping (categorized as noisy, clean re-entry, or acrobatic flip) were recorded during each interval (Markowitz 2004). Beginning in 1998, we supplemented this data collection with the use of digital recording to more accurately document rapid sequences of sexual activity.

Social–sexual contact behaviors examined in this study are shown in Figure 8.1. Social rubbing (Connor et al. 2000a, defined as any touching of the body), ventral presentation (or two dolphins swimming "belly-to-belly"), and sexual approach (approaching another dolphin with the penis out), often preceded mating. Copulation was noted only if intromission was confirmed. Social contact behaviors were the most difficult behaviors to observe at a distance; however, within a typical observational radius for small groups of roughly 20 m, all near-surface social–sexual contact could reliably be documented. Only small groups (<50 dolphins) were classified as mating groups if either sexual approach and ventral presentation or intromission (confirmed copulation) were observed. Interactive behaviors were recorded during all intervals (Markowitz 2004), including spyhopping (eye-outs), speed bursts (white-water slicing), porpoising (lateral, traveling leaps), chasing conspecifics (rapid follow with changes in directions), and inverted swimming ("belly up").

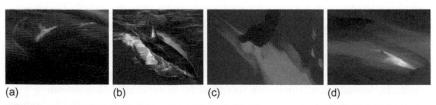

(a)   (b)   (c)   (d)

FIGURE 8.1   Social–sexual contact behaviors examined in this study were: (a) social rub (contact with body/rostrum), (b) ventral presentation ("belly-to-belly swim"), (c) sexual approach (with penis out), and (d) intromission (confirmed copulation).

Digital data recording

To learn more about the dusky dolphin mating system, behavioral observations were combined with the use of digital video, digital photography, and digital audio, documenting rapid sequences of sexual activity. These devices greatly increased the quantity and quality of data we could collect per unit time while monitoring high-speed social–sexual interactions. Time-stamped observations and images collected in rapid succession in real time could be exhaustively and extensively reviewed in post-hoc frame-by-frame, moment-to-moment analyses. This allowed for more detailed analyses of behavioral events, sequences, durations, latencies, and interactions, as well as identification and sexing of individuals that could not be accomplished immediately in the field.

Digital video

A Canon Optura digital video camera was used to record a total of 6.87 hours of above-water video of mating activity in 26 small mating groups (McOmber 1999). Each tape included a data code with the date, time of day, tape counter, and tape time remaining. This data code allowed frame-by-frame analysis of events, each frame covering 1/30 of a second. Taping continued throughout encounters with dolphin groups, except during brief periods when the dolphins were out of range. These data were used to examine mating group size, composition, and the involvement of particular individuals in mating activities. In addition to capturing the visual record, the videographer provided a real-time narration of the events recorded on the tape, with observed behaviors called out by members of the research team. The narration focused on several particularly conspicuous behaviors (Figure 8.1), including inverted swimming, ventral presentation, re-entry leaping, speed burst, social rub, conspecific chasing, sexual approach, and intromission. This audio record proved to be every bit as valuable as the video record.

Digital video records, in combination with film photography of dorsal fins, were used to identify individuals engaged in mating activities, and their roles in sexual encounters. In some cases, relatively subtle identifying features, insufficient for long-term mark–recapture, could be used to keep track of individual roles in small groups of mating dolphins (McOmber 1999). A length of 5 minutes was chosen as the minimum videotape time needed to obtain useful information about a mating group. Of the 26 mating groups videotaped, 20 mating groups were included in this analysis. Data from the six remaining groups were not useful because short video length prevented tracking the behavior of individual dolphins over time. The median digital video sample length per mating group was 15.27 minutes, ranging from 5.0 to 57.52 minutes. Film identification photos taken concurrently were scanned into a computer using a Nikon Coolscan III (LS-30) 35 mm film scanner. Scanning the images allowed for easier comparison between the photos and video images when making individual identifications. Initially, behavioral sequences were reviewed on a color television monitor frame-by-frame. Once the frames containing useful information were determined, they were downloaded to a Silicon Graphics computer for subsequent analyses of individual frames (McOmber 1999).

As much as possible, the number of females and males in each group, and their position with respect to sexual and social interactions, were documented in real time as well as in post-hoc analyses. Since dusky dolphins show no marked sexual dimorphism at any age (Cipriano 1992), one must be able to see the genital area on the dolphin's ventral surface to determine the sex of an individual. The simplest method of determining gender is the observation of an erect penis. This was the method most often used to sex individuals in the field. Dolphins have an anal opening and genital groove on the ventral surface (Figure 8.2). The genital groove in males contains a fibroelastic penis (similar to that found in ruminants) which narrows to a small tip at the end (Schroeder 1990). In males, the anal opening occurs several centimeters posterior to the genital slit. In females, however, the ano-genital slit is continuous (Schroeder 1990). Females also show small slits on either side of the genital slit which contain mammary glands. Gender identification of the dolphins from the digital video was accomplished by examining sequences of frames on the computer with the ventral surface showing (McOmber 1999). We have continued to use such ano-genital digital records (AGDRs) to assess the sex of individuals by the relative location of the genital groove and anal opening (Figure 8.2).

Digital still photography

Although at first glance it may seem like a step back from digital video, the advent of high-speed, high-resolution digital SLR photography greatly increased the quality and quantity of information we could collect regarding individuals (Markowitz et al. 2003a, b). Beginning in 2000, we used high-speed, high-resolution digital still photography for individual photo-identification.

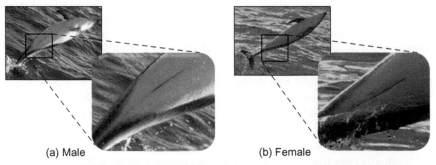

(a) Male   (b) Female

FIGURE 8.2 Dolphins were sexed using ano-genital digital records (AGDRs), either frames captured from digital video sequences or digital still photographs (as shown). (a) Males were identified by the distance between the genital slit and anus. (b) Females were identified by the lack of separation between the genital slit and anus, and by mammary slits.

Since that time, digital SLR cameras have improved (both in speed and resolution), enabling us to document sequences of events (individual position relative to sexual activity), and to sex individuals using AGDRs (Figure 8.2) as well as identify them with rapid series of high-resolution still digital images. In this effort, we have used a variety of Nikon digital SLR camera bodies, including the D1, D1H, D90, and D200 models, with variable focal length (zoom) lenses of various sizes, and found all to be successful in capturing images for individual identification and sexing. Photographs were stored as fine-resolution jpeg files on compact flash memory cards and downloaded to computers for onscreen review and analysis (Markowitz et al. 2003a). Images were sorted, cropped, and renamed using ACDSee and Adobe Photoshop software, and then catalogued using Finscan software (Hillman et al. 2003). Associations between individuals were subsequently examined using SOCPROG (Whitehead 1997, Whitehead and Dufault 1999).

Digital audio

Beginning in 2007–2008, we have been pairing continuous digital audio records of dusky dolphin behavior with rapid sequences of digital still photography. Focal group continuous behavioral sampling is initiated upon each group encounter. All behavioral events (including all social–sexual, interactive, high-speed, and leaping activities) are recorded continuously using an Olympus VN-2100PC digital voice recorder with noise-reducing headset. Using this method, we document behavioral state, group dispersion, heading, and number of individuals associated with the research vessel (bow riding) whenever these parameters change and at each 2-minute interval. In sexually active groups, additional information recorded includes visual and/or photographic identification (frame number) of individuals with position (dorsal or ventral presentation) during chase or mating sequences and sex of individuals (if discernible).

For mating groups monitored continuously via digital audio with concurrent digital photo-identification records, the minimum sample duration was 7.7 minutes and the maximum was 25.2 minutes. Digital recordings of focal group follows were downloaded to a PC using Olympus Digital Wave Player software. These recordings were subsequently played back and entered into JWatcher V.1.0 data files for quantitative analysis (Blumstein and Daniel 2007). Re-entry leaps, group dives (time from the last animal at the surface to the first animal re-surfacing) and above and below surface chase sequences were then analyzed by group for frequency, duration, and synchrony. Using this technique, it was possible to assess rates and durations of copulation, as well as inter-copulatory intervals.

EFFECT OF THE BREEDING SEASON ON SEXUAL BEHAVIOR

All marine mammals exhibit seasonal breeding patterns, which in turn can influence feeding (including fasting) and movement (including long-range migration) patterns (reviewed by Boyd et al. 1999). For species that breed seasonally, endocrine profiles tied to photo-period influence sexual and parenting behavior (Ketterson and Nolan 1994), which can be important for raising young in environmental conditions favorable for development. Some animals have annual biological clocks, which can run freely without exogenous input but can be entrained, or calibrated, by appropriate environmental stimuli related to the seasons (Pengelley and Asmundson 1971). Post-mortem examinations of net-caught dusky dolphins in Kaikoura show seasonal peaks in adult male testis size and calving (Cipriano 1992), indicating seasonal reproduction. The timing is such that calves are born into the warmest part of the year in spring and summer (Würsig et al. 1997), when nearshore mean surface water temperatures are roughly double those of winter (Markowitz 2004).

Although dolphins do not undergo long-distance seasonal migrations for breeding as do baleen whales (Wells et al. 1999), all dolphins studied to date show seasonal mating and calving, with gestation lengths generally ranging from 10 to 16 months (Perrin and Reilly 1984). Post-mortem studies of dusky dolphins indicate that birthing occurs during only one period of the year (Cipriano 1992, Van Waerebeek and Read 1994). In New Zealand, neonatal calves were observed at the end of the austral spring and early summer (late October–mid-January, Cipriano 1992, Markowitz 2004). In Peru, most newborns were found in the late austral winter (August–October, Van Waerebeek and Read 1994), with a sudden onset of calving followed by tapering off at the end of the calving season (Van Waerebeek and Read 1994). Würsig and Würsig (1980) also note that most births of dusky dolphins in Argentina occur during the austral summer. While mating behavior occurs during all parts of the year in both New Zealand and Argentina (Cipriano 1992, Würsig and Würsig 1980), these results suggest that mating serves a reproductive function for only a limited amount of time.

Post-mortem analyses show that female dusky dolphins are seasonally polyestrous (Cipriano 1992, Van Waerebeek and Read 1994). The data provided by these post-mortem analyses roughly coincide with information gathered about many other cetaceans, including bottlenose (*Tursiops truncatus*; Schroeder 1990) and spinner dolphins (*Stenella longirostris*; Wells and Norris 1994, Östman 1994). Hormone profiles of captive bottlenose dolphins have demonstrated peaks of progesterone during several spontaneous ovulations in addition to two to three ovulations during the breeding season (Schroeder 1990). Spinner dolphins are also seasonally polyestrous (Wells and Norris 1994).

The hormonal changes that indicate a period of estrous in cetaceans are not manifested in an outward, observable, manner. Several suggestions for the detection of estrous in dolphins include a slightly pink coloration of the belly or a subtle increase in the tactile behavior of an individual. However, in a study of captive bottlenose dolphins, neither indicator strongly correlated with confirmed estrous (determined using laboratory tests, Schroeder 1990). Connor et al. (1996) report patterns of attractiveness of free-ranging, female bottlenose dolphins from the Indian Ocean which may be associated with multiple estrous cycles. In their study, alliances of 2–3 males herd females throughout the year, and herding events tend to be longer during the breeding season (Connor et al. 1996). Herding events are often associated with reproduction and generally occur when the female has a calf that is approximately 2–2.5 years old, or 1–2 weeks after losing an infant (Connor et al. 1996).

Testis size in male dusky dolphins varies considerably at different times throughout the year, although adult males from Peru had at least some seminal fluid in the epididymes year-round (Van Waerebeek and Read 1994). Maximum testis size coincided with the recorded period of peak ovulation and conception (calving season) in both Peru and New Zealand (Cipriano 1992, Van Waerebeek and Read 1994). Therefore, while testicular activity does not cease during the year, it peaks during the breeding season (Figure 8.3). Studies of captive male dolphin hormone profiles show similar seasonality to the post-mortem studies. Testosterone levels of one captive male spinner dolphin changed 60-fold throughout the course of 14 months, peaking during a brief season (Wells and Norris 1994). Heterosexual genital–genital contact only occurred when the male's testosterone level was high, but no clear relationship was found between genital–genital contact and female estradiol or progesterone levels (Wells and Norris 1994).

Although dusky dolphin sexual activity occurs throughout the year off Kaikoura, it peaks during austral summer in all dolphin group types (Figure 8.3). Mating groups were seen consistently throughout the spring and summer

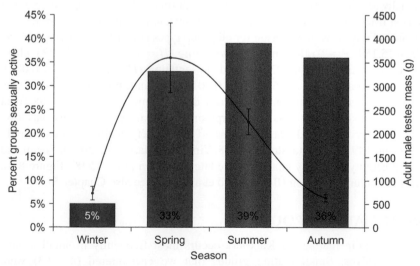

FIGURE 8.3 The percent of all dusky dolphin groups encountered off Kaikoura, New Zealand in which sexual activity (defined as sexual approach with ventral presentation and/or confirmed intromission) was documented is compared by season (bars, left axis, $n = 289$ group encounters), with mean ($\pm$SE) combined testicular mass of adult males also shown by season (scatter plot, right axis, from post-mortem measurements by Cipriano 1992, $n = 9$ adult males).

and into the autumn. These sightings correspond roughly to the timing of the breeding season reported by Cipriano (1992). Although mean testicular mass based on a limited sample ($n = 9$ adult males) peaked somewhat earlier in the spring than sexual activity ($n = 289$ groups, Figure 8.3), observations of neonatal calves into mid-January (Markowitz 2004) and a gestation length of 11.4 months (Cipriano 1992) suggest that conception occurs throughout much of the summer (as late as February) as well as in the late spring (beginning in November). Combining data collected from these sightings with reports that the calving season starts suddenly and then tapers off (Van Waerebeek and Read 1994), it seems likely that this pattern holds for both the mating and the calving portions of the dusky dolphin breeding season.

SOCIAL–SEXUAL ACTIVITY BY TIME OF DAY

In the deep coastal waters off Kaikoura, New Zealand dusky dolphins feed mostly at night on small squid and lanternfishes associated with the deep scattering layer (DSL) (Cipriano 1992). This leaves greater time during the day for resting and social–sexual activities. This pattern differs from that in the shallow bays of Patagonia in Argentina, the Marlborough Sounds in New Zealand, and the west coast of South Africa where dusky dolphins forage much of the day and engage in social–sexual activity following a successful foraging event (Würsig and Würsig 1980, Würsig et al. 1997, Markowitz et al. 2004, Vaughn et al. 2007, see also Chapter 6). Although we observed small groups of sexually active dolphins (mating groups) at all times of day during the breeding season (McOmber 1999), we found social–sexual activity at Kaikoura to vary by time of day, as well as season for all group types (Figure 8.4). Social rubbing ($H = 15.159$, $P = 0.004$), inverted swimming ($H = 12.660$, $P = 0.013$), chasing ($H = 16.638$, $P = 0.002$), ventral presentation ($H = 25.300$, $P < 0.001$), sexual approach ($H = 27.628$, $P < 0.001$), and intromission ($H = 16.276$, $P = 0.003$) varied significantly between diurnal time periods ($n = 298$ focal groups, Kruskal–Wallis tests). Not surprisingly, social–sexual activity was lowest at midday, the period when dolphins off Kaikoura typically rest most heavily (Cipriano 1992, Barr and Slooten 1998, Yin 1999, Markowitz 2004). Social and sexual activity generally peaked in the late afternoon (after 15:00, Figure 8.4), prior to nocturnal feeding (Benoit-Bird et al. 2004, see also Chapter 5).

SMALL MATING GROUPS

Dusky dolphin mating off Kaikoura occurs most frequently in small groups without calves. Small mating groups that we encountered ($n = 53$) were composed mainly of adult animals (median = 7, range = 3–36), with fewer juveniles (median = 0.5, range = 0–6). While not all individuals could be sexed (McOmber 1999), we can infer from our observations (Markowitz 2004) that most groups were composed of <10 males and a single female

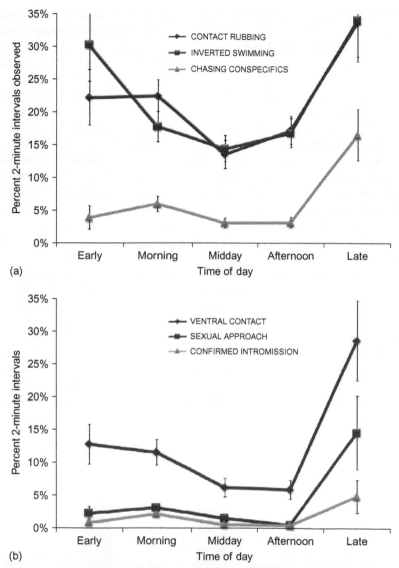

FIGURE 8.4 Social (a) and sexual (b) behaviors are compared by time of day, divided into early (<09:00), morning (09:00–10:59), midday (11:00–12:59), afternoon (13:00–15:00), and late (>15:00). Points represent the mean (±SE) proportion of 2-minute intervals during which social and sexual behaviors were observed.

(median: males = 6, female = 1). Among mating groups, the percentage of 2-minute intervals with sexual activity was generally highest in smaller groups (<10 dolphins), although there was considerable variability in sexual activity among all mating groups (Spearman's correlation, $R^2 = 0.123$, $P < 0.03$).

Mating groups are often found near the shore, but occur in both shallow and deep water. Following the general ranging patterns of other dusky dolphin groups off Kaikoura (Markowitz 2004), mating groups are found further offshore later in the day than earlier in the day (McOmber 1999). They typically form as satellite groups in the vicinity of large groups of hundreds of dolphins, and frequently join and split apart from larger social assemblages, in fission–fusion societal fashion.

In these small and sexually active groups, the mating period (period of highest sexual activity during the focal observation) ranged from 4 to 34 minutes (median = 13 minutes). During mating periods, confirmed intromission occurred rapidly, often with multiple partners (McOmber 1999, Markowitz 2004). We observed both males and females to have multiple sexual partners over these short periods of time; however, single females more often mated with multiple males. In one exceptional case, a single male mated exclusively with two females over the course of one of these brief mating periods (Markowitz 2004). Inter-copulatory intervals ranged from just over half a minute to over 5 minutes, averaging less than 1.5 minute (Table 8.1). In our most recent study (2007–2008), mean ($\pm$SE) duration of intromission was 6.8 ± 4.59 seconds.

The rate of social and contact behavior in mating groups was markedly different from other social groups. The percentage of intervals with social rubs ($H = 46.194$, $P < 0.001$), inverted swimming ($H = 42.556$, $P < 0.001$), and ventral presentation ($H = 76.166$, $P < 0.001$) differed significantly across social group types (small mating groups, mother–calf nurseries, and non-sexually active adult groups, Kruskal–Wallis tests). Social contact behaviors occurred in 4–20 times as many intervals in mating groups as in other social group types (Figure 8.5).

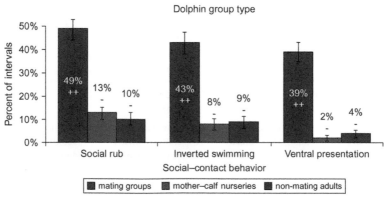

FIGURE 8.5 The percent of 2-minute intervals during which social rubbing, inverted swimming ("belly up"), and ventral presentation ("belly-to-belly" swimming) were observed is compared for small mating groups, mother–calf nurseries, and non-sexually active adult groups. Bars show mean values with standard errors. Significant differences are indicated by + and − symbols (Kruskal–Wallis, Bonferroni post-hoc, $P < 0.05$).

TABLE 8.1	Inter-copulatory intervals in mating groups				
Study years	Seasons	Digital method	Inter-copulatory intervals (s)		
			Low	High	Median
1998–99	Spring–summer	Video	34	126	49
2007–08	Summer–fall	Audio + photo	40	353	78

MATING OF THE QUICKEST: MALE–MALE COMPETITION

If you've ever watched a high-speed car chase scene, then you have a pretty good image of the sexual antics of dusky dolphins in the breeding season. The males race and swerve after the female, each vying to be the one that catches up to her first. This contest is not just about size, strength or speed; it is about physical and mental agility, awareness, and anticipation. In a word, it's all about "quickness."

Introducing a concept he called "sexual selection," Darwin (1859) noted that, "generally, the most vigorous males, those which are best fitted for their places in nature, will leave the most progeny." Among mammals, male–male competition for access to mates often takes the form of aggressive confrontation, in which mating success is largely determined by the outcome of either violent or ritualized combat (Eibl-Eibesfeldt 1961). Physical size and/or age strongly influence the outcomes of male–male conflicts over mates in many species, including red deer (*Cervus elephus*; Clutton-Brock et al. 1979), African elephants (*Loxodonta africana*; Poole 1989), rhesus macques (*Macaca mulatta*; Manson 1996), and elephant seals (*Mirounga angustirostris*; LeBoeuf and Kaza 1981; *Mirounga leonina*; McCann 1981, Modig 1996). In some species, weapons play an important role in male–male competition for access to females, sometimes leading to the evolution of extremely hypertrophic male characteristics that contribute to male success in aggressive encounters (Lorenz 1966). Examples include the tusks of African elephants (Poole 1989) and narwhals (*Monodon monoceros*; Gerson and Hickie 1985, Brear et al. 1993), the antlers of elk (Leslie and Jenkins 1985), and the horns of a number of African antelopes (Packer 1983).

Hormonal state can also influence outcomes of male–male contests over access to females (e.g., African elephants in musth, Poole 1989; red deer in rut, Lincoln et al. 1972). Among primates, social dominance rank is generally correlated with mating success, but this does not mean that only the most dominant males successfully mate or that they necessarily have the greatest mating success (reviewed by Huntingford and Turner 1987).

Non-aggressive competition between males for access to mates may take the form of sperm competition (Trivers 1985). In primates, testes mass relative to body mass is generally larger in species with multi-male groups than in monogamous species or single-male groups, indicating sperm competition (Harcourt et al. 1981). Among dolphins, the testis size relative to standard length is especially high in the genus *Lagenorhynchus*, suggesting sperm competition, but smaller body size might also favor maneuverability or quickness in mating success (Connor et al. 2000a).

Dusky dolphin mating chases appear to involve male–male competition, but quickness may be more important than size or aggression in reproductive success. Among small groups of dolphins, high-speed activities occurred most frequently in mating groups (Figure 8.6). The percentage of intervals with speed bursts (Kruskal–Wallis $H = 6.670$, $P = 0.04$), lateral porpoising (Kruskal–Wallis $H = 10.146$, $P = 0.006$), and chasing (Kruskal–Wallis $H = 34.882$, $P < 0.001$) varied significantly between social group types, with mating groups outpacing mother–calf nurseries and non-sexually active adult groups in all categories (Figure 8.6). Mean ($\pm$ SE) duration of surface chases in mating groups was 20.6 ± 3.6 seconds.

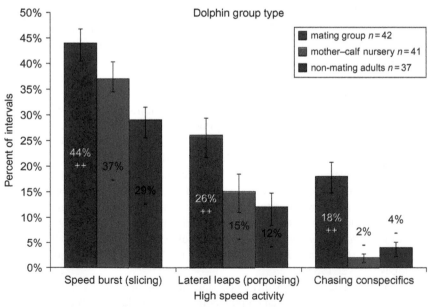

FIGURE 8.6   The prevalence of high-speed behaviors (speed bursts, lateral leaps and chasing) is compared for small groups (<50 dolphins) monitored off Kaikoura. Bars represent the mean ($\pm$SE) percent of 2-minute intervals during which high-speed activities were documented. Significant differences are indicated by $+$ and $-$ symbols (Kruskal–Wallis, Bonferroni post-hoc, $P < 0.05$).

Further evidence for rapid high-speed chasing competition was encountered in the variability of directional heading and lack of clear group swimming formations in mating groups. Mating groups exhibited less linear and less coordinated movement and formation than other dolphin groups (Figure 8.7). Heading variability, defined as the proportion of instantaneous samples at which group members were pointed in different directions (>45° different in the horizontal plane by compass bearing), differed significantly by social group type (Kruskal–Wallis $H = 12.186$, $P = 0.002$), with mating groups twice as likely to swim in an uncoordinated fashion as other small groups (Bonferroni, $P < 0.05$, Figure 8.7a). Swimming formation varied between small groups (Figure 8.7b), with mating groups most often having no swimming formation (Kruskal–Wallis $H = 7.087$, $P = 0.03$). Mating group formation was less often parallel than other small groups (Kruskal–Wallis $H = 20.510$, $P < 0.001$, Figure 8.7b), indicating that the dolphins were less likely to be swimming "with" (alongside) and more likely to be focused "toward" a central point (or individual, i.e., the

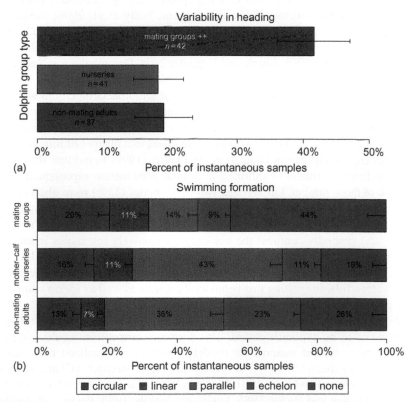

FIGURE 8.7   The degree of coordinated movement in small (<50 dolphins) mating groups, mother–calf nurseries, and non-mating adult groups is compared with respect to: (a) variability in heading and (b) swimming formation. Bars represent the mean (±SE) percent of instantaneous samples at which variable headings and swimming formations were documented.

female). Mating groups also showed the tightest inter-individual proximity (Kruskal–Wallis $H = 52.129$, $P < 0.001$) as the males clustered around the female(s), with a mode inter-individual distance of <1 body length (<1.8 m) 75% of the time (Markowitz 2004).

Among male dusky dolphins, intra-sexual competition appears to take on two main forms: sperm competition and physical chasing (Cipriano 1992, Markowitz 2004, Van Waerebeek and Read 1994). Unlike bottlenose dolphins (Connor et al. 2000a), male dusky dolphins appear unable, either singly or cooperatively, to monopolize females (although see Chapter 9). The absence of size dimorphism between sexes in dusky dolphins (Cipriano 1992, Van Waerebeek 1993a) suggests that physical strength is relatively unimportant in the context of intra-sexual competition.

Scramble competition for access to mates is common among many polygynous animals, including wasps (*Pleistodontes froggatti*; Bean and Cook 2001, Alcock and Kemp 2005), bees (*Anthophora plumipes*; Stone 1995), fireflies (*Photinus* spp.; Vencl 2004), orb-weaving spiders (*Argiope aurantia*; Foellmer and Fairbairn 2005), wetas (*Deinacrida rugosa*; Kelly et al. 2008), treefrogs (*Osteopilus septentrionalis*; Salinas 2006), crustaceans (Bertin and Cezilly 2003), rodents (Fleming and Nicolson 2004; Spritzer et al. 2005a, 2005b), and primates (Matsubara 2003, Eberle and Kappeler 2004, Dammhahn and Kappeler 2005, Eberle et al. 2007). Although it is possible that the rapid mating chases by dusky dolphins are simply another instance of scramble competition with no particular individual males favored, our observations suggest that male prowess in the form of quickness strongly influences the outcome of this mating competition. While females may mate with multiple males in a manner of minutes, not all males engage equally in copulation. McOmber (1999) found that the number of individuals (males) confirmed in ventral presentation represented only 22–75% of those present. On average, two of six males (33%) were observed to successfully copulate with a female during the time we were monitoring them (Markowitz 2004). More recent research (Markowitz and Markowitz, unpublished data) generally confirms the tendency for relatively few males in mating groups to successfully copulate with the female (although due to our limited observation times we cannot rule out the possibility that all the males are eventually successful), suggesting that behavioral contests as well as sperm competition apparently play an important role in male reproductive success.

We propose that in the chase, quick males may be highly favored, although this remains to be explicitly tested. This form of competition would contrast with the more typical mammalian model of violent or ritualized male–male combat (Eibl-Eibesfeldt 1961) in which size, age, strength (Clutton-Brock et al. 1979, LeBoeuf and Kaza 1981, McCann 1981, Manson 1996, Modig 1996), weapons (Gerson and Hickie 1985, Leslie and Jenkins 1985, Brear et al. 1993), and/or hormonal state (Lincoln et al. 1972, Poole 1989) determine male reproductive success. This is not to say that male dusky dolphins engaged in mating chases never show signs of overt aggression. Rather, our observations suggest that agonistic interactions are rare, and that males generally appear unable to completely dominate mating contests or sequester females. Apparent agonistic behaviors we have witnessed include jostling for position (<20% of groups), mid-air collisions (in three groups), biting (on one occasion), open mouth threats (on two occasions), and a "rolling over" behavior in which one male rolled over the top of a copulating pair of dolphins, forcing the coupled pair to split apart (on six occasions). Low rates of agonistic interactions may partly explain the lower mark rate of dusky dolphin dorsal fins compared to those of bottlenose dolphins (Markowitz et al. 2003a).

TEAM MATES TO GET MATES: ASSOCIATIONS AMONG MALES

Basketball coaches often emphasize "hustle" when encouraging their team before a big game. In a competition involving two well-matched teams, the one which out-hustles the other is usually the winner. Being the quickest team involves far more than being quick in a one-on-one situation, and on the basketball court, great teams usually prevail against one-on-one superstars. Success is measured in assists as well as scores. Although at first glance, dusky dolphin sexual competition appears to be a free-for-all, closer inspection reveals something perhaps more organized among the males in their "fast break" to get to the female.

While "cooperating to compete" sounds contradictory, it is a common feature of many social mammals (De Waal and Harcourt 1992). Whether they are temporary (coalitions) or lasting (alliances), cooperative relationships based on reciprocal altruism (Trivers 1971) or by-product mutualism (Dugatkin et al. 1992) can benefit social partners in conflicts with other conspecifics. Cooperation among males can be an important factor in the outcome of aggressive male–male competition for mates. Male lions (*Panthera leo*) form egalitarian alliances with both related and unrelated males to take over and maintain prides (Grinnell et al. 1995). Male savanna baboons (*Papio anubis*) that form coalitions overthrow more dominant males and gain access to females (Packer 1977, Smuts and Watanabe 1990, Noë 1994). Extra-troop male langurs (*Semnopithecus* spp.) cooperate in raids to take over troops held by other males (Hrdy 1977). Male coalitions and alliances among chimpanzees (*Pan troglodytes*) and bonobos (*Pan paniscus*) play a key role in agonistic contests and mating success (chimpanzees: Watts 2000, 2002, 2004, Arnold and Whiten 2003, Newton-Fisher 2006, Gilby and Wrangham 2008; bonobos: Stanford 2000, Hohmann and Fruth 2003).

The level of cooperation exhibited by dolphins competing for access to mates can rival that of primates. For example, male bottlenose dolphins form coalitions and alliances, roughly analogous to those observed in baboons or chimpanzees, to obtain mates (Connor et al. 1992a). Male coalitions to gain access to mates have also been noted in Atlantic spotted dolphins (*Stenella frontalis*, Herzing 1996). Social intelligence related to complex cooperative relationships has been proposed as one factor promoting the evolution of large brain size in dolphins (Connor 2007; see also Chapter 16).

Our studies indicate that some dusky dolphin males apparently cooperate to compete, as do male bottlenose dolphins (Connor et al. 1992a), with long-term associations between males in mating and coordinated foraging groups (Markowitz et al. 2004). Males photographed in mating groups in Kaikoura during the breeding season (spring–summer) have been re-sighted in Admiralty Bay hundreds of kilometers to the north in the non-breeding season (winter), where they form small coordinated foraging groups with consistent associations within and between years (Markowitz and Würsig 2004). In one case, four males photographed chasing a female together in a small mating group off Kaikoura during the breeding season were subsequently re-sighted in Admiralty Bay where they repeatedly foraged together the following winter 2000 (Figure 8.8). This group was consistently observed in the same part of the bay and there

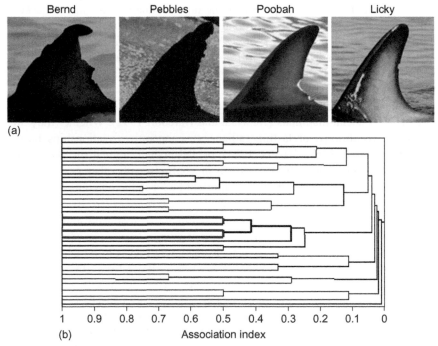

FIGURE 8.8   (a) Dorsal fin photographs of four individuals identified in Kaikoura chasing a female in a mating group during the breeding season and re-identified in Admiralty Bay chasing fish together, from left to right named "Bernd," "Pebbles," "Poobah," and "Licky." (b) This diagram, based on simple ratio coefficients of association (the proportion of the number of times individual dolphins were photographed together to the total number of times they were photographed), for individuals photographed in Admiralty Bay foraging groups during winter 2000, shows the extent to which associations were clustered. Bold lines show the group "Tribe Kahuna," consistently observed feeding together. Bold vertical lines at the 1 level indicate constant companions (dolphins always photographed together).

was no overlap between this group and other foraging groups, suggesting the possibility of territoriality (Markowitz 2004). In Kaikoura, male strategies in mating groups include pushing the female up against the surface of the water (swimming ventrum to ventrum), much as they herd fish. It is unclear how important male–male cooperation is in determining male reproductive success. Given the length of these associations, across seasons and between habitat types (Markowitz 2004), we suspect that dusky dolphin males may form long-term cooperative alliances, cooperating in different ecological and social contexts.

We have few direct observations of cooperative or apparently coordinated efforts by males in mating groups, perhaps due to the high speed of the action, but we have witnessed some suggestive interactions. For example, in our most recent field season, we observed the highest intromission rate (1.5 per minute) and maximum intromission duration (22 seconds) in a small group with just two males and a single female. Both males copulated with the female repeatedly. The two males consistently performed synchronous ventral chases with copulations by both males in quick succession. The female in this group was observed playing with kelp and resting at the surface during the focal group follow. The long copulation was performed by the first male in the sequence, which could indicate sperm competition in what appeared otherwise to be a relatively non-competitive interaction.

FEMALE CHOICE AND FEMALE STRATEGIES

There's something very special, very different, about swimming with a group of mating dusky dolphins. If you happen to plunk into the sea with them during a lull in the high-speed action, and if you swim as hard as you can, you may just be able to keep up. Unlike other times, dusky dolphins having sex aren't interested in playing with you. At best, they tolerate you. They are . . . otherwise occupied. And if you manage to swim along upside down, underneath the action, then you are really in for a show, because then the most astounding thing happens. The female suddenly takes off, streaking away at high speed in an impressive display of quickness and agility. And of course, being only human, your chance of keeping up is zero. All you can do is watch in amazement as the last of the female's suitors disappear ahead of you, in the wake of the great race.

Female choice often plays an important role in reproductive contests (Trivers 1985). Evolution favors discriminating females, choosing males based on reliable indicators of male fitness (Fisher 1930). Showy characters can represent handicaps, reliably demonstrating to females the fitness of males (Zahavi and Zahavi 1997), for example resistance to parasites (Hamilton and Zuk 1982). Courtship displays among lekking birds provide females an arena in which to select mates (Kruijt and Hogan 1967). Similar lekking occurs in a number of ungulates, including topi antelope (*Damaliscus lunatus*; Bro-Jørgensen 2002, 2003, 2008, Bro-Jørgensen and Durant 2003), kob antelope (*Kobus kob kob*;

Fischer and Linsenmair 1999), blackbuck (*Antilope cervicapra*; Isvaran and Jhala 2000, Isvaran 2005), and Kafue lechwe (*Kobus leche kafuensis*; Nefdt and Thirgood 1997). Among cetaceans, the mating system of humpback whales (*Megaptera novaeangliae*) shares features with that of lekking birds and mammals, with some evidence for female choice (Clapham 2000).

For dusky dolphins, extended chases provide a mechanism for females to exercise choice in sexual partners (Whitehead and Mann 2000), a competitive arena that encourages fitness of their mates. Females may opt to evade males that fail to demonstrate requisite vigor or social skill. Although females might mate with multiple males in a manner of minutes, not all males engage equally in copulation. On average, we witnessed successful copulation by roughly one-third of males present in mating groups (McOmber 1999, Markowitz 2004) and relatively little overt aggression, suggesting that successful males were not those that could defend a female, but rather those that could keep up with her.

Our studies to date indicate that the female is always the one on top during copulation. Several observations from our work support this generalization (McOmber 1999, Markowitz 2004; Markowitz and Markowitz, unpublished data). During ventral presentation all dolphins that we have sexed in the ventral-up position were males. In mating groups, usually one (rarely two) individuals were consistently in the ventral-down position (on top) during intromission. These same individuals were consistently chased, and those we managed to sex were always females. In the context of the chase, being on top gives the female an inherent advantage; she can remain swimming at speed, while surfacing to breathe.

Although not involving flagrantly showy features, the requirement that males swim upside down while keeping up with the chase could represent a form of "handicap" (Zahavi and Zahavi 1997), providing a reliable indicator of the quickness, agility, and physical condition of candidate males with whom a female might copulate. According to sociobiological theory, mate choice can promote "good genes" which in turn increase the probability that the females' own offspring inherit these advantageous phenotypic traits such as quickness, social skills, or other attractive characters (Trivers 1985). Unlike some showy features of males favored by sexual selection that might otherwise be maladaptive (Lorenz 1966), the above male attributes represent qualities that reliably demonstrate fitness in a variety of contexts. For example, quickness and social skills are likely to be important in foraging (see Chapter 6) and may be the difference between life and death when interacting with predators (see Chapter 7).

Our observations indicate that females are not helpless or passive participants in dusky dolphin mating activity. Females employ a number of tactics to evade males, effectively extending the chase. Apparent avoidance strategies by females include rapid changes in direction (quickness and agility), re-entry leaps (which allow the female to dive below chasing males), and head-up braking maneuvers (female stops suddenly during a chase causing males to "overshoot" the female).

Rapid changes in direction

As females evade males and males race to keep up, mating group movement patterns are relatively non-linear, with groups often doubling back on themselves and making frequent changes in heading. In the most recent mating season (2007–2008), the average linearity (net distance/total distance) estimated from GPS tracks at 1-minute intervals was just 0.55 (or 55% linear, Table 8.2). In other words, the net distance, or "distance made good," by dolphins in mating groups was just a bit over one-half the total distance traveled. Calculated values in Table 8.2 probably underestimate actual swimming speeds and overestimate linearity due to the limited maneuverability of the vessel relative to the dolphins, so actual zig-zag movement may even be greater.

Mating groups changed heading regularly at a mean reorientation rate of 36°/min (Table 8.2). For those groups with detailed information on confirmed copulation rates (intromissions per minute), copulation rate was negatively correlated with reorientation rate (Kendall's tau, $R^2 = 1.0$, $P = 0.04$). In other

TABLE 8.2 Movement patterns of small mating groups off Kaikoura (2007–2008)

Group size	% Linearity (total/net km)	Reorientation rate (degrees/min)	Leg speed (km/h)		
			Mean	Top	Variance
3	50%	18.9	4.6	12.5	7.5
5	58%	66.6	4.1	8.3	10.3
5	55%	35.5	3.6	13.0	13.7
5	50%	72.3	2.6	4.7	3.3
6	33%	42.3	1.7	12.0	4.8
6	89%	12.9	3.7	8.6	2.6
6	10%	10.8	3.9	19.5	13.2
7	75%	39.6	6.0	10.7	6.9
8	72%	34.7	2.3	7.9	5.6
9	86%	16.5	7.2	27.0	22.7
9	27%	44.4	1.8	2.6	0.4
Min	10%	10.8	1.7	2.6	0.4
Mean	55%	35.9	3.8	11.5	8.3
Max	89%	72.3	7.2	27.0	22.7

words, as the dolphins' rate of reorientation increased, the rate of intromission decreased. This indicates that a chase with many rapid changes in direction reduces the frequency of copulations. Frequent changes in direction by the female may extend the chase, and consequently the length of time between copulations (inter-copulatory interval).

Three-dimensional chases: re-entry leaps

These data above do not address movement in three dimensions, and it is clear that chasing includes a three-dimensional component. One tactic females can employ to extend mating chases in three dimensions is re-entry leaping. Head-first re-entry leaps, which occur at a high rate in mating groups, allow a dolphin to catch a breath and dive efficiently back down through the water. Females apparently use re-entry leaps as an evasion tactic, and the typical response of males is to follow the female with re-entry leaps of their own (Figure 8.9a). Mating groups vary from other small groups (mother–calf nurseries and non-mating adult groups) not only in the number of leaps counted but also in their prevalence, with much higher rates of clean head-first re-entry leaps (Kruskal–Wallis, $H = 25.619$, $P < 0.001$) than other small social groups (Figure 8.9b).

In 2007–2008, the mean ($\pm$ SE) number of re-entry leaps per individual per minute was 0.4 ± 0.1 in mating groups. High rates of re-entry leaping (maximum $= 0.71$ leaps per individual per minute) were associated with low rates of copulation (minimum $= 0.167$ intromissions per minute), indicating that re-entry leaps are an effective female evasive tactic for prolonging pre-copulatory male–male competition. Synchronous surfacing and re-entry leaping occurred in all mating groups monitored in 2007–2008, with a mean ($\pm$ SE) of 2.5 ± 0.2

(a)   (b)

FIGURE 8.9 (a) A female dusky dolphin vaults up out of the water in a clean head-first re-entry leap with a male dolphin close behind. (b) The frequency of clean head-first re-entry leaping is compared between mating groups, mother–calf nurseries, and non-sexually active adult groups at Kaikoura. Bars show the mean ($\pm$SE) number of leaps per 2 minutes. Data labels show the percentage of 2-minute intervals with leaping.

individuals surfacing synchronously and 2.5 ± 0.3 re-entry leaping synchronously. In 13 cases where more than two individuals performed synchronous re-entry leaps, only one led to subsequent intromission, again suggesting that when competition between males is greater (with three or more leaping after the female at once) successful intromission is less likely to immediately follow. Most intromissions were observed after re-entry leaping by just one or two individuals (perhaps indicating reduced competition).

We found re-entry leaps to be commonly associated with surface chases, although the correlation between chasing and re-entry leaping was not a very strong one (Spearman's correlation, $R^2 = 0.182$, $P < 0.01$). Head-first re-entry leaps, which also occur frequently in feeding groups when dolphins are chasing fish from below (Markowitz et al. 2004), likely play a role in male pursuit of females as well as females evading males. Below-surface chasing behavior was also observed to immediately follow re-entry leaping in mating groups, although the focus of the chase was conspecifics. In this context, subsurface chases associated with re-entry leaps were considered to begin when the last leaping dolphin re-entered the water and end when any dolphin in the mating group broke the surface to take a breath. As only the initiation and portions of the chase near the surface could be observed (mainly near the end of the chase), it is possible that subsurface chases were not always continuous. The mean ($\pm$ SE) duration of these chases was 11.5 ± 2.6 seconds.

Typically, re-entry leaps are seen in a context something like this: A female (or dolphin being chased we guess is female but haven't been able to sex) races along, zigzagging rapidly at the surface, followed closely by several males slicing through the water at high speed. One or more have turned upside down and are rapidly approaching her to copulate. But before they can reach her, she vaults high into the air, catches a breath and plunges down below her avid pursuers. The males follow her, leaping up into the air, often in close quarters and sometimes in unison, diving back in after the female.

Males generally follow females closely, and the speed of these chases during the breeding season clearly demonstrates male fitness in terms of quickness, agility, and motivation. Within mating groups, this high-speed chasing activity, including rapid changes in direction, is associated with sexual activity, although both sexual activity and chasing, which are generally elevated in mating groups, show considerable variability (Spearman's correlation, $R^2 = 0.127$, $P < 0.05$, $n = 40$).

Head-up braking maneuver

Eye-outs (spyhopping) were more common in mating groups than in non-sexually active groups (Kruskal–Wallis $H = 16.280$, $P < 0.001$). In mating chases, these generally take the form of a braking maneuver. In the heat of the chase, with several males in hot pursuit, the female suddenly raises her head and lowers her lower body down through the water, stopping suddenly. Just as in a high-speed car chase, the result is that the pursuers (the males in this case) "overshoot" the female. Typically, she turns quickly and heads in the opposite direction.

Morphological adaptations

Female dusky dolphins may also have morphological features promoting female choice. Pseudocervix structures in the upper vagina occur in a number of cetaceans, including the closely related Pacific white-sided dolphin (*Lagenorhynchus obliquidens*: Schroeder 1990). It is likely that dusky dolphins have a similar pseudocervix structure, the folds of which create a section of the vagina between the cervix and pseudocervix into which sperm may be deposited. A certain degree of sexual stimulation may be required for muscular contractions of the pseudocervix that protect the sperm from seawater (Schroeder 1990). At any rate, the pseudocervix can serve as a sperm depository, and likely allows sperm from multiple males to be kept from washing out during a copulation session, thus promoting sperm competition by volume of sperm alone.

INTERACTIONS BETWEEN MATING GROUPS AND NURSERIES

Among dusky dolphins, mating occurs most frequently in small groups without calves. Mothers with calves appear to avoid mating group activities. We have observed mothers with calves in mating groups on just three occasions. In each case, a lone female with a very young calf was the object of the chase. These chases were particularly rapid, with both the mothers and the calves showing signs of distress. In one case, calf tossing was observed (Markowitz 2004). In another case, the female and calf were escorted away from mating activity by two other adults, apparently from their nursery group.

Observations of sexual harassment of mothers with calves suggest that aggressive mate coercion (Huntingford and Turner 1987, Connor et al. 2000a) may be employed by male dusky dolphins in some cases. The proximate cause and/or function of this behavior on the part of the males is unknown, but based on observations of it occurring with very young calves present we speculate that it may involve chemosensory or behavioral signals related to female postpartum hormonal state. Protection of calves from conspecific harassment may be an important factor contributing to the formation of nursery groups, and we have noted apparent avoidance of mating groups by nursery groups. For more information on interactions between mother–calf pairs and other dolphins, see Chapter 9.

PROCREATION AND RECREATION: THE ROLE OF NON-REPRODUCTIVE SEX IN DOLPHIN SOCIAL LIVES

"Sex is an antisocial force in evolution. Bonds are formed between individuals in spite of sex and not because of it." These words open the chapter on "Sex and society," in the classic book *Sociobiology* by E.O. Wilson (1975), emphasizing the divisive, competitive nature of sexual reproduction in general. Here, we examine the role of non-reproductive sex in the social lives of dusky dolphins, and explore the possibility that, at least in this instance, sex reinforces important social bonds.

Like bottlenose dolphins (Connor et al. 2000), dusky dolphins apparently engage in sexual behavior for social reasons as well as reproduction. Sexual behavior may play a role in communication (Wells and Norris 1994), or in greeting ceremonies (Würsig and Würsig 1980). Although female dusky dolphins are seasonally polyestrous, with conception only possible during a limited breeding season, mating occurs in all seasons (Figure 8.3, Cipriano 1992, Markowitz 2004). Social–sexual behavior outside the breeding season appears to be much more relaxed, lacking the high-speed competitive chasing we have described during the breeding season. Homosexual activity between males has been noted in both Kaikoura and the Marlborough Sounds (Cipriano 1992, Markowitz 2004). In a foraging context, social–sexual activity may be a form of celebration and/or strengthen social bonds between foraging companions (Würsig and Würsig 1980, Markowitz 2004).

Dusky dolphins also mate with other species. Off Kaikoura, dusky dolphins were frequently observed mating with common dolphins (*Delphinus delphis*; 28.6% of mixed groups encountered). In the Marlborough Sounds, social–sexual activity between common dolphins and dusky dolphins was observed in two of five mixed species groups encountered (Markowitz 2004). Apparent hybrids between the two species have been noted (Reyes 1996, Würsig et al. 1997). Hybrids have also been noted between other cetaceans, including Dall's porpoise and harbor porpoise (*Phocenoides dalli* and *Phocoena phocoena*; Willis et al. 2004), and bottlenose and common dolphins (Zornetzer and Duffield 2003).

SUMMARY AND CONCLUSIONS

Although dusky dolphins are clearly polygynandrous (what is termed promiscuous by many) breeders with sperm competition, this should not be misconstrued as random breeding without variability in mating success or mate choice. Our vessel-based focal group research, utilizing digital data recorders to document social–sexual interactions continuously while identifying and sexing individuals, has revealed social mechanisms we believe result in nonrandom breeding. Dusky dolphins breed seasonally, but mate year-round for social as well as reproductive reasons. Off Kaikoura, sexual activity peaks in the late afternoon, prior to nocturnal feeding. During the breeding season, mating occurs most frequently in small groups without calves. These groups are generally composed of about six males chasing a single female, and appear to function as competitive breeding arenas.

Quickness and agility, rather than physical strength, seem to be the main factors determining male reproductive success. Reproductive chases are extremely rapid, involving frequent changes in direction, male chasing tactics and female "escaping" tactics. This role of quickness in male–male competition is either unusual among mammals or has not been emphasized in the literature. Re-sights of males in mating groups and coordinated foraging groups in different seasons and habitats provide some evidence for alliances that function in different social and ecological contexts (social partners that chase females and chase fish together).

Our observations suggest a role for female choice as well as male–male competition in dusky dolphin reproduction. Rather than runaway selection for an ecologically irrelevant showy feature, quickness is a male characteristic with obvious fitness consequences in terms of survival (foraging and avoiding predation) as well as reproduction. Therefore, we propose that it represents a reliable indicator of male quality to choosy females. With females apparently always on top during copulation, males must chase from an inverted position, a "handicap" in the context of the chase which could help facilitate female choice of the most vigorous males. Females utilize a number of apparently effective tactics, including rapid changes in direction, re-entry leaps, and head-up braking maneuvers, to evade males and prolong the chase.

On rare occasions, males chase females with calves. These chases can be especially rapid, and may cause some distress to calves and their mothers. The source of attraction is unknown. Mother–calf nurseries may partly function to protect females and their calves from marauding males as well as other dangers. Dusky dolphins have sex for social as well as reproductive reasons. Evidence for social sex includes mating outside the breeding season, homosexual activity, and mating with other species. Clearly, sex is an important component of the rich social lives of dusky dolphins.

ACKNOWLEDGMENTS

Research in Kaikoura was accomplished with the help of many hard-working full-time field assistants, including Tracy McKeown, Joana Castro, Cathryn Owens, Heidi Petersen, Kim Andrews, and Anita Murray. Additional assistance was provided by many capable, enthusiastic Earthwatch volunteers. Jack Van Berkel at the University of Canterbury Edward Percival Field Station supplied lab space and logistical support. Field research was financially supported by the New Zealand Department of Conservation, the Earthwatch Institute, the National Geographic Society, and Texas A&M University. Boat maintenance was supplied by John Hocking, Murray Green, and Steve Howie. Bernd and Melany Würsig provided guidance and support throughout the work.

Chapter 9

Dusky Dolphin Calf Rearing

Jody Weir,[a,b] Sierra Deutsch,[b] and Heidi C. Pearson[b]

[a]Sidney, British Columbia, Canada
[b]Marine Mammal Research Program, Texas A&M University, Galveston, Texas, USA

It's a bright sunny day off the coast of New Zealand, the albatrosses are sitting patiently on the surface waiting for the winds to pick up, and the tourist boats are surrounded by a group of over 400 flipping, tail slapping, belly flopping dusky dolphins. As we motor along on our survey, we find a nursery group. This group is not leaping or making big splashes like the bigger "main" group. It's a small tight group, and if we hadn't been scouring the waters for small gray fins against the background of the rugged coastline, we would likely have passed them by. As we approach, we notice that there are at least two distinct sizes of animals, adults and calves. The young calves are bobbing like corks, somewhat over-shooting the surface. We keep a fair distance; these nursery groups are easily disturbed. The camera and data sheets come out, and we begin a focal follow of the group. As we see the small calves beating their flukes rapidly beside their mothers and other adults, we ourselves feel like we are taking small fluke strokes towards unravelling the social lives of these unique subgroups. We were not the first to notice these groups. Bernd and Mel Würsig recorded dusky dolphin nursery groups in Argentina, and Frank Cipriano's and the Würsigs' early work in New Zealand describes nursery groups generally staying closer to shore off Kaikoura. However, studies of calf rearing and development in dusky dolphins are still in their infancy, and here we explore what we have managed to decipher thus far.

Jody Weir, Kaikoura, New Zealand, December 21, 2008

INTRODUCTION

Dolphins are one of the few carnivorous mammals considered precocious at parturition. They are able to swim and come to the surface to breathe almost immediately after being born, yet they spend an extended amount of time with their mothers after birth. For example, Indian Ocean bottlenose dolphins (*Tursiops* sp.) in Shark Bay, Australia, nurse their calves for 3–8 years (Mann et al. 2000). Given the high energetic cost of lactation to females, it is at first somewhat puzzling that dolphins invest so much time and energy in young who are born relatively mature and mobile, until we realize that the mother–calf bond may be critically important for social learning and general development.

In this chapter, we describe some of the behaviors of mother–calf pairs from our work with dusky dolphins (*Lagenorhynchus obscurus*) off Kaikoura, New Zealand. Here, mother–calf pairs separate from the larger groups to associate with other females with calves of roughly the same age (Figure 9.1). Similar grouping behaviors and patterns have been documented in other species of delphinid, including Indian Ocean bottlenose dolphins (Mann et al. 2000, Gibson and Mann 2008a, 2008b), marine tucuxis (*Sotalia fluviatilis*; Azevedo et al. 2005), spotted dolphins (*Stenella attenuata*; Pryor and Shallenberger 1991), common bottlenose dolphins (*T. truncatus*; Wells 1991), and humpback dolphins (*Sousa chinensis*; Karczmarski 1999).

Long-term studies of bottlenose dolphins in Shark Bay provide a foundation for what is known about wild delphinid calf rearing and development (Mann and Smuts 1998, 1999, Mann et al. 2000, Mann and Sargeant 2003, Mann and Watson-Capps 2005, Gibson and Mann 2008a, 2008b). Although much less is known about dusky dolphin calf rearing, we describe mother–calf grouping behaviors that are notably different from those in Shark Bay. For example, bottlenose dolphin mother–calf pairs in Shark Bay are alone nearly 50% of the time (Gibson and Mann 2008b), which may be a result of female foraging specializations such as sponging or beach-hunting (Mann and Sargeant 2003, Sargeant et al. 2005). Dusky dolphin mothers and calves are usually seen with at least one other mother–calf pair, and are often part of a "nursery group" composed of at least two calves per six non-calves (Weir et al. 2008; see elaboration below).

Delphinid nursery groups exhibit intra- and inter-specific variation in group size. Off Kaikoura, dusky dolphin nursery group size may range from 4 to 100 individuals (2–50 mother–calf pairs), with a median group size of 14 (Weir 2007). In Golfo Nuevo, Argentina all but one dusky dolphin mother–calf group was composed of <20 individuals (Degrati et al. 2008). Earlier work by Würsig and Würsig (1980) in Golfo San José, Argentina described dusky dolphin nursery groups composed of 8–20 mother–calf pairs. In Shark Bay, Australia, when

FIGURE 9.1   Dusky dolphin mother and calf pair.

bottlenose dolphin mothers and calves do associate with other individuals, they are usually found in small groups of 3–7 individuals (Mann et al. 2000). In Algoa Bay, South Africa humpback dolphin groups with calves had a mean size of 10 (Karczmarski 1999).

Delphinid mothers and calves may also occur in groups of varying composition. Dusky dolphin mothers and calves off Kaikoura may be found amongst the larger, mixed sex and age groups. However, 18% of dusky dolphin groups encountered during systematic surveys off Kaikoura were nursery groups, composed "entirely" of mothers and calves (Weir et al. 2008). Similarly, in Golfo Nuevo, Argentina dusky dolphin mother–calf groups were composed of >80% mother–calf pairs (Degrati et al. 2008). In contrast, humpback dolphin calves made up just 14% of the individuals found in nursery groups (Karczmarski 1999).

Here, we describe basic reproductive parameters of dusky dolphins, provide an overview of influences on nursery group formation, and describe our methods for conducting behavioral observations of nursery groups. Next, we present some of the dangers associated with calf rearing and how nursery group formation may minimize these dangers. We then give examples of opportunities that may be granted to females and calves when nursery groups are formed, and discuss some possible benefits to mothers and calves. Throughout the chapter, we place dusky dolphin mother and calf behavior in a comparative context by providing examples from other species. Many examples are provided from bottlenose dolphins in Shark Bay, due to the long-term studies of mother and calf behavior in this population.

FEMALE DUSKY DOLPHIN REPRODUCTIVE PARAMETERS

Female dusky dolphins off New Zealand first reproduce at about 7–8 years of age (Cipriano 1992). Off Argentina, a small sample of females was sexually mature at about 6–7 years (Dans et al. 1997b), but we do not know whether this slight difference is due to sampling biases. Gestation length is approximately 11 months off New Zealand and up to 13 months off Peru (Cipriano 1992, Van Waerebeek and Read 1994).

Off New Zealand and Peru, calving primarily occurs in late spring and summer, while off Argentina, calving appears to occur slightly earlier in the year (Würsig and Würsig 1980). However, calving also occurs during other seasons, and many investigators have recorded newborn calves in fall and winter off Kaikoura.

Cetacean milk has a fat content of 22–30%, which is higher than that of any group of mammals other than pinnipeds (Whitehead and Mann 2000). Lactation is energetically costly to mothers, and female odontocetes therefore increase their caloric intake by about 50% while nursing (Oftedal 1984). From examination of dusky dolphin calves captured incidentally in fishing nets, lactation is estimated to last at least 18 months off New Zealand and 12 months off Peru (Leatherwood and Reeves 1983, Cipriano 1992, Van Waerebeek and Read 1994).

OVERVIEW OF INFLUENCES ON NURSERY GROUP FORMATION

There are different explanations for why dusky dolphin mother–calf pairs may associate to form nursery groups separate from the main group. As described below, nursery groups may provide protection to calves from dangers such as predation and harassment from male conspecifics (Weir et al. 2008; see also Chapter 7), while simultaneously increasing opportunities for foraging, resting, allomaternal care, social bond formation, and social learning by the calf (Weir 2007, Deutsch 2008, see also Chapter 16). Ultimately, nursery group formation may increase female reproductive success. For example, female bottlenose dolphins in Sarasota, Florida who raised calves in larger, more stable groups had increased reproductive success (Wells 2003).

An alternative explanation for nursery group formation may be that lactating females form nursery groups simply by adopting a "falling out" strategy from other types of social groupings, due to the physical demands of calf "carrying." As there is no place to hide an infant in the three-dimensional ocean environment, dolphin mothers protect their calves by "carrying" them. Through the Bernoulli effect, a calf traveling in close proximity to its mother becomes sucked in towards her body and can ride her pressure wave (Weihs 2004). However, while calves may get a somewhat "free ride," this carrying behavior increases the drag force on the mother, reducing her mean maximum swim speed by nearly 25% (Noren 2008). In dusky dolphins, mothers who are slowed down by calf carrying may be unable to keep up with the fast-paced swimming found in large groups, and thus may form groups with other females facing similar swim speed limitations.

Calves may be carried in either the echelon position (to the side of the mother by her dorsal fin), or the infant position (just ventral and to one side of the mother near her mammary slits). Like bottlenose dolphin calves (Cockcroft and Ross 1990, Gubbins et al. 1999, Mann and Smuts 1999), dusky dolphin calves are more likely to be in echelon position after parturition, but are increasingly found in infant position as they become older (Deutsch 2008). However, dusky dolphin calves of all ages appear to revert to echelon position when traveling rapidly (Deutsch 2008).

BEHAVIORAL OBSERVATIONS

In 2005 and 2006, we conducted boat-based surveys off Kaikoura to locate nursery groups and conduct behavioral studies. For each nursery group that was encountered, a 30-minute group focal follow was attempted. At 5-minute intervals, the behavioral state of the group was determined via scan sampling (Altmann 1974). Behavioral state was classified as traveling, resting, socializing, or foraging. During focal follows, we also recorded all instances of displays (slapping a body part such as a belly, tail, dorsal surface, or pectoral flipper on the surface of the water), head-first re-entries (leap out of the water that involves the entire body of the dolphin, and re-entering with a clean, or minimal splash, dive), and coordinated leaps. We recorded whether displays and head-first re-entries were by adult or calf.

Age classification

We defined four main dusky dolphin age groups: young calves, yearlings, juveniles, and adults. Yearlings, like younger calves, are still dependent on maternal care; therefore, both age groups are collectively referred to as "calves" in most of our work.

Young calves

Young calves are <1 year old. While there is not much information on length at birth for New Zealand dusky dolphins, Cipriano (1992) obtained lengths of 97–102 cm from three small animals believed to be newborns. However, Peruvian dusky dolphins are a slightly larger morphotype, measuring approximately 91 cm long and weighing approximately 9.6 kg at birth (Van Waerebeek and Read 1994). Therefore, Cipriano's data may have been biased by calves that were slightly older than newborns. Newborns are recognizable by fetal folds (Figure 9.2a) and floppy dorsal fins (Cockcroft and Ross 1990, Cipriano 1992, Mann and Smuts 1999). At 1–2 weeks of age, the dorsal fin becomes rigid (McBride and Kritzler 1951), fetal folds smooth out, and the infant is left with fetal lines for 1–3 months (Figure 9.2b; Cockcroft and Ross 1990, Mann and Smuts 1999). Young calves range from about one-fifth to one-third the size of an adult and have a yellow hue that fades as they grow. They exhibit "cork-like" breathing (Figure 9.3) which is most pronounced in the first few months (McBride and Kritzler 1951) as they tend to overshoot the surface for each breath instead of exhibiting the smooth rolling pattern of older youngsters and adults. Young calves are also characterized by their almost constant close proximity to adults. During the first few weeks of life, young calves rarely leave their mother's side. They begin taking small ventures on their own at 2–3 months, and these increase with age, lasting several minutes at 5–6 months.

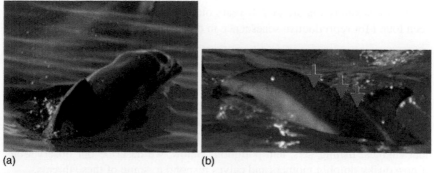

(a)   (b)

FIGURE 9.2 Fetal folds (a) and fetal lines (b) in dusky dolphin calves.

FIGURE 9.3   Dusky dolphin calf exhibiting "cork-like" surfacing behavior.

Yearlings

Yearlings are 1–2 years old, and are physically similar to adults, but roughly three-fourths to four-fifths the length of an adult (Cipriano 1992). The dorsal fin of a yearling is slightly more triangular than the falcate (concave trailing edge) fin of a mature adult. The dorsal fin is also more uniformly dark, while an adult fin shows a mottled graduation of darker to lighter skin from the leading to trailing edges. Yearling dorsal fins also tend to be unmarked by splotches and trailing edge marks that are more characteristic of older animals. Yearlings spend most of their time in close proximity to adults, but swim independently more often than young calves, and rarely exhibit cork-like breathing.

Juveniles

Juveniles are 2–8 years old, and become progressively more difficult to decipher from adults as they age. Juveniles are not yet sexually mature, although they are probably weaned and are not greatly dependent on their mothers (Cipriano 1992).

Adults

Adults are sexually mature at 7–8 years old (Cipriano 1992). No evidence has been found for reproductive senescence in dusky dolphins.

DANGERS IN CALF REARING

For dusky dolphin mothers off Kaikoura, it is no easy task to raise a calf around a deep submarine canyon. There are potential predators such as sharks and killer whales to avoid (see also Chapter 7), potentially aggressive male conspecifics searching for females, and humans which bring potential hazards such as fast boat engines, fishing nets, and pollution. Below, we describe our observations of how dusky dolphin mothers and calves respond to some of these threats.

Predation

While newborn cetaceans are precocial, calves swim more slowly than adults (Noren et al. 2006), cannot dive for as long as adults (Szabo and Duffus 2008), and are dependent on their mothers for food (Mann and Smuts 1999). Females with young also need to spend a considerable amount of time nursing. Thus, the "flight" or escape response of mother–calf pairs is assumed to be less efficient than that of adults, making them particularly susceptible to predation.

Killer whales have been documented killing and feeding on dusky dolphins off Kaikoura (Constantine et al. 1998), and kills have occasionally been witnessed by tour operators. To a killer whale, a dusky dolphin calf may be an easy target, and some dusky dolphin calves bear scars from probable (but not proved) interactions with killer whales (Figure 9.4). When killer whales are present, dusky dolphins are more commonly seen close to shore, sometimes even in tide pools or beaching themselves (Würsig et al. 1997, Constantine et al. 1998).

In the two main areas where dusky dolphin behaviors have been studied, Argentina and New Zealand, similar patterns of nursery group distribution emerge. Off Golfo Nuevo, Argentina, mothers and calves occur in shallow waters (<60m) close to shore (<3km), while mixed and larger groups occur in deeper water farther from shore (Garaffo et al. 2007). Off Kaikoura, New Zealand, 77% of nursery groups are found in waters <20m deep (Weir et al. 2008). While these depths may not appear to be shallow when compared to dolphin habitat in other regions (e.g., Shark Bay where maximum depth is 12.5m; Mann and Watson-Capps 2005), water reaches depths of 184m off Golfo Nuevo, Argentina (Garaffo et al. 2007) and >1400m off Kaikoura (Lewis and Barnes 1999). Increased use of shallow water habitat by females and calves may be a predator avoidance strategy, as killer whales are typically found in deeper water off Kaikoura (Table 9.1). Similarly, in Shark Bay, Australia mother–calf pairs that used shallower waters had a higher chance of calf survival past age 3, likely due to increased predator detection and avoidance in shallow water (Mann et al. 2000).

Over three field seasons, we (personally) did not witness an attack by a shark or killer whale on a nursery group off Kaikoura. When killer whales were sighted in the area, we were generally unable to find nursery groups. During one focal follow of a dusky dolphin nursery group when killer whales were reported in the area, the latter clearly influenced dolphin behavior. The nursery group had been foraging in a tight "pack" formation very close to shore. At 28 minutes into the focal follow, another small group of dolphins traveled rapidly towards the nursery group we were following. When the two nursery groups were <10m from one another, both groups merged and began burst swimming in parallel formation towards the shoreline. At this time, a tour boat operator radioed that a group of 20 killer whales had just been sighted in the area. After this, we were unable to approach within 50m of any dolphin group for the rest of the day, as the dolphins were skittish and avoided us.

FIGURE 9.4   Dusky dolphin calf with severe injuries, possibly incurred by a killer whale (*Orcinus orca*) bite.

TABLE 9.1   Number of dusky dolphin nursery groups and killer whales (*Orcinus orca*) encountered per hour of search effort during 2006 in waters of depths <20 m and >20 m

Variable	<20 m deep	>20 m deep	Total
Nursery groups	1.00 (*n* = 21)	0.17 (*n* = 7)	0.45 (*N* = 28)
Killer whale groups	0.05 (*n* = 1)	0.19 (*n* = 8)	0.15 (*N* = 9)

Weir 2007.

We also observed one particularly extreme incidence of presumed stress to a young calf, likely as a result of killer whale presence. We had returned to the boat ramp at the end of a field day to find a solitary dusky dolphin calf (later measured to be 119 cm in length) swimming within the shallow harbor. Killer whales had been recorded in the area earlier that day and locals reported first seeing a calf close to shore <1 hour after the killer whales left the area. By late afternoon, the dolphin had already been within the shallow harbor for 3–4 hours. We recorded the dolphin's breathing patterns and other behaviors until losing daylight at 21:04. During this time, the dolphin frequently came within meters of the shoreline, often coming ≤1 m of onlookers standing in the water. The calf frequently made loud squeaks, whistles, and clicks that were heard above water. On three occasions, the calf engaged in a head-first re-entry leap in slightly deeper waters. Occasionally, it came close to the exit of the harbor, although it did not leave the harbor during daylight hours.

The following morning, the dolphin was further west along the shoreline, in a very shallow (<1 m) tide pool. It was still alone and now had numerous scratches and marks on its body, and was bleeding at the rostrum tip. Since the dolphin's condition appeared to be deteriorating, the decision was made by the local authorities to capture the calf and transport it by boat to the large group of dusky dolphins which, on that day, was approximately 20 km south of the boat ramp. The dolphin was quick to rejoin the other dolphins and appeared strong as it swam away with an adult within seconds of being released.

This particular incident presented a rare opportunity to study the behavior of a dusky dolphin calf under presumed stress from killer whales. The dolphin did not leave the nearshore area to search out its group. It is possible that, during the fast swim towards shore, it was separated from its group and mother or became disorientated. It is also possible that the dolphin's mother was injured or killed by the killer whales.

Harassment from conspecifics

Mating often involves high-energy chases and leaps by both sexes, and males appear to aggressively herd single females (see also Chapter 8). This may create a dangerous situation for a calf, as calves may become separated from their mothers during chases and may themselves become the target of male harassment. The apparent chaos caused by a high speed chase of a female by a male or group of males may leave the calf disorientated, exhausted, and possibly more vulnerable to predation.

Formation of nursery groups may make the group less detectable (encounter effect), and reduce the likelihood of male harassment. This possibility may be supported by the fact that nursery groups are smaller in size in the earlier part of the calving season (Deutsch 2008), which coincides with enlarged testes in males (Cipriano 1992) and heightened mating behavior (Markowitz 2004). Mothers also keep their calves closer to them during this time, regardless of age (Deutsch 2008), which appears to be an additional strategy to protect calves from agonistic interactions with males. However, males have been observed chasing mothers in nursery groups; in such cases, the nursery group rapidly dissipates, leaving 1–2 mothers and their calves to become the sole targets of male harassment.

In one extreme example of harassment, eight large adults, probably males, arrived while we were following a nursery group of eight adults and eight calves. Coinciding with the initial arrival of the outside individuals, there were aggressive chases of nursery group members, and the probable males continually surged their bodies between mothers and calves in what appeared to be attempts to break up the group. The calves appeared to be struggling, with labored breathing and random disoriented movements. The water in the localized area was churned and wavy even though it was a calm day. Females attempted to evade the probable males by rolling on their sides and rising vertically out of the water. On at least four occasions, a female and calf pair moved away from the group, only to slowly return to the group, apparently intent on staying with their nursery group. This very high-energy behavior continued for >90 minutes.

In another example of apparent harassment, the males seemed to employ an alternative strategy. We had been following a nursery group with five adults and four calves. For the first 10 minutes of the follow, the group was resting until a group of five probable males arrived. The group then began fast swimming. Instead of chasing the entire nursery group, the probable males separated one female and her young calf from the group, and spent over 20 minutes chasing her from both sides and from below. The female was unable to flee for more than a few meters before the group of males would stop her. The female frequently rolled on her side and occasionally onto her back to avoid the males, but often seemed to be struggling to get her head out of the water to breathe. During this encounter, the rest of the nursery group had fled towards the shoreline and the female's calf stayed close (within 7–10 m) but on the outside of the tightly packed mating group consisting of its mother and five presumed males. While there is currently no strong evidence for male alliance formation in dusky dolphins, these interactions seem quite like those described for bottlenose dolphin male alliances which "kidnap" a female and force copulations on her (e.g., Connor et al. 1999).

Human interactions

An animal's response to humans may resemble an anti-predator response (Bejder et al. 1999, Frid and Dill 2002, Beale and Monaghan 2004a, 2004b). Off Kaikoura, young dusky dolphin calves and their mothers are likely to be more vulnerable to human disturbance than other sex or age groups. The preferred shallow habitat for calf rearing overlaps with areas of high human use, such as recreational boating (Figure 9.5) and commercial fishing (Table 9.2). Additionally, nets set in shallow nearshore habitats threaten dusky dolphins and other species such as Hector's dolphins (*Cephalorhynchus hectori*; Bräger et al. 2003) and New Zealand fur seals (*Arctocephalus forsteri*; Boren et al. 2006). Protecting known calf-rearing habitats from human harm by restricting vessel access or fishing activities may thus minimize risk to nursery groups.

Calf-rearing females may exhibit situation specific behavioral reactions to humans. Often, mothers allow their calves to interact with boats, while on other occasions mothers will herd their young away from boats. Calves tend to keep their distance from humans that are interacting via swim-with-dolphin operations, and appear to rarely approach humans. There are exceptions, however. On one occasion, while observing a large group that included several mothers and their calves, two calves swam past the swimmer and circled around for another look. One of the two calves seemed particularly interested in the swimmer, while the other kept its distance, but remained close enough to observe the interaction. The interested calf darted in and out of view and circled around the boat. Within several minutes, an adult dusky dolphin (presumably the mother) came into view and began to try to herd the calf away from the swimmer. The calf resisted and turned away to come back towards the swimmer several times before it finally disappeared with the adult; the "herding-away" interaction lasted about 3 minutes.

FIGURE 9.5 Human activities in the preferred nearshore habitat of dusky dolphin nursery groups may pose a risk to calves and adults.

TABLE 9.2 Number of dusky dolphin nursery groups and boat types encountered per hour of search effort during 2006

Variable	<20 m	>20 m	Total
Nursery groups	1.0 ($n = 21$)	0.2 ($n = 7$)	0.5 ($N = 28$)
Private recreational	2.3 ($n = 49$)	1.2 ($n = 48$)	1.6 ($n = 97$)
Commercial tour	2.5 ($n = 52$)	2.1 ($n = 86$)	2.2 ($n = 138$)
Commercial fishing boats	0.1 ($n = 3$)	0.5 ($n = 20$)	0.4 ($n = 23$)
All boats	5.0 ($n = 104$)	3.7 ($n = 154$)	4.2 ($N = 258$)

Weir 2007.

Such "herding" behavior has been noted in bottlenose dolphins (McBride and Kritzler 1951, Tavolga and Essapian 1957, Mann and Smuts 1999) and can be interpreted in at least two ways. The probable mother may be actively protecting her young from a perceived threat until the calf becomes less curious and more wary of its surroundings, or she may be teaching it to stay away from humans. Teaching is an extremely difficult behavior to quantify in non-human animals, but there is strong evidence for teaching in killer whales (Guinet and Bouvier 1995) and chimpanzees (*Pan troglodytes*; Boesch and Boesch-Acherman 2000), two species which exhibit prolonged maternal care. Thus, it is possible that one of the functions of the extended mother–offspring bond is to teach young how to respond to certain threats in the environment.

Hoyt (2005) recommends that areas used by cetaceans for calving, nursing, and raising calves be classified as critical habitat. Critical habitat may be defined as "those parts of a cetacean's range that are essential for day-to-day survival, as well as for maintaining a healthy population growth rate" (Hoyt 2005); such areas usually include areas used for feeding, breeding, and raising calves. Once critical habitat is identified, managers can find ways to mitigate the effects of humans on that habitat and establish species protection measures. We suggest that human disturbance to dusky dolphin populations could be mitigated by focusing on protecting critical habitat, and reducing human interactions in places where dusky dolphin nursery groups occur.

OPPORTUNITIES IN CALF REARING

While nursery groups may be vulnerable to predation, harassment by males, and negative interactions with human activities, there are benefits for mothers and calves which form nursery groups. These include increased time for resting and foraging, potential for allomaternal care, and opportunities for calf socialization and social learning.

Resting

The formation of nursery groups may grant mothers and calves increased time for resting. Time to rest is important for growing calves, and for females which face increased energetic constraints due to lactation. For example, bottlenose dolphin mothers and dependent calves spend the greatest percentage of their time resting (Mann and Smuts 1999). Off Kaikoura, we conducted 56 focal follows of nursery groups with durations ≥30 minutes. Forty-four (79%) of these follows had a predominant group behavior, meaning the group was engaged in the same behavioral state for at least two-thirds of the focal follow. For groups exhibiting a predominant group behavior, 68% were resting, 18% were traveling, 9% were socializing, and 5% were foraging. While the behavior of nursery groups differed by month, resting was the predominant behavior during most months (Figure 9.6).

Foraging

In many mammals, pregnant and lactating females spend an increased amount of time foraging (Ruckstuhl et al. 2003, Breed et al. 2006). Thus, it may be advantageous for females to form nursery groups composed of individuals which share similar energetic constraints and patterns of behavior. Pregnant and lactating females may also feed on different prey species than other individuals. For example, in one study, lactating spotted dolphins fed primarily on flying fish (family Exocetidae) while pregnant females and other individuals fed primarily on squid (family Ommastrephida; Bernard and Hohn 1989).

We observed members of nursery groups chasing fish, suggesting that nursery groups may provide foraging opportunities for mothers and calves. Off Kaikoura,

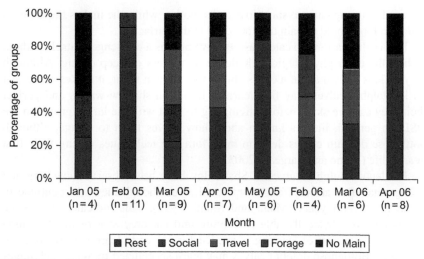

FIGURE 9.6 Predominant behavioral state of dusky dolphin nursery groups during focal follows in 2005 ($n = 37$) and 2006 ($n = 18$) by month. A category for January 2006 is not displayed here due to small sample size ($n = 1$, no main behavior) (from Weir 2007).

water clarity does not allow us to see >2 m deep, even on the clearest and calmest days. However, we observed behaviors that were indicative of foraging at depth, including tail-out dives and loud breathing upon resurfacing. Of 56 focal follows conducted in 2005 and 2006, two groups (4%) predominantly foraged, while five other groups (9%) foraged during ≥1 scan of the focal follow. During one focal follow in February 2006, an adult (presumed to be a mother based on her close proximity to a calf at all times) surfaced ≤2 m of our research boat with a fish about 30 cm long held cross-wise in her mouth. On another occasion, a group composed of five adults, one infant, and three yearlings was observed chasing fish. At one point, an adult captured a fish and subsequently spit it back out. The fish was then immediately pursued by a yearling, presumably the offspring of the adult. While this was an isolated incident, it is possible but certainly unproved that this may have been an example of a mother teaching her calf how to forage.

While our observations of dusky dolphin foraging in nursery groups are limited, nursery group formation in shallow nearshore waters may enable individuals to exploit prey species other than those found within the deep scattering layer (DSL). Most dusky dolphins off Kaikoura feed on myctophids and squid within the DSL, which rises towards the surface at night (Chapter 5). However, deep waters at night are likely a dangerous place for calves to learn to forage. Additionally, it may be dangerous for females with dependent calves to leave the nursery group, travel offshore to meet the DSL, and make long foraging dives. Furthermore, a calf may not have the swimming and diving skills, or the required lung capacity, to stay close to its mother while she makes these dives, and therefore would remain unsupervised at the surface.

For a calf, learning to catch fast, evasive prey is a challenge in itself without adding the extra physiological challenge and stress of deep diving. Although very little is known about a calf's ability to forage at night, it makes sense that dusky dolphin calves may first learn to forage in shallow waters and perfect their prey capture skills before attempting to feed with the larger group on the DSL. In general, little is known about how calves learn to forage, although bottlenose dolphin calves develop most foraging techniques during their first year of life (Mann and Sargeant 2003).

Bottlenose dolphin calves in Shark Bay have been observed catching and eating small fish at only four months old, even though they still continue to nurse for several years (Mann and Smuts 1999). This indicates that while calves may be physically able to capture and eat prey at a relatively young age, a longer period of learning is needed for calves to learn certain foraging strategies. Bottlenose dolphin calves may learn specialized foraging techniques (e.g., sponging) through vertical social learning (Krützen et al. 2005, Sargeant et al. 2005), while less specialized techniques may be acquired through horizontal, oblique, or individual learning (Sargeant et al. 2005). A calf-rearing strategy that allows opportunities for calves to learn how to capture and handle difficult prey before they are weaned is likely beneficial to both mother and offspring (Whitehead and Mann 2000). Specialized foraging techniques and individually distinct female foraging repertoires have not been reported for dusky dolphins. However, continued behavioral observations of females in nursery groups may reveal the presence of foraging specializations.

Socializing

Social behavior in nursery groups means that individuals in the group have their attention focused on one another. When socializing, dolphins make frequent direction changes and exhibit contact behaviors such as pectoral fin rubs. Quite often, it is the calves that are doing most of the obvious or "high-energy" socializing, chasing each other within and around nursery groups and occasionally attempting less than perfected leaps.

Weir (2007) counted the number of displays and head-first re-entries per calf and non-calf (juveniles and adults) during each 30-minute focal follow (Figure 9.7a,b). Overall, the mean number of high-energy behaviors by calves was significantly greater than for non-calves (Table 9.3). Calves also performed significantly more displays than non-calves, and there was a trend for calves to perform more head-first re-entries than non-calves.

Young dusky dolphins apparently are not born with the ability to perform leaps and must learn to master each one. Based on our observations, it appears that calves are able to independently learn the physical formation of leaps as their motor skills develop. Calves appear to first learn how to perform noisy leaps, followed

(a) (b)

FIGURE 9.7 Dusky dolphin calf attempting a head-first re-entry (a) and adult dusky dolphin with a perfected head-first re-entry (b).

TABLE 9.3 Mean number of displays, head-first re-entries and total high energy behaviors per calf and per non-calf during 30-minute focal follows (*n* = 46)

	Per calf	Per non-calf (SE)	*P*-value
Mean no. displays	0.11 (0.03)	0.02 (0.00)	0.01
Mean no. head-first re-entries	0.09 (0.02)	0.04 (0.01)	0.05
Total high-energy behaviors	0.20 (0.04)	0.05 (0.01)	0.02

Focal follows for which high-energy behaviors were not recorded due to the large size of the group are excluded from this analysis (n = 10).

Weir 2007.

by head-first re-entries, coordinated leaps, and finally acrobatic leaps (Table 9.4). However, unlike adult dusky dolphins which perform different leap types in different contexts (Würsig and Würsig 1980, Würsig 2002, Markowitz 2004, Pearson 2008), leap type does not appear to be context-dependent for calves (Deutsch 2008). Thus, while calves may independently learn *how* to perform leaps, calves may learn *when* to perform each leap type through interactions with conspecifics.

Offspring socialization is another benefit of nursery group formation. By interacting with other individuals of roughly the same age, calves may obtain important information about future social and mating partners. Long-term social bonds may be formed among calves in nursery groups, as exhibited by bottlenose dolphins in Shark Bay where calves that socialize together tend to stay in close association with each other in their adult lives (Mann et al. 2000). For marine and

TABLE 9.4 Months and estimated ages for first performance of specific leap types in dusky dolphin calves

Leap type	Month	Estimated age
Noisy leap	Early November	1 month
Clean leap	Mid December	2.5 months
Coordinated leap with adult	Late December or early January	3 months
Acrobatic leap	Early January	3.5 months
Coordinated leap with peer	Mid January	3.5 months

Dusky dolphins were considered to be 0 months old in October.
Deutsch 2008.

terrestrial species in which males form alliances (e.g., bottlenose dolphins, Wells et al. 1987, Connor et al. 1999; chimpanzees, Goodall 1986), nursery groups may also provide opportunities for young males to learn about future alliance partners.

Nursery group formation may also facilitate allomaternal care or "baby-sitting." While it is possible that dusky dolphins exhibit allomaternal care, there is currently no direct evidence for this behavior. However, allomaternal care has been well-documented in other odontocetes such as sperm whales (*Physeter macrocephalus*; Whitehead 1996) and bottlenose dolphins (Shane 1990, Mann and Smuts 1998). Allomothering may benefit the mother, allomother, and the calf. Mothers are released from caregiving activities and granted increased time to forage. Allomothers may benefit by "learning to parent" and forging social bonds with the calf's mother. Calves may benefit by receiving protective care, learning about different types of prey items, and expanding their social network.

SUMMARY

Nursery group formation play an important role in the rearing of dusky dolphin calves. These groups likely provide protection to calves from dangers such as killer whales and male conspecifics, while simultaneously increasing opportunities for resting, foraging, socialization, and social learning. Dusky dolphin calves spend an extended period of time with their mothers; however, there are no clear empirical data on the length of the mother–calf bond. Continued long-term research into mother–calf sociality and nursery group behavior will further illuminate specific calf-rearing behaviors and tactics employed by dusky dolphins, and how they compare with other marine and terrestrial mammals.

ACKNOWLEDGMENTS

We thank the New Zealand Department of Conservation, Earthwatch Institute, National Geographic Society, Texas A&M University, the Mooney Foundation, and Encounter Kaikoura for providing funding for this research. We thank Jane Packard, Bernd Würsig, Patty Edwards, Robin Vaughn, and Adrian Dahood for statistical advice and Nicholas Duprey for assistance and support in most aspects of this research. We thank the many volunteers and dedicated research assistants in the field and the lab: Jennifer Bennett, Trashon Boudreaux, Keltie Dienes, Thomas Egli, Marie Fournier, Sarah-Lyn Kirkman, Carrie Skorcz, Mridula Srinivisan, Lauren Wilson, and the Earthwatch volunteers. We thank Bernd and Mel Würsig for their continued support throughout the project, and Encounter Kaikoura for field support. NIWA generously provided bathymetry data for the study area. This research was conducted under permits from Texas A&M University and the New Zealand Department of Conservation.

Chapter 10

Sexual Segregation and Genetic Relatedness in New Zealand

Deborah E. Shelton,[a] April D. Harlin-Cognato,[b] Rodney L. Honeycutt,[c] and Tim M. Markowitz[d]

[a]*Ecology and Evolutionary Biology, University of Arizona, Tucson, Arizona, USA*
[b]*Department of Zoology, Michigan State University, East Lansing, Michigan, USA*
[c]*Natural Science Division, Pepperdine University, Malibu, California, USA*
[d]*School of Biological Sciences, University of Canterbury, New Zealand*

It is morning, and we are in a small boat among dusky dolphins, not far south of Kaikoura, New Zealand. Our best glimpses of them are when they surface to breathe, but we can see them underwater too, weaving among each other. The water is just clear enough for us to follow an individual dolphin among the group for a few moments before he (or she) descends out of sight. The dolphin reappears and swims rhythmically and slowly, sleepily, as if it was getting ready to wake up but is not quite there. An anthropomorphic interpretation, but perhaps a little anthropomorphism is warranted. We are both mammals, after all, and it is not so farfetched to think that we could share some common ground—a basic set of mammalian feelings and behaviors. On the other hand, the air–water interface is a dramatic and abrupt divide, and we humans are also divided from dolphins by the 80+ million years that have passed since our ancestors began following distinct evolutionary paths.

The dolphin we are watching likely has a stomach full of squid and lanternfishes from foraging farther offshore the previous night. Within sight of this focal individual is a small cluster of more active animals. Perhaps they are young males. Perhaps they are eager to socialize, to learn how they fit in among others their own age. Or perhaps the boisterous group of associates is familial—half-siblings confirming ties that will be important for the future of their shared genes. From the boat, all we can do is speculate. Later, in the lab and behind computer screens, we hope to discover whether our ideas about the ecology and evolution of dolphin behavior are supported by more detailed analyses.

The focal dolphin surely sees the nearby cluster of active animals. He does not join them, however. He is near the periphery of the pod, so perhaps his outward-facing eye registers the stark empty image of open water. Perhaps he is ready for a rest but is reluctant to stray too far. Could the safety of the group draw him in among others of his kind who are not interested in resting just now? That is, do the benefits of group-living keep him from separating and following his own agenda? What other costs must he endure to be a part of a society and how do the costs and benefits of group membership play out over evolutionary time?

We see another distinct subgroup up ahead and slightly more separated from the hundreds of dolphins in the "main pod." Even from a distance, we see that this group is not very active. The dolphins in it are also swimming slowly and maintaining a steady heading. Will the individual we have been watching join with this similarly behaving subgroup?

The dolphin disappears below us, into a physical and social world full of information to be perceived and decisions to be made. We on the boat are left with our clipboards and dictation recorders and video and still cameras, hoping that these and other tools, put to use methodically, will yield some answers about how and why individuals choose to group.

Deborah Shelton, reminiscing while in Tucson, Arizona, September 2008

GROUP COMPOSITION

Many factors could influence animal group composition, as it is a property that arises from different individuals deciding to either join or to be alone, to stay together or to separate from the group. Animal group composition is often structured according to characteristics such as age, body size, sex, and relatedness (Krause and Ruxton 2002). Of these characteristics, sex and relatedness have important implications for understanding the evolution of social patterns (Hamilton 1964a, 1964b, 1970, Ruckstuhl and Neuhaus 2000). When selection is distinct for males and females, the sexes tend to separate along behavioral and ecological lines, and segregation occurs at one or more spatiotemporal levels in a variety of animals (Parmelee and Guyer 1995, Main et al. 1996, Gonzalez-Solis et al. 2000, Bon et al. 2001, Sims et al. 2001, Encarnacao et al. 2005, Michaud 2005, Wolf et al. 2005, Croft et al. 2006). Kin-based social structure has been well documented for several species of terrestrial mammals that show group formation involving closely related females and their offspring (Armitage and Schwartz 2000, Gurnell et al. 2001, Kaminski et al. 2005).

In societies with individualized social bonds, an individual's decision to join or leave a particular group can be based on the relationships among individuals. Non-random group composition can result from social preferences, which are often manifestations of differential allocation of benefits among individuals (Connor 2000). For example, altruists may prefer to partner with close relatives (Hamilton 1964a, 1964b), and reciprocal altruists are expected to value reliable, long-term social partners differentially based on qualities such as tendency to reciprocate or gullibility (Trivers 1971). In "by-product" mutualisms, individual variation in the ability to either dispense or utilize by-product benefits could also lead to individualized partner preferences (Wrangham 1982). Thus, in addition to the factors that can create structured groups in the absence of individualized relationships, group composition can also be influenced by the types of interaction that occur.

In spite of the complexity behind the factors that could affect animal grouping decisions, researchers have learned much about the composition of groups in different animal societies. Comparing patterns of group composition among different organisms is a powerful method for testing general ideas about how these properties evolve, but this strategy requires detailed knowledge of diverse systems. Cetaceans have been evolving independently from terrestrial mammals in response to a drastically different habitat, playing out a roughly 50 million-year-long natural experiment in the ecology of social evolution (Bromham et al. 1999). Yet, compared with terrestrial mammals, little is known about the structure of cetacean societies. Detailed studies of social life in cetaceans indicate societies that are structured in terms of both sex and relatedness, much like those seen in terrestrial mammals. Some of these best-studied examples include matrilineal groups in sperm whales, *Physeter macrocephalus* (Whitehead and Weilgart 2000), lifetime family pod structure in killer whales, *Orcinus orca* (Baird 2000), "bands" of related female common bottlenose dolphins, *Tursiops truncatus* (Wells et al. 1987, Duffield and Wells 1991), and alliances among male Indian Ocean bottlenose dolphins, *Tursiops aduncus* (Connor et al. 1992a, Krützen et al. 2003). For dolphins, current knowledge comes largely from studies of species and populations that are alike in forming relatively small and stable inshore societies (Connor 2000), so detailed research on a wider range of systems is needed.

One area that needs further study is the sociality of dolphin species that live in large groups, use "semi-pelagic" as well as inshore habitats, and routinely travel long distances. The dusky dolphin (*Lagenorhynchus obscurus*) has these characteristics. Groups of dusky dolphins occur off two areas of the South Island of New Zealand: Kaikoura on the east coast and Admiralty Bay in the Marlborough Sounds on the north coast, among many other areas. In the Kaikoura area, dusky dolphins occur year-round, whereas in the Marlborough Sounds they are present primarily in winter months (Cipriano 1992, Markowitz 2004; see also Chapter 6). Photo-identification data show movement of some dolphins between Kaikoura and Admiralty Bay, and marked changes in patterns of grouping and behavior (Markowitz 2004). In Kaikoura, dusky dolphins feed nocturnally, exploiting the shallower night-time depths of fishes and squid associated with a deep sound-scattering layer (DSL; Cipriano et al. 1989, Würsig et al. 1989, Cipriano 1992, Benoit-Bird et al. 2004; see also Chapter 5). In sharp contrast, dusky dolphins in Admiralty Bay feed diurnally by cooperatively herding small schooling fishes, with groups spending much of the day searching for food and feeding (McFadden 2003, Benoit-Bird et al. 2004, Markowitz 2004, Markowitz et al. 2004; see also Chapter 6).

In this chapter, we describe recent work on the group composition in terms of sex and relatedness of New Zealand dusky dolphins. First, sex-specific genetic markers are used to determine whether or not the composition of groups changes with respect to sex in different parts of the dusky dolphin's range. Some of the same individuals have been sighted in both Kaikoura and Admiralty Bay, yet behavioral patterns at the two sites are different, suggesting that individuals of one sex may have more or less incentive to use a particular site or to travel between sites. Therefore, by comparing the overall sex ratio observed at each site and by determining whether individuals group based on sex within a site, we can shed light on ecological and behavioral factors that may influence migration and grouping decisions. Our results suggest that each sex is making distinct decisions about migrating and grouping. We consider these results in the context of selective pressures that are distinct for males and females. Second, both mitochondrial and nuclear genetic markers are used to examine intra-group relatedness in multiple social and ecological contexts, such as small nursery, mating, adult, and feeding groups off Kaikoura and Admiralty Bay. Technical and logistic challenges preclude strong conclusions about relatedness within groups. However, the results so far support two themes that have emerged from studies of other animals: matrilineal relationships are particularly important in mammalian societies, and cooperative behavior need not arise only among highly related individuals.

METHODS AND RESULTS

Field methods

Skin swabs (Harlin et al. 1999) were collected from bow-riding dusky dolphins off Kaikoura (February 1998 and May 2000) and in Admiralty Bay (June 1999 and July 2001). Tissue samples from 200 individuals (118 in Kaikoura; 82 in Admiralty Bay) were collected, fixed either in dimethylsulfoxide (DMSO) saturated with salt or in 95% ethanol, and stored in a standard freezer. These samples were obtained from 54 distinct small groups of four types: mating, nursery, adult, and feeding (Table 10.1). Calves were not sampled. We identified distinctive marks of sampled individuals and moved the vessel position within the group to avoid sampling the same individual more than once within a sampling bout.

Dolphin groups were characterized by size, age classes present, and behavior following the methods of Markowitz (2004). Groups were defined by spatial proximity according to the "10-m chain rule" (Smolker et al. 1992), meaning that dolphins were considered part of the same group if they were within 10 m of another dolphin. Group size was estimated by noting the maximum number of animals simultaneously observed at or near the surface and also by taking into account distinctly marked individuals that were not seen during the largest simultaneous surfacing (Markowitz 2004). Groups were broadly categorized as "small" (<50 individuals) and "large" (≥50 individuals). Three distinct types of small groups were identified in Kaikoura: (1) mating groups, which were defined based on confirmed sexual activity, (2) nursery groups, which consisted of adults with calves (neonatal up to about one year old), and (3) adult groups, which consisted of adult and subadult group members without sexual behaviors. Calves were identified as smaller individuals that consistently swam in close proximity to an adult. Fetal folds (white, vertical markings on the sides of small dolphins)

TABLE 10.1 Samples of dusky dolphin (*Lagenorhynchus obscurus*) tissue collected for molecular analyses, by behavioral context

Study site	Group type	Number sampled	Number sexed	Number genotyped, nuclear markers	Number haplotyped, mitochondrial markers
Admiralty Bay	Feeding	118	82	45 (24 groups)	63 (26 groups)
Kaikoura	Mating	37	8	9 (6 groups)	13 (3 groups)
Kaikoura	Nursery	12	1	2 (1 group)	9 (5 groups)
Kaikoura	Adult	33	9	5 (4 groups)	7 (4 groups)
Admiralty Bay or Kaikoura	Other	131	7	3	105

The total number of sampled individuals exceeds the sample sizes for particular analyses due to issues associated with template quantity and quality. The "other" category encompasses samples taken from large groups as well as samples taken from groups that were not characterized as one of the listed group types.

distinguished neonatal calves (see also Chapter 9). In Admiralty Bay, samples were collected predominantly from a fourth group type, feeding groups, defined as such if individuals were observed "pursuing fish or holding fish in their mouths" (Acevedo-Gutiérrez and Parker 2000, Markowitz 2004).

Segregation by sex

Sex was determined for 107 dusky dolphins by the simultaneous polymerase chain reaction (PCR) amplification of the zinc-finger gene on the X-chromosome (ZFX) and the sex-determining region on the Y-chromosome (SRY) with primers and PCR protocols for either mammals in general (Banks et al. 1995), or specifically for odontocetes (Rosel 2003).

A total of 20 females and 87 males were identified at both locations. The overall sex ratio at Kaikoura and Admiralty Bay was addressed by calculating 95% confidence intervals for the proportion of females identified at each site (i.e., if sampling were repeated, the population parameter—that is, the proportion of dolphins that are female—would fall within the confidence interval 95% of the time), regardless of small group membership (Ott and Longnecker 2001). For Kaikoura, but not for Admiralty Bay, the overall habitat–scale sex ratio appeared to be relatively even (Figure 10.1). The Kaikoura estimate lacked precision due to small sample size, while the Admiralty Bay estimate clearly indicated that males primarily use this area.

Within Admiralty Bay, the distribution of the sexes among groups within a group type was investigated for feeding groups (Figure 10.2), and patterns of sexual composition within a group type were examined. Due to sample size limitations, only Admiralty Bay feeding groups were included in this analysis. For each group, the number of males and females determined by molecular sex identification was plotted in conjunction with an estimate of the number of unsexed individuals, inferred by subtraction from field estimates of group size (Figure 10.2). A randomization test (Manly 2001), specifically the "category membership" procedure in the program PERM version 1.0 (Duchesne et al. 2006), was employed to address the question of whether groups are structured by either attraction among same-sex individuals or (equivalently) avoidance among different-sex individuals. PERM was used to calculate homogeneity statistics based on the number of specimens belonging to the most frequently identified category within

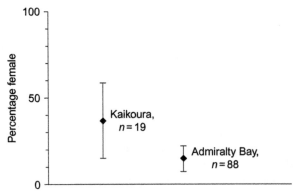

FIGURE 10.1   Estimated percentage of females identified, with individuals pooled by sampling location. The 95% confidence interval for the percentage of females in Admiralty Bay ($n = 88$) is 7.4–22.2; for the percentage of females in Kaikoura ($n = 19$), it is 15.2–58.5.

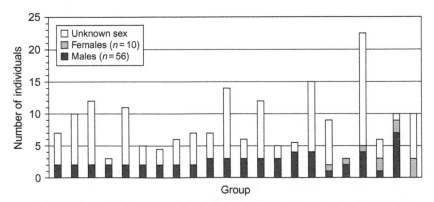

FIGURE 10.2   Sexual composition of Admiralty Bay feeding groups (22 groups, 66 sexed individuals). The null hypothesis that individuals group randomly with respect to sex was tested with a randomization test (Duchesne et al. 2006) and was rejected ($P = 0.024$).

each group. P-values were estimated by comparing the observed homogeneity statistics with a null distribution of values generated from 10 iterations of 1000 random permutations of the data (i.e., the observed grouping of the sexes), in which the observed number and sizes of groups were kept constant (Duchesne et al. 2006). This procedure tests the degree to which groups are structured by sex relative to the overall sex ratio for the feeding group type. The randomization test indicated that feeding groups were more likely to consist predominantly of one sex, more so than would be expected if individuals were grouping without regard for the sex of other group members ($P = 0.024$).

Relatedness in social groups

Microsatellites

To estimate genetic relatedness, individuals were genotyped for 11 micro-satellite loci (Tables 10.2 and 10.3). Microsatellites, segments of DNA consisting of

TABLE 10.2   Summary of primer information

Reference	Target species	Locus	Repeat sequence	Annealing temperature (°C)
Shinohara et al. 1997	*Tursiops truncatus*	D14	$(AC)_{16}$	56
		D28	$(CA)_{24}$	57
Buchanan et al. 1996	*Delphinapterus leucas*	DlrFBC11	$(CA)_8(CC)(CA)_{15}$	56
This study	*Lagenorhynchus obscurus*	Lo105	$(AC)_{12}$	57
		Lo514	$(CA)_{10}(GT)_5(CA)_{23}$	56
		Lo6	$(CA)_5(GA)(CA)_4(TG)_5(CA)_{24}$	60
Cassens et al. 2005	*Lagenorhynchus obscurus*	Lobs_Di7.1	$(TG)_3/(TG)_4/(TG)_2/(TG)_{19}$[a]	60
		Lobs_Di9	$(TG)_{16}$	56
		Lobs_Di19	$(CA)_{11}$	57
		Lobs_Di21	$(TG)_{15}$	57
		Lobs_Di24	$(GT)_9(GA)_{10}$	60

[a]*The sequence cloned by Cassens et al. (2005) also included unique sequences that intervened among the repeated elements, the locations of which are indicated by forward slashes. See Cassens et al. (2005) for details.*

TABLE 10.3　Summary of microsatellite alleles observed in this study

Locus	Number of samples	Fragment size (base pairs)	Number of alleles	Expected heterozygosity	Observed heterozygosity
D14	72	119–137	10	0.77	0.58
D28	66	131–165	16	0.88	0.79
DlrFBC11	87	99–133	9	0.81	0.76
Lo105	85	232–250	8	0.80	0.58
Lo514	96	231–239	5	0.64	0.49
Lo6	86	145–211	22	0.92	0.74
Lobs_Di7.1	83	123–153	14	0.87	0.71
Lobs_Di9	87	86–114	12	0.87	0.76
Lobs_Di19	81	94–124	15	0.86	0.80
Lobs_Di21	86	99–119	11	0.85	0.72
Lobs_Di24	91	106–132	12	0.83	0.71

tandem repeats of short (2–6 base pairs long) nucleotide motifs, are well-suited as molecular markers for relatedness estimation (Queller et al. 1993). Microsatellite alleles differ in the number of times the motif is repeated, and in many wild out-bred populations, microsatellites are highly polymorphic. For this study, microsatellite primers were developed in the laboratory of Dr. Mohammed Noor, Duke University, using standard protocols (Maniatis et al. 1982). Microsatellite loci were amplified in multiplexed PCR reactions performed by the core genetics facility Nevada Genomics Center (University of Nevada, Reno, Nevada, USA) with fluorescent dye-labeled primers and primer-specific annealing temperatures (Table 10.2). Fluorescently labeled PCR products were visualized on a 3730 automated sequencer (Applied Biosystems). Allele sizes were determined by analysis of electropherograms with the program GENEMAPPER (Applied Biosystems, Foster City, California, USA), and the resulting allele calls were exported into a Microsoft Excel spreadsheet for subsequent analyses.

Due to the quality and quantity of DNA extracted from skin swabs (Taberlet et al. 1999, Pompanon et al. 2005), amplification success was low. Ninety-seven of 188 individuals were genotyped for at least seven of the 11 microsatellite loci; only these samples were considered in subsequent analyses. Size range, number of alleles, and expected heterozygosity were calculated (Table 10.3). Expected heterozygosity (H_E) was calculated with the program ML-RELATE (Kalinowski et al. 2006), which employs an unbiased H_E estimate (Nei 1978). A web-based (http://wbiomed.curtin.edu.au/genepop) version of GENEPOP (Raymond and Rousset 1995) was used to test the null hypothesis of random union of gametes (Haldane 1954, Guo and Thompson 1992) and to estimate the value of F_{ST} (Weir and Cockerham 1984). A general excess of homozygote genotypes (in all allele size classes) indicates that null alleles may be affecting the dataset (Van Oosterhout et al. 2004). A deficiency of heterozygote genotypes with alleles differing by one repeat (in this case, two base pairs) is likely caused by mis-scoring due to artifacts (e.g., stuttering patterns) on the electropherograms. Both of these possibilities were examined for each locus using the program MICRO-CHECKER (Van Oosterhout et al. 2004). Loci were tested for linkage using the Fisher exact test implemented by GENEPOP (Raymond and Rousset 1995). P-values were considered significant at the $\alpha = 0.05$ level and, when necessary, were analyzed using the sequential Bonferroni technique to allow for multiple comparisons (Rice 1989).

Eight independently sampled dolphins were suspected to be recaptures of an individual based on a high level of genetic similarity to another individual. Of the 93 unique individuals sampled, 66 were sampled in Admiralty Bay and 27 in Kaikoura. The F_{ST} value for the two areas was 0.0055, indicating that the samples collected would not support partitioning of the two geographic areas. Samples were pooled while testing for Hardy–Weinberg proportions. After correcting for multiple comparisons, half of the loci showed deviations from Hardy–Weinberg proportions. This result is not altogether surprising, considering that the male bias in samples from Admiralty Bay indicates that these individuals are not representative of the overall breeding population. However, because all of the Hardy–Weinberg deviations were due to an excess of homozygotes, null alleles are a potential concern. Additionally, three loci showed evidence of mis-scoring due to electropherogram artifacts (stuttering). Three of the pairwise comparisons of loci showed evidence (at the $\alpha = 0.05$ level, corrected for multiple comparisons) of linkage (Van Oosterhout et al. 2004). Evidence for linkage could be due to a sampling effect.

Checking for Hardy–Weinberg proportions and linkage disequilibrium was repeated with a restricted dataset in an attempt to use a dataset that meets the assumptions necessary for relatedness analyses. In the restricted dataset, four loci (D14, L0105, L0514, and L06) were eliminated based on high error rate, low proportion of individuals scored, highly significant deviation from Hardy–Weinberg proportions, or a combination of these factors. Further, individuals not scored for at least six of the remaining seven loci were eliminated, leaving 47 individuals from Admiralty Bay and 17 from Kaikoura. The F_{ST} for the two locations in this dataset was 0.0078, so the samples were assumed again to come from a genetically mixed population and were pooled for further analyses. Two loci (Lobs_Di7.1 and Lobs_Di9) still showed deviation from Hardy–Weinberg proportions in the restricted dataset. None of the loci showed evidence of linkage. The restricted dataset (7 loci and 64 individuals) was used in subsequent relatedness analyses.

Maximum-likelihood values for the relatedness coefficient, r, were calculated for every pair in the dataset using the program ML-RELATE (Kalinowski et al. 2006). Allele frequencies for Lobs_Di7.1 and Lobs_Di9 were adjusted to accommodate for null alleles (Kalinowski et al. 2006). This resulted in pairwise r-values for 2016 pairs, 51 of which were collected from the same group. The r-values ranged from 0.00 to 0.80, with a median value of 0.00 and a mean of 0.06. Relatedness within groups was estimated for 46 Admiralty Bay feeding group pairs, three Kaikoura mating group pairs ($r = 0$, 0.0764, 0.0845), one Kaikoura nursery group pair ($r = 0.6163$), and one Kaikoura adult group pair ($r = 0$). Of the 2016 pairs (64 individuals) examined from both locations, the third-highest r-value belonged to the only nursery group pair examined. Because few genotyped individuals were sampled in Kaikoura, grouping patterns and relatedness were not analyzed further for this site.

Using the "Groups: pairwise relationship test" in the program PERM (Duchesne et al. 2006), the question of whether groups are structured by relatedness was addressed for Admiralty Bay feeding groups. This program sums pairwise r-values within each observed group. The r-values were then re-arranged 1000 times into random groups of the same sizes as those observed, and this process was repeated 10 times. Significance of the observed sum of within-group r-values was determined by comparison to the distribution of random values. The observed intra-group sum of r-values was 2.840. Comparison to randomly generated groups produced a high P-value ($P = 0.404$), indicating that the intragroup sum of r-values in "random" groups was comparable to the observed value. That is, individuals in Admiralty Bay appear to be joining and leaving feeding groups without regard to their relatedness to other group members, though it is possible that this result may be an artifact of the problems associated with our microsatellite genotype dataset or of the relatively low sample size.

Mitochondrial DNA

To investigate patterns in maternal relatedness, a 472-base pair fragment from the control region of the mitochondrial genome was amplified by PCR following the protocol and primers of Harlin et al. (1999). PCR products were purified with Qiaquick spin columns (Qiagen, Valencia, California, USA) and were sequenced with Big Dye version 1 dye-terminator chemistry (Applied Biosystems) on a model 377 or 3100 automated sequencer (Applied Biosystems). Sequences were edited and aligned using the program SEQUENCHER (Gene Codes Corporation) and haplotypes were identified using the program DnaSP, version 4.10.8 (www.ub.es/dnasp/; Rozas and Rozas 1999).

For the mitochondrial DNA haplotypes, the probability that any two samples drawn at random would have the same haplotype was calculated using the formula $\sum_i (p_i)^2$, where p is the frequency of the ith haplotype in the dataset (Ott and Longnecker 2001). This probability was multiplied by the number of within-group pairs to obtain the expected number of within-group haplotype

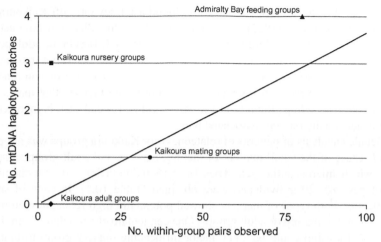

FIGURE 10.3 mtDNA haplotype matches by group type. The expected number of matches per within-group pair was calculated based on the frequency of haplotypes in the dataset and is indicated by the line.

matches for each type of group. The expected number of matches was compared with the number of matches observed in the dataset.

The control region was sequenced for 197 individual dusky dolphins in Kaikoura and Admiralty Bay. Seventy-three distinct haplotypes were identified. The probability that two individuals, drawn at random from those sampled, would have a matching haplotype is 0.036. The number of within-group haplotype matches that would be expected if the dolphins were grouping randomly with respect to maternal relatedness was compared to the number of matches observed in the dataset (Figure 10.3). Kaikoura nursery groups ($n = 9$ individuals in five groups) showed more matches than would be expected by chance, while haplotype matches within Kaikoura mating ($n = 13$ individuals in three groups), Kaikoura adult ($n = 7$ individuals in four groups), and Admiralty Bay feeding groups ($n = 63$ individuals in 26 groups) were close to expected values of randomness.

GROUPING PATTERNS OFF NEW ZEALAND

Kaikoura group composition

Overall, several sources indicate that the dusky dolphin population off Kaikoura has a relatively even sex ratio. Based on a small sample size ($n = 19$), we calculated a 95% confidence interval of 15–58% females in Kaikoura (Figure 10.1). This result is based on samples that were collected from small "satellite" groups not expected to be representative of the general population of dolphins present. For example, nursery groups are suspected to be female-biased and mating groups are male-biased (Markowitz 2004). In examining samples from the presumably more representative larger groups, Harlin (2004) found a 1:1 sex ratio off Kaikoura. In the same area, Cipriano (1992) sexed individuals that were killed in fishing nets or found dead on the beach. Results indicated a potential male bias in the adult population (4 adult females and 9 adult males), though this difference is not statistically testable and may well be an artifact of either small sample size or non-random sampling. A relatively even sex ratio in Kaikoura would indicate that males and females are responding similarly to the ecological and social factors involved in feeding nocturnally on DSL-associated prey.

Detailed analysis of patterns of relatedness for Kaikoura groups was not possible because of the small number of mating, nursery, and adult group samples from which microsatellite genotypes ($n = 16$ individuals) and mitochondrial haplotypes ($n = 29$ individuals) were obtained (Table 10.1). In spite of small sample size, both types of markers showed potentially high relatedness in nurseries, but not in mating or adult groups. One can tentatively conclude from these results that kinship could be a key factor influencing nursery group formation, while it is less likely to be important in mating or adult group formation.

The above tentative conclusions resonate with current knowledge of dusky and other dolphin species. For instance, adult female bottlenose dolphins are often known to associate with relatives in their natal "band" (Wells et al. 1987, Duffield and Wells 1991). Maternal care of young and matriarchal social organization is common in mammals, and in some systems alloparental care has also been documented (e.g., Whitehead 1996, Mann and Smuts 1998). For dusky dolphins, associations between mothers and sub-adult or adult offspring or between maternal half-siblings could create high relatedness in nursery groups. High relatedness in nursery groups could promote alloparental care in dusky dolphins as well, though this possibility remains to be investigated. Morphological and behavioral lines of evidence suggest that dusky dolphins are highly polygynandrous, or promiscuous (Cipriano 1992, Markowitz 2004; see also Chapter 8). Therefore, lack of discrimination according to any factor (except sex) in the formation of mating groups is reasonable to expect. Information relative to the formation of adult groups, in which individuals tend to rest (Markowitz 2004), is generally lacking. One plausible scenario is that segregation of individuals from the main pod into adult groups occurs not on the basis of social relationships with kin, but rather on the basis of the coincidentally overlapping resting needs of individuals throughout the day.

Admiralty Bay group composition

In contrast to the situation in Kaikoura, the data reported here clearly indicate habitat-level sexual segregation in Admiralty Bay, where the population of dusky dolphins was estimated to consist of only $15 \pm 7\%$ females (95% confidence interval, $n = 88$, Figure 10.1). The seasonality of dusky dolphin breeding (Cipriano 1992) indicates that males do not suffer reduced opportunities to sire offspring by choosing to winter with (mostly) other males in Admiralty Bay. Because dusky dolphins are only slightly sexually dimorphic for size (Cipriano 1992) and because females invest heavily in parental care, the "reproductive strategy/predation risk" and "social factor" hypothesis categories (Bon et al. 2001) encompass most of the potential adaptive explanations for dusky dolphin sexual segregation. The pattern of dusky dolphin habitat-level sexual segregation could be the result of sex differences in the costs and benefits influencing seasonal migration to Admiralty Bay, or could be more directly related to the distinct feeding (and associated social) behavior of dusky dolphins while in Admiralty Bay.

Within the male-dominated population of Admiralty Bay, sexual segregation was further investigated at the social level. A randomization test indicates that the few females observed in Admiralty Bay ($n = 13$) tended to co-occur within a group more often than one would expect if sex were not a factor by which the dolphins were grouping. Würsig and Würsig (1980) also reported social-level sexual segregation in conjunction with dusky dolphin feeding groups in Golfo San José, Argentina, where dusky dolphin mother–calf pairs tended to separate from fish-herding groups. They noted that two aspects of fish herding—boisterous or competitive interactions among conspecifics and the conspicuousness of such groups to dusky dolphin predators—could drive mother–calf pairs to separate. In Admiralty Bay, social-level sexual segregation is more subtle and does not primarily involve mother–calf pairs. Nevertheless, similar social factors, such as sex differences in motivation to socialize or in tolerance for boisterous interactions, could be important in Admiralty Bay.

A bias in the "catchability" of the sexes during sample collection from bow-riding dolphins is a potential concern that should be considered when interpreting these data. However, the uneven sex ratio observed in Admiralty Bay is supported by Harlin's (2004) observations of a 1:1 sex ratio of bow-riding dusky dolphins (off Kaikoura) and two primarily female dusky dolphin populations (off Otago and off the west coast of the South Island). Additionally, even if males in Admiralty Bay were three times as likely as females to be sampled, the conservative conclusion that most (i.e., >50%) of the dolphins using Admiralty Bay in the winter are males would be supported by the data reported here.

A permutation test indicated that Admiralty Bay dusky dolphin feeding groups typically consisted of unrelated individuals ($P = 0.404$) and that they had approximately the same number of within-group mitochondrial DNA haplotype matches as would be expected by chance (Figure 10.3). Preliminary analysis of relatedness and association patterns among a small number of repeatedly photographed individuals ($n = 13$) seems to confirm that relatedness does not strongly influence grouping decisions for Admiralty Bay dusky dolphins (Shelton 2006) but larger sample sizes are needed to confirm this tentative conclusion.

The social relationships within a large society, like that of the New Zealand dusky dolphin, are not strictly kin-based. Lack of within-group relatedness for dusky dolphins is consistent with current ideas concerning the potential for the benefits of living in larger groups to outweigh the benefits of grouping based on kin relationships (Avilés et al. 2004, Lukas et al. 2005). However, the picture is complicated for this system by the fact that the typical group size in Admiralty Bay is relatively small and only a subset of individuals from a much larger dusky dolphin society use this location. Additionally, lack of grouping by kinship does not rule out the possibility that within-group behavior is kin-biased. In spite of these complications, our results suggest that potential conflicting interests among Admiralty Bay dusky dolphins during coordinated fish herding are not mediated by high relatedness and that the long-term relationships documented by Markowitz and Würsig (2004) are not primarily kin-based.

Conclusions and future directions

Overall, our data support several conclusions about dusky dolphin grouping patterns. The observations that females are relatively rare in Admiralty Bay and that individuals appear to group preferentially with others of their own sex indicate that males and females are responding to distinct selective pressures. Yet, the relatively even sex ratio off Kaikoura indicates that dusky dolphin society is not always strongly separated by sex. In terms of within-group relatedness, our data hint that relatedness could be high within small nursery groups. Our data also suggest that foraging partners tend to be unrelated. High relatedness among interacting individuals is a mechanism that can promote pro-social behaviors (i.e., cooperation or altruism). Whether social organization with respect to relatedness is an important determinant of behavior in dusky dolphin behaviors is yet unknown.

Our results begin to paint a picture of how dusky dolphins group, but they also highlight some challenges of characterizing group composition. We used the "less-invasive" sampling method of swabbing skin from bow-riding dolphins (Harlin et al. 1999) rather than a biopsy dart (Krützen et al. 2002). The dolphins' reaction to this type of sampling was generally mild, and the method avoids creating a wound. The lowest possible level of disruption to the animals' lives is clearly desirable, especially for studies in which behavioral data are collected in addition to physical samples (however, it is arguable whether skin swabbing is truly less invasive than modern biopsy darting). Harlin et al. (1999) found that the skin swabbing method provides samples that are suitable for analyses based on amplification of mitochondrial DNA. However, we found that skin swabbing was more problematic when using nuclear markers (i.e., microsatellites). Epithelial cells in cetaceans often do not have a nucleus, so even when a substantial amount of skin is collected, the yield of nuclear DNA can be low and degraded. For microsatellite data, low quantity or quality of DNA can cause genotyping failures and errors. A conservative approach to address the reliability of genotypes is to repeat each DNA amplification independently for each locus several times (i.e., the "multi-tubes approach"). This approach can yield reliable results, but is expensive, time-consuming, and can be difficult to execute with low amounts of DNA (Taberlet et al. 1999). An additional approach that could help to produce better results in future studies would be to focus on designing and optimizing more dusky dolphin-specific microsatellite primers.

A major challenge in cetacean sociobiology is to move beyond "the gambit of the group" (i.e., the assumption that grouped individuals are interacting; Whitehead and Dufault 1999) to uncover the content and quality of social interactions (Hinde 1976). This level of detail is difficult to achieve, yet is essential if social behavior is to be understood in an evolutionary context. In this chapter, we contributed detail to the current picture of dusky dolphin sociality by investigating the sex and relatedness of interacting individuals across different habitats and different behavioral contexts. The same features that make dusky dolphins such a challenging system for addressing social behavior questions (e.g., marine habitat, large group sizes, "semi-pelagic") also make knowledge of dusky dolphin social life important from a comparative perspective. Future work on the composition and behavior of dusky dolphin groups will continue to make the system more relevant to evolutionary questions about social behavior.

ACKNOWLEDGMENTS

Data were collected and analyzed with monetary support from the Marlborough District Council, the New Zealand Department of Conservation, the National Fish and Wildlife Service, the National Geographic Society, the Earthwatch Institute, and Texas A&M Department of Marine Biology. Tissues and behavioral observations from free-living dolphins were collected pursuant to the Marine Mammal Protection Act of New Zealand with permission from the New Zealand Department of Conservation, and with approval of Animal Use Protocol committees from the University of Auckland, New Zealand, and Texas A&M University, USA. Dr. Sharon Gursky, Texas A&M Department of Anthropology, provided advice and support for this project. Dr. Adam Jones, Texas A&M Department of Biology, deserves special thanks for being so generous with lab space, equipment, and advice. We also thank Dr. Terry Thomas, Texas A&M Department of Biology, for the use of lab equipment. D. Shelton was supported by graduate fellowships from Texas A&M University and Science Foundation Arizona. We thank J.R. Alvarado Bremer and one anonymous reviewer for helpful comments on earlier drafts of this manuscript.

Chapter 11

Human Interactions with Dusky Dolphins: Harvest, Fisheries, Habitat Alteration, and Tourism

Tim M. Markowitz,[a] Silvana L. Dans,[b] Enrique A. Crespo,[c] David J. Lundquist,[d] and Nicholas M.T. Duprey[e]

[a]School of Biological Sciences, University of Canterbury, New Zealand
[b]Centro Nacional Patagónico, CONICET, Chubut, Argentina
[c]Facultad de Ciencias Naturales, Universidad Nacional de Patagonia, Chubut, Argentina
[d]Department of Anatomy and Structural Biology, University of Otago, Dunedin, New Zealand
[e]Marine Mammal Research Program, Texas A&M University, Galveston, Texas, USA

If you ever get the chance to lean out over the front of a boat and watch a group of bow-riding dusky dolphins, you will likely see something truly remarkable, as they roll over on their sides to look right back at you. Then, just as you are reveling in the experience and wondering what they are thinking, with a sudden leap and a splash, the dolphins disappear in the time it takes you to wipe the saltwater from your eyes. Interactions with wild dolphins are often like that, so tantalizing they leave us totally, utterly stunned and amazed.

Across many cultures and social divides, dolphins hold a very special place in the human psyche. We sing, dream, paint, and write poetry about dolphins. And we have done so for at least as long as our written history. One could say that wild dolphins have a heck of a lot to do with what it means to be human. The proliferation of myths about dolphins has resulted in people treating them as magical beings, sort of like "floating hobbits" (Pryor and Norris 1991). Yet, even as such touchy-feely "dolphin hugger" attitudes have proliferated across the world, people continued to chop dusky dolphins up by the thousands for human food or fish bait. Dolphins are worth saving for their own sake, and for all that they have to offer humanity, not because they are mythical beings with supernatural, otherworldly powers. Dusky dolphins are truly remarkable creatures, and we are cautiously optimistic that people around the world are beginning to form a greater awareness, understanding and appreciation of these aquatic acrobats on their own terms.

Tim Markowitz, Edward Percival Field Station, Kaikoura, New Zealand, 2008

INTRODUCTION

The International Union for Conservation of Nature (IUCN) lists the dusky dolphin (*Lagenorhynchus obscurus*) as a species for which currently available data are insufficient to assign conservation status (IUCN 2008). Limited information on dusky dolphin abundance worldwide, combined with documented high levels of direct harvest and incidental fisheries by-catch in some locations, have made it challenging to assess the current conservation status for this species. We begin by examining available information on dusky dolphin mortality through direct harvest and fisheries information. We then examine the more subtle potential effects of habitat alteration and degradation, and conclude with an assessment of dusky dolphin–tourism interactions.

DIRECT HARVEST OF DUSKY DOLPHINS

It is, of course, a misnomer to call the hunting of dolphins a "fishery." Dolphins are nothing like fish with respect to important reproductive and life history parameters critical to assessing fisheries sustainability (Perrin 1999). Dolphins are long-lived, slow-growing, and slow to reproduce. Although dolphins are cetaceans managed under various marine mammal protection acts and monitored by the International Whaling Commission, they are also not great whales, which results in different problems for management (Gambell 1999). Far greater numbers of dolphins than whales must be harvested to fill a ship's hold. Despite this, direct harvest of dolphins often garners considerably less critical attention than harvest of great whales (Perrin 1999). Dolphins are highly social, meaning that harvests have great potential not only to affect population viability but also to disrupt dolphin societies; cultural inheritance as well as genotypes can be lost due to overharvesting (Whitehead 2002). Because of these unique life history and societal traits, there is no reason to expect that models and procedures for managing fisheries and whaling should be effective or appropriate for managing dolphin harvests. We open this chapter by examining the effects of direct harvest on dusky dolphins, examining the history and, as much as possible, the current status of dusky dolphin hunting.

For over three decades, dusky dolphins have been taken directly in the multi-species small cetacean fisheries off Peru. Although most dusky dolphins have been taken in directed net fisheries, they have also been taken by a harpoon fishery (Brownell and Cipriano 1999). Dusky dolphins are caught along with long-beaked common dolphins (*Delphinus capensis*), Burmeister porpoises (*Phocoena spinipinnis*), and common bottlenose dolphins (*Tursiops truncatus*). Most animals are caught in medium-sized multi-filament nylon drift gill nets with stretched mesh size up to 20 cm. Nets are usually set at dawn and recovered in the morning by artisanal fishermen in small open boats (Van Waerebeek et al. 1997). Other target species include several species of sharks and rays, such as the blue shark (*Prionace glauca*), shortfin mako shark (*Isurus oxyrhynchus*), hammerhead sharks (*Sphyrna* spp.), eagle rays (*Miliobatis* spp.), and other schooling fish, for example bonito (*Sarda chiliensis*). Dolphin meat is used for human consumption and fishing bait.

Cetaceans were first reported in the scientific literature for sale in fish markets at Peruvian fishing ports, during the 1960s (Clarke 1962, Grimwood 1969, reviewed by Read et al. 1988). Dusky dolphins taken during this time were likely landed mostly as by-catch, incidental to the primary fisheries targets. An expanded and more directed harvest of dolphins and porpoises apparently started in Peru following the demise of the anchoveta (*Engraulis ringens*) fishery in 1972. Large catches of anchoveta in 1972 combined with a very strong ENSO (El Niño Southern Oscillation) event led to collapse of the fishery, which in turn led fishermen to switch from fishing for anchoveta with some incidental catch of dolphins, to a direct dolphin harvest (Read et al. 1988, Vidal 1993).

Although direct harvest of dusky dolphins in Peruvian fisheries apparently began in the early 1970s, the first systematic studies of cetacean harvests off Peru were not conducted until the mid 1980s. Large catches of small cetaceans (approximately 10000) were reported from the coastal waters of central Peru in 1985, with dusky dolphins constituting over 90% of that number (Read et al. 1988). Van Waerebeek and Reyes (1990) reported an increase in catches of dusky dolphins in 1987 at Pucusana Port. There was also an increased use of alternative harvesting methods such as harpooning or setting dynamite charges to stampede dolphins into set nets, which confirmed that this was a directed fishery. Numbers of dolphins harvested off Pucusana Port continued to grow in 1988–1989, and reached a peak in 1989 (Van Waerebeek and Reyes 1994a). Harpooning was largely eliminated at Pucusana Port at this time via a local prohibition, but returned on a small scale in 1990 (Van Waerebeek and Reyes 1994a). The highest number of landings occurred in winter, coinciding with the dusky dolphin reproductive peak and resulting in the catch of large numbers of pregnant females and mothers with calves (Van Waerebeek and Reyes 1994a).

In 1990, the Peruvian government issued a ban on the harvest, processing, and trade of small cetaceans; however, because enforcement of the ban was the responsibility of regional and national government, it was largely unenforced and harvest continued (Vidal 1993, Van Waerebeek and Reyes 1994b). Catches of dusky dolphins remained high during the early 1990s. From 1991 to 1993, an estimated 7000 dusky dolphins were caught per year, a level thought to be unsustainable. It has been estimated that more than 700 dusky dolphins were landed per year at just one port in the early 1990s. These dolphins were sold for food in local markets, and were taken both incidentally and as deliberate targets (Jefferson et al. 1993). The percentage of dusky dolphins caught dropped significantly between 1985–1990 and 1991–1993 (Van Waerebeek 1994). At the same time, the relative percentage of common dolphins (*Delphinus* spp.) landed rose, which might have been a natural cycle or a sign of a decrease in abundance of dusky dolphins due to high levels of harvest (Van Waerebeek 1994). Dusky dolphins continued to be taken by direct harvest in large numbers in Peru, at least until new regulations were enacted following a meeting of the International Whaling Commission (IWC) in 1994. In January–August 1994, over 87 days, 82.7% of 722 cetaceans landed at Cerro Azul, central Peru were dusky dolphins, captured mostly in multi-filament gill nets. The total kill estimate for a seven-month period, stratified by month, was 1567 cetaceans. Data collected at 16 other ports showed that high levels of dolphin and porpoise mortality persisted in coastal Peru at least until August 1994.

The number of animals harvested during the late 1980s and early 1990s was deemed by the IWC (1994) as probably not sustainable. Subsequent legislation in 1994 placed the responsibility for enforcing the cetacean harvest ban on district and municipal authorities (Van Waerebeek and Reyes 1994b). Further pressure was applied by conservation non-governmental organizations (NGOs) and public education efforts were implemented, resulting in partially reduced harvests, but enforcement varied from port to port depending on local factors (Van Waerebeek et al. 1997).

It is believed, but not confirmed, that the level of exploitation declined in part because dolphin hunting was completely banned by law in 1996, but mostly due to the depletion of dolphin numbers (Van Waerebeek and Würsig 2009). Circumstantial evidence suggests that after 1996 increasing enforcement reduced direct takes and illegal trade in meat, but also hampered monitoring. The absence of abundance data precludes any assessment of the impact of direct harvest or indirect takes on dolphin populations (Van Waerebeek et al. 1997). For the period 1999–2001 a minimum of 471 small cetaceans (310 identified to species) were found close to ports and landing beaches. Species composition of identifiable specimens included: Burmeister's porpoise (42.6%), long-beaked common dolphin (24.2%), dusky dolphin (20.6%), and bottlenose dolphin (12.6%). The number of specimens tallied often included the visible fraction of animals landed that day plus remains of other animals butchered on earlier days; therefore, no per diem landing rates could be deduced.

The new legislation and increased enforcement resulted in dramatic changes in landing procedures for dolphins taken in fisheries off Peru, with new covert dolphin harvesting methods employed to conceal illegal hunting operations. New practices began in the early 1990s, shortly after the ban was initially passed, and included butchering captured specimens at sea and landing concealed, filleted meat (Van Waerebeek and Reyes 1994b). Depending on the port, entire cetacean carcasses were very rarely (Cerro Azul, Pucusana), or somewhat infrequently, landed due to conservation regulations. These "standard" practices that serve to conceal illegal dolphin hunting still go on. As before, the meat is used predominantly for human consumption (mostly in fresh form) and as bait in fisheries for large elasmobranchs (sharks and rays) using both long lines and gill nets.

Direct harvest of dusky dolphins in Peru apparently still continues today (Van Bressem et al. 2007b). Enforcement of legislation continues to be an issue, and there is a market for cheap dolphin meat among lower income Peruvians

(Van Waerebeek et al. 2002). It is estimated that roughly 1000 dolphins and other small whales are still being caught annually by fishermen to supply bait for the shark fishery. There are a complex set of economic, social, cultural and moral factors which must be acknowledged and prioritized in order to effect change within this fishery and the associated fishermen.

Direct harvest of dusky dolphins also occurred historically off the coast of southern Chile. Dusky dolphins were included in the group of species taken for bait in the king crab (*Lithodes santolla*) fishery in the Fueguian Channels in the extreme south of South America (Goodall and Cameron 1980, Goodall et al. 1994). The Chilean king crab fishery had its largest impact on marine mammals in the 1970s. Dolphin takes had already declined by the 1990s when a management program was initiated by the Chilean Government to curtail dolphin harvests.

DUSKY DOLPHIN INTERACTIONS WITH FISHERIES

Incidental by-catch of dusky dolphins in fisheries off Argentina

As in many other parts of the world, fisheries have increased off Patagonia, Argentina during the past 30 years, and became especially important for the regional and national economy during the 1990s. Trawling, jigging, and long-line fisheries were all developed and increased during this period. The main targets for these fisheries include Argentine hake (*Merluccius hubbsi*), Argentine red shrimp (*Pleoticus muelleri*), and Argentine shortfin squid (*Illex argentinus*). The actual extraction of hake exceeded the maximum allowable catch during the 1990s (Bezzi et al. 1995, Aubone et al. 1999). Over-fishing led to the collapse of the hake fishery in the late 1990s, and illustrated a clear need for new fisheries policies.

The first investigations of direct and indirect effects of fisheries on wildlife in Patagonia began in the early 1990s. The focus of this research was to evaluate the effects of fisheries on top predators, especially marine mammals. This research program has documented species involved in the interactions, calculated mortality rates due to incidental by-catch, examined diet and overlap in the use of common resources, and developed population and community models to assess whether fisheries interactions are affecting dolphin populations in a significant or unsustainable manner (Crespo et al. 1997b, 2000, Dans et al. 1997a).

Several species of birds, marine mammals, rays, and sharks become entangled and are discarded from the hauls (Crespo et al. 1994, 1997b). Although high-seas fisheries target relatively few species (e.g., hake, squid, and shrimp), more than 70 species are taken incidentally as by-catch. Target species, as well as most of the by-catch species, represent items in the diet of top predators. Therefore, direct effects of incidental mortality are combined with indirect effects of fisheries on marine mammal prey.

Dusky dolphins' preference for pelagic fish and squid makes them prone to be caught in midwater or purse seine nets targeting these species (Crespo et al.

2000). Dusky dolphins not only feed on pelagic species but they do so in large groups, increasing the potential for entanglement of large numbers of individuals (Crespo et al. 2000).

The highest rates of incidental catch of dusky dolphins off the Patagonian coast occur in midwater trawling for shrimp at night (Crespo et al. 1997b). Mortality of dusky dolphins interacting with this fishery was high during the 1980s, with by-catch occurring at a possibly unsustainable rate for the population (Dans et al. 1997a, 2003a,b). At the peak of the Patagonian midwater shrimp trawling fishery in 1984, the number of dolphins caught was estimated to be 442–560 individuals per year. Thus, incidental mortality during 1984–1986 would have led to a maximum annual mortality close to 8% of the present estimated population size. The effect on the dusky dolphin population may have been even more severe than these numbers indicate, considering that the catches affected mostly reproductive females (Dans et al. 1997a). Since 1994 there has been greater regulation of nets used in the shrimp fishery and incidental catch has declined consistently. Subsequently, dolphin mortality estimates for 1994 dropped to as low as 36 dolphins per year, mostly females and young adults.

After 1997, with the depletion of hake stocks, the fishing effort in Patagonia shifted to other species, including southern anchovy (*Engraulis anchoita*), which are caught with midwater nets. This increased the potential for entanglement of dusky dolphins and other pelagic predators. The anchovy fishery has been somewhat more erratic than the hake fishery, and in terms of both biomass and market price, cannot completely replace the hake fishery. At present, dusky dolphins are still caught incidentally in the anchovy fishery, but there are no current estimates of mortality rates due to lack of monitoring in this fishery.

Where available, information on catch rates and abundance estimates, together with demographic data, allow the comparison of annual mortality with critical values (Dans et al. 2003b). These comparisons show that while annual catch rates of dusky dolphins may be low, they could be unsustainable, making dusky dolphins off Argentina vulnerable to population declines as a consequence of incidental fisheries by-catch. The sample of dolphins collected from Patagonian fisheries was biased toward females (70% of the catch) with ages ranging between 0 and 11 years old (Dans et al. 2003b). Among the females, 44% were mature, and half of these were pregnant. Given that the average age of the females was 6 years old, it is likely the fishery was taking the most sensitive sex and age classes of the population in terms of reproductive potential.

In addition to these Patagonian fisheries, there are other fisheries affecting dusky dolphins along the Argentine coast. For example, dusky and common dolphins are taken incidentally in purse seine fisheries for anchovy and mackerel in the northern fishing grounds of Buenos Aires Province (39°S; Corcuera et al. 1994, Crespo et al. 1994). Fishermen use the activity of dolphin herds to find anchovy and mackerel, and then set the purse seines around the herd, which in turn causes the death of individual dolphins by stress and suffocation. This fishery, similar to the Patagonian anchovy fishery, has not been monitored by any fishery agency or research or conservation group in the past 15 years. The total annual take of dusky dolphins was estimated in the early 1990s to be roughly 100 individuals at a single port (Necochea), but the same fishery operates out of other harbors (e.g., Mar del Plata). As the purse seine fishery has not been properly evaluated and never monitored, the total take of dolphins and by-catch trends over time are uncertain.

The effects of fisheries in different areas may be additive in the case of the dusky dolphins, as there is some evidence that the same population may occur in Buenos Aires Province and Patagonia. Dusky dolphins are highly mobile, and two individuals marked in Península Valdés, Patagonia were found still together 7 years later 780 km to the north in Mar del Plata, Buenos Aires Province (Würsig and Bastida 1986). Similar long-range movements have been documented for several dusky dolphins over relatively short time-frames in New Zealand (Markowitz 2004), suggesting this is not a rare occurrence.

Current management of fisheries in Argentina is based on mono-specific models of the target species. Marine mammals, together with other biological, economic, and social components, are not seriously considered as part of the equation when fishery management is discussed (Crespo and Hall 2001). Mortality rates of marine mammals, the number of jobs created or eliminated, and the regional socio-economic impact of the management measures are largely omitted from fisheries management plans. We hope that these important issues will one day take their rightful place in the fishery management agenda.

Interactions with fisheries in New Zealand and South Africa

Although high levels of fisheries by-catch have been shown to impact Hector's dolphin populations (Dawson 1991, Slooten et al. 2000, Dawson and Slooten 2005, Slooten 2007), relatively few dusky dolphins in New Zealand become entangled in gill nets (Cipriano 1997). The estimated annual incidental kill of dusky dolphins in fishing gear around New Zealand was within the range of 50–150 during the mid 1980s (Würsig et al. 1997). Current rates of incidental take for dusky dolphins in New Zealand waters are unknown, but presumed to be sustainable given the large, robust population (Harlin et al. 2003, Markowitz 2004). Necropsy reports suggest a low level of dusky dolphin by-catch. Only four females and two males were reported as incidentally taken in commercial fisheries during 1997–2002 (Duignan et al. 2003, 2004). A considerably greater number are likely to have been taken and not reported, but it seems unlikely that unsustainable levels of dusky dolphin by-catch in New Zealand would go undetected. However, there is also potential for indirect interactions with pilchard and squid fisheries, as these favorite prey species of dusky dolphins have been targeted in New Zealand fisheries (see Chapters 3, 5, and 6). Off South Africa, some dusky dolphins are taken as by-catch in beach seines and purse seines, but these numbers are not thought to be large, and likely have a negligible effect on the population (reviewed in Chapter 14).

ANTHROPOGENIC CHANGES TO DUSKY DOLPHIN HABITAT

Although more difficult to measure than intentional and incidental take, habitat loss and degradation are two of the major threats to wild cetacean populations (Whitehead et al. 2000). Competition with humans for aquatic resources can impact aquatic mammals, and may exert a greater influence on their populations than either directed hunting or incidental catch due to fishing (Crespo and Hall 2001). In addition to direct or indirect competition for resources, human-made structures such as those associated with aquaculture may compete with marine mammals for space in the coastal environment (Würsig and Gailey 2002). Another form of habitat alteration that can affect marine mammals is environmental contamination (O'Shea and Tanabe 2003). We review below the potential effects of habitat alteration associated with the aquaculture industry and environmental contamination on dusky dolphins.

Effects of aquaculture on dusky dolphin habitat

Mussel farming is the largest aquaculture industry in New Zealand and a substantial source of revenue for the Marlborough Sounds region (see review in Chapter 12). In 2000, green-lipped mussels (*Perna canaliculus*) were grown in 520 farms around New Zealand, 87.5% of which were located in the Marlborough Sounds (Gall et al. 2000). In the early 1990s it was recognized that more information on potential impacts from this growing industry would be needed to ensure sustainable development (Smaal 1991). While significant advances have been made in some areas, concerns surrounding potential impacts on marine mammal populations have only recently received attention, and as such our understanding of these ecological interactions remain somewhat limited.

Aquaculture can affect dolphin habitat use in a number of ways. Objects at the surface and lines in the water column may impede dolphin movements, impacting the animals' ranging and foraging patterns (Würsig and Gailey 2002, McFadden 2003, Markowitz et al. 2004). For example, an oyster farm established in Shark Bay, Australia excluded mother–calf pairs of Indian Ocean bottlenose dolphins (*Tursiops aduncus*) from the farm area (Mann 1999b). Consequently, the long-term ranging patterns of adult female bottlenose dolphins were changed, and the dolphins used the area within farms less often than an adjacent ecologically similar area (Watson-Capps and Mann 2005). When cetacean habitat use and marine farming are both concentrated in the same small areas, increased vessel traffic due to aquaculture activity can increase noise levels and disturb cetaceans. Chilean dolphins (*Cephalorhynchus eutropia*) in an area of intensive mussel farming were found to respond to vessels by bunching, increasing speed, and increasing reorientation rate (Ribeiro et al. 2005).

In the course of research on dusky dolphin distribution in New Zealand waters, it became clear that the dolphins might be affected by habitat alteration due to aquaculture developments in the Marlborough Sounds region. Studies of dusky dolphin habitat showed extensive overlap between dolphin winter foraging habitat and proposed aquaculture development in the region (Markowitz et al. 2004). The area in and around Admiralty Bay, heavily utilized by dusky dolphins in winter months, has also been an area proposed for a substantial increase in green-lipped mussel farms (Figure 11.1, top). Sighting rates per survey effort were much higher in inner Admiralty Bay than in adjacent areas, suggesting that this area where substantial aquaculture development has been proposed might be particularly important winter foraging habitat for dusky dolphins (Markowitz et al. 2004).

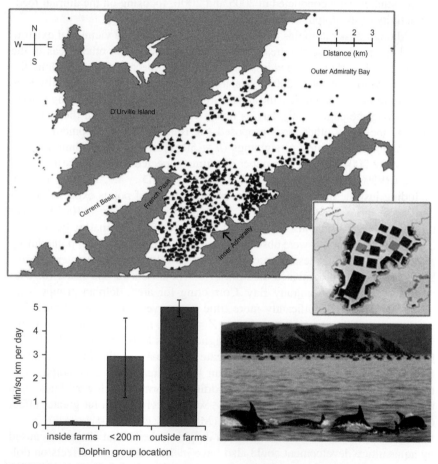

FIGURE 11.1 Positions where dusky dolphin (*Lagenorhynchus obscurus*) groups were encountered during winter surveys of the Marlborough Sounds, New Zealand (top) in 2001 (pentagons, $n = 303$), 2002 (triangles, $n = 250$), 2003 (squares, $n = 189$), and 2004 (circles, $n = 68$) showed a high degree of overlap with proposed marine farms (inset). Large, dark gray boxes in the middle of the bay (inset) show mussel farms that were proposed. Existing farms can be seen as small boxes in light gray along the coast. Dusky dolphins often forage near existing mussel farms (bottom) in Admiralty Bay during the winter months, but rarely inside the farms.

Concerns over the potential impacts on dusky dolphins of further marine farm developments in the nearshore environment (Würsig and Gailey 2002) prompted further investigation of long-term variation and trends in the distribution and abundance of dusky dolphins in the Marlborough Sounds. We report here findings from research on dusky dolphin use of Admiralty Bay and surrounding areas conducted during seven consecutive winters (1998–2004), with 713 hours of boat-based effort on 135 days producing focal follow data from 948 dusky dolphin group encounters (Markowitz and Würsig 2004). Survey methods and other details can be found in Markowitz (2004). Subsequently, further studies were conducted in 2005 and 2006, focusing on the diurnal feeding activity of the dolphins (see Chapter 6).

Although dusky dolphins often are found foraging in the vicinity of existing mussel farms, they use the areas within the farms less than other areas in the bay (Figure 11.1, bottom), and less than would be expected given a random distribution (Markowitz et al. 2004). Despite extensive use of inner Admiralty Bay, dusky dolphins did not utilize the areas within the boundaries of existing marine farms along the edges of the bay as much as adjacent areas and the center of the bay, including other areas proposed for future farm development. Based on a random distribution of sightings ($n = 50$ replications) in Admiralty Bay (Markowitz 2004), the mean number of points expected to fall in marine farms was 20 (95% confidence interval (CI) = 18–21, SD = 4.7, range = 11–30). Including just the nearshore zone, the mean number of points expected to fall within the farms was 16 (95% CI = 15–17, SD = 3.2, range = 10–23). Over the course of the study, just one encounter occurred within a marine farm.

A total of 17 groups were observed entering marine farms at any point during a focal follow (16 of these were encountered outside farms, then entered farms briefly), spending a combined duration of 0.7 hours out of 208 hours (0.35%) of focal follows in Admiralty Bay. Correcting for area, dolphin groups were observed spending significantly more time per survey day outside farms than inside farms (Wilcoxon signed ranks test $Z = 5.777$, $P < 0.001$), and near farms (<200 m) than inside farms (Wilcoxon signed ranks test $Z = 2.934$, $P = 0.003$, Figure 11.1 bottom). These findings indicate that the 44 existing inner Admiralty Bay farms, occupying a total area less than $1\,km^2$, already influence dolphin distribution and habitat use. The proposed additional farms, which would take up a much greater amount of space in the bay, would likely have a far greater effect on dusky dolphin winter foraging habitat (Markowitz et al. 2004).

In addition to direct interactions with dolphins, the habitat alteration caused by aquaculture development could also have indirect ecosystem effects on dolphins. Shellfish farming can introduce new species or pathogens into marine ecosystems, cause biodeposition, and result in faunal changes (Buschmann et al. 1996). The biochemical effects of accumulated faeces and pseudofaeces under mussel farms (Grant et al. 1995) can alter benthic communities, increase aerobic heterotrophic bacteria (La Rosa et al. 2001), and decrease meiofauna (Mirto et al. 2000). Mussels increase nitrogen levels (La Rosa et al. 2002), and deplete chlorophyll *a* levels (Grange and Cole 1997, Ogilvie et al. 2000) within and around farms, although shellfish farming appears to influence water column biochemistry less than fin fish farming. At present, indirect trophic effects of mussel farms on dusky dolphins, other apex predators, and their prey in the Marlborough Sounds are unknown. For these reasons alone, caution should be taken in placing a high density of mussel farms in one geographical location.

A high proportion of dusky dolphin groups encountered in Admiralty Bay were observed actively feeding (Markowitz et al. 2004), and sighting rates in the area were highest in years when a greater proportion of dolphin groups were feeding (Figure 11.2). If the proportion of groups feeding is an indication of prey availability, which seems likely, then these data indicate that the observed inter-annual variation in dolphin abundance in Admiralty Bay is linked to fluctuation in prey abundance. In 2002, estimates of prey density using backscatter intensity from an echosounder during small vessel surveys indicated that dolphin distribution in inner Admiralty Bay closely mirrored that of schooling fishes (Benoit-Bird et al. 2004). Not surprisingly, the dolphins went where the fish were most abundant. Assuming this rule of thumb holds on a broader spatial and temporal scale, it follows that a greater number of dolphins use Admiralty Bay in years when a greater number of fish are present, and that the dolphins disperse over a broader area to search for food in years when fewer prey are available. This suggests that dusky dolphins may be indicators of local fish abundance, and, consequently, ecosystem health. Further aquaculture development with very high densities of farms could not only displace dolphins but also reduce local fish abundance, as mussels feed on plankton that provides the basis for the pelagic food web in aquatic ecosystems.

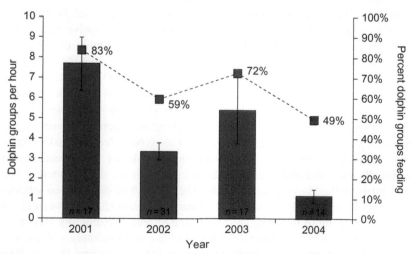

FIGURE 11.2 Dolphin groups encountered per hour of systematic survey effort in Inner Admiralty Bay, New Zealand (left axis, bars) and percent of groups feeding (right axis, line) are compared by year with sample size (*n* surveys) shown for each winter season (2001–2004). Bars represent mean values with standard errors.

As in Golfo San José, Argentina (Würsig and Würsig 1980), dusky dolphins apparently coordinate their diurnal foraging activities in Admiralty Bay (see Chapter 6), feeding in small groups on schooling fish such as pilchard (*Sardinops neopilchardus*) that are locally abundant and may be concentrated in certain areas by tides and currents (Baker 1972a). This coordinated fish-herding activity apparently improves prey accessibility for a number of other predators, including other dolphins, New Zealand fur seals (*Arctocephalus forsteri*), predatory fishes, small sharks, and a number of seabirds (McFadden 2003, Markowitz 2004, Vaughn et al. 2007). Therefore, in addition to being indicators of fish abundance, dusky dolphins may fill the role of a keystone species in this region. Consequently, the effects of anthropogenic changes to dusky dolphin habitat may have greater ecological consequences than are immediately apparent.

Abundance estimates show that large numbers of dusky dolphins use the relatively small area of inner Admiralty Bay each winter, the same area proposed for a greater increase in aquaculture development than any other location in the Marlborough Sounds. Photo-identification data demonstrate that, despite considerable inter-annual variation, over 100 dusky dolphins utilize Admiralty Bay each winter (Figure 11.3). From 1998 to 2004, a computer catalog of 530 marked individuals was developed (Araabi et al. 2000, Hillman et al. 2003, Markowitz et al. 2003a) from a sample of 21 355 dorsal fin photographs taken during the seven consecutive winters (mark rate = 75%). Abundance estimates were calculated in SOCPROG version 2.2, with the best fit model selected based on Akaike Information Criterion (AIC) values (Markowitz and Würsig 2004). The estimated total abundance of dusky dolphins (marked and unmarked) using Admiralty Bay as winter foraging habitat over this 7-year period was 711 (95% CI = 608–844, based on a 1-year interval, best fit model mortality + trend). Over this time period, there was a negative trend (trend: −0.25, 95% CI = −0.12 to −0.35), suggesting dusky dolphin use of the bay may have been declining overall, but there is also considerable inter-annual variability in use of this habitat (Figure 11.3). During 2000–2004, single-season estimates ranged from 122 to 272 (Figure 11.3, based on a one-week sample interval, best fit model closed-Schnabel in 2000, mortality in all other years).

Preliminary comparison of photo-identification records between regions showed regular seasonal migrations of at least some New Zealand dusky dolphins ($n = 37$), with dolphins spending summers in the Kaikoura area, where they feed nocturnally on squid and lanternfish associated with the deep scattering layer, and winters in the Marlborough Sounds, where they feed diurnally on schooling fishes (Markowitz 2004). Re-sightings of particular individuals between seasons indicate that many of the same animals return to feed in Admiralty Bay in subsequent winters (81%), and that dolphins associate non-randomly with preferred foraging companions across multiple seasons (Markowitz and Würsig 2004). Such a foraging tradition among close associates suggests a cultural component to winter foraging by dusky dolphins in Admiralty Bay (Whitehead et al. 2004). We have argued that this winter habitat and dolphin foraging culture are worth

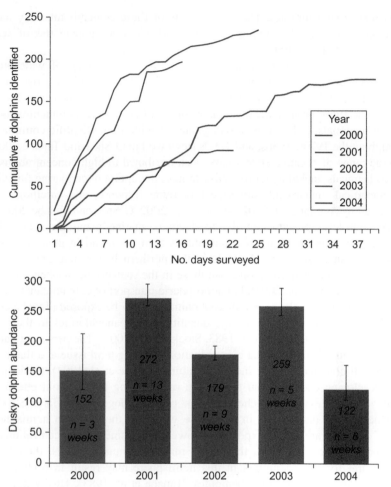

FIGURE 11.3 Discovery curves of the number of marked individuals photographed by survey day (top) and mark-recapture abundance estimates with 95% confidence intervals (bottom) are shown for dusky dolphins encountered in Admiralty Bay, New Zealand during five winter seasons, 2000–2004.

preserving (Markowitz et al. 2004), and the local regional council has agreed, declining a number of applications for additional marine farms. However, the issue remains unresolved, as court appeals have been lodged by companies that are intent on expanding mussel farming operations in Admiralty Bay.

Effects of environmental contaminants on dusky dolphins

As top-level predators, dolphins are exposed to high concentrations of environmental contaminants produced by humans, ranging from heavy metals to persistent organochlorines. The toxic effects of these contaminants may hamper responses of dolphin immune systems and result in a greater risk of death due to disease or infection.

The end of the Second World War marked the beginning of widespread industrial use of organochlorine pesticides and PCBs, as well as a rise in the introduction of heavy metals into marine environments (O'Shea and Tanabe 2003). The 1960s brought the first studies of the effects of these contaminants on marine mammals, followed by a rising interest within the scientific community throughout the 1970s, 1980s, and 1990s (reviewed in O'Shea and Tanabe 2003). A wide range of negative effects have been attributed to high concentrations of these chemicals, including reproductive failure, disruption of endocrine function, adrenocortical impairment, suppressed immune response, and carcinogenesis (reviewed by Aguilar et al. 2002, Moore et al. 2002, O'Shea and Tanabe 2003).

As high-level predators in the food chain, marine mammals accumulate much higher concentrations of contaminants than are found in their surroundings (Koeman et al. 1975). Animals in the northern hemisphere are typically exposed to greater concentrations than those in the southern hemisphere (Tanabe et al. 1983, Moore et al. 2002), but atmospheric transport of contaminants means that animals far from the source of contamination may be exposed (Tanabe et al. 1994). Mothers may also pass large quantities of organochlorines to their offspring via lactation (Tanabe et al. 1983, Stockin et al. 2007). The persistent nature of these chemicals ensures that contamination is a long-term issue and that toxic effects will continue into the indefinite future (Tanabe et al. 1994), particularly near developing countries that are unable or unwilling to phase-out cheap but effective pesticides and other chemicals that pose a biological risk (Tanabe 2002).

Due to their distribution in temperate waters of the southern hemisphere, dusky dolphins are likely exposed to lower environmental concentrations of organochlorine contaminants than other marine mammal species, though the agricultural focus of countries in this hemisphere may result in higher relative levels of DDT and other pesticides (Tanabe et al. 1983). Dusky dolphins have been reported with elevated (but still low) levels of DDT and PCBs in the southern Pacific (Tanabe et al. 1983) and off the coast of South Africa (de Kock et al. 1994). Elevated PCBs and other organochlorine contaminants have also been found in dusky dolphins off New Zealand (Jones et al. 1999). Scarce data exist for contaminant levels in dusky dolphins around South America, though Aguilar et al. (2002) reported that organochlorine concentrations in marine mammals off South America were generally low relative to those in the northern hemisphere. Local variations may exist near sources of high pollution, such as ports and cities, as noted by Van Bressem et al. (2007a), though this is more likely to affect inshore species rather than dusky dolphins.

Exposure to high levels of toxins has been shown to result in a wide range of deleterious effects on marine mammals. Disruption of normal endocrine function, sterility, abortion, and stillbirths have all been attributed to contaminants (reviewed in Ross et al. 1996b, Tanabe 2002). Sexually transmitted diseases in dusky dolphins (Van Bressem et al. 1998) occur naturally, but may be more easily transmitted to animals suffering from reduced fitness due to environmental toxins (Van Bressem et al. 2007b). Increased concentrations of PCBs and DDT have been shown to result in lower immune response in many marine mammals, including bottlenose dolphins (Lahvis et al. 1995), harbor porpoises (*Phocoena phocoena*; Ross et al. 1996a), and harbor seals (*Phoca vitulina*; Ross et al. 1996b). Toxin-compromised immunity is suspected as a factor for several virus-related mass mortality events among European marine mammals, including harbor seals, gray seals (*Halichoerus grypus*), and striped dolphins (*Stenella coeruleoalba*; Ross et al. 1996b).

Immunotoxicity can be an important indirect effect of environmental contaminants, as lowered immune response may result in population-level effects due to increased parasitism, widespread lethal infection, or reduced reproductive success (reviewed in Van Bressem et al. 2007b). Ross (2002) and Harvell et al. (1999) have suggested that high contaminant levels may facilitate the emergence of new diseases by lowering the collective immunity of a population and allowing pathogens to persist and even be passed to new populations or species (also see Barrett et al. 1995). A toxic algal bloom may have also facilitated the transmission of morbillivirus from dolphins to Mediterranean monk seals (*Monachus monachus*; Harvell et al. 1999).

Morbilliviruses were responsible for several of the above-mentioned mass mortalities, and subsequently were found in dusky dolphins in Peru (Van Bressem et al. 1998) and other species worldwide (Van Bressem et al. 2001). Because large mortality events have not been seen among Peruvian dusky dolphins, it was suggested that the virus may have a long-established presence among the dolphins. The virus would then primarily infect newborns and calves without immunity, and survivors would retain immunity for the rest of their lives (Van Bressem et al. 1999). The same suggestion was made regarding poxviruses in Peruvian dusky dolphins (Van Bressem et al. 1999). Lower immune response due to environmental toxins would likely result in lower calf survivorship rather than mass mortality events. Van Bressem et al. (1999) also cautioned that reducing the number of animals in a population (such as via direct harvest, which is the situation in Peru) may result in such infections no longer being self-sustaining within that population, setting the stage for recurrent mass mortality events such as those seen in other species. Thus, subtle and indirect effects of environmental contaminants can have dramatic consequences when interacting with other anthropogenic impacts.

DUSKY DOLPHIN INTERACTIONS WITH TOURISM

Given their special place in human cultures, dolphins have the potential to act as "flagship species," ambassadors for conservation at a time when the earth is in the midst of widespread environmental devastation and a global extinction crisis. Dolphin tourism can also provide economic incentives for maintaining healthy dolphin populations to communities. Thus, well-managed dolphin tourism can potentially provide conservation as well as economic benefits. However, this does not mean that all dolphin "eco-tourism" is necessarily a good thing for the animals, and even well-managed tourism can always be improved.

Marine mammal tourism worldwide

Strong public interest in seeing and interacting with the natural world around them has spawned a large industry in nature-based tourism. Eco-tourism has been increasing worldwide for several decades and tourism focusing on interacting with marine mammals is no exception. Cetacean-watching activities have grown considerably in the last two decades, with annual revenues of greater than US$1 billion from more than 9 million participants in 1998 (Hoyt 2001). Between 1991 and 1998, the number of people participating in whale watching increased 12.1% per year, and the amount of revenue from whale watching increased by 21.4% per year, providing incentive for coastal communities to shift from consumptive uses of marine resources to tourism (Hoyt 2001).

Marine mammal tourism can include a variety of activities, ranging from viewing marine mammals from vessels, planes, helicopters, or shoreline vantage points, to entering the water and swimming with pinnipeds or cetaceans. The continued growth of this industry has generated much interest within the scientific community in examining the effects of tourism on marine mammals. Although marine mammal tourism is relatively sustainable when compared with certain activities such as direct harvest, there remain concerns regarding the potential effects of interactions with tours on the behavior of the animals in their natural environment (Constantine 1999). Tourism has been shown to affect the behavior of a number of free-ranging marine mammals, including polar bears (*Ursus maritimus*; Dyck and Baydack 2004), Florida manatees (*Trichechus manatus latirostris*; Nowacek et al. 2004, Sorice et al. 2006), harbor seals (Johnson and Acevedo-Gutiérrez 2007), common bottlenose dolphins (Samuels and Bejder 2004), Hawaiian spinner dolphins (*Stenella longirostris longirostris*; Timmell 2005, Courbis 2007), killer whales (*Orcinus orca*; Williams et al. 2002a, 2002b), gray whales (*Eschrichtius robustus*; Ollervides 2001), and southern right whales (*Eubalaena australis*; Lundquist 2007). Behavioral responses specifically described relative to swim-with-cetacean tourism include avoidance of swimmers (Constantine et al. 2004), increased speed and reorientation rates (Lundquist 2007), and increased communication and echolocation (Scarpaci et al. 2000). In addition to concerns about harassment of the animals, in some cases there may also be a risk of injury to either the human participants or the animals (Samuels and Bejder 2004).

Assessing the effects of tourism on wild dolphins

Assessing the negative effects of tourism on dolphin population status is a difficult goal, and the way to appropriately evaluate biologically significant impacts is in continuous debate among researchers. The most crucial task is to determine how short-term responses, which are the easier to handle and measure, may be related to long-term changes and biologically significant consequences (Bejder and Samuels 2003). Recently, there appears to be much confusion in the literature between short-term statistically significant behavioral responses and long-term biologically significant impacts. Such confusion can be misleading to resource managers, stakeholders, and the general public, particularly when any behavioral change is called an "impact." Here, we reserve the term "impact" for biologically significant population changes, and use the term "response" for statistically significant behavioral changes between a control (no tour vessel) and treatment (tour vessel) condition. Further, we will endeavor to discriminate between behavioral responses indicative of stress, potentially affecting the welfare of individuals (Dawkins 1980), and apparently affiliative responses such as approaching and bow-riding vessels.

There are some studies of dolphin interactions with tourism that indicate biologically significant impacts. For example, bottlenose dolphins in Shark Bay, Western Australia shifted their distribution in response to tourism (Bejder et al. 2006), with dolphin abundance decreased in areas with high boat traffic. It was also suggested that more sensitive dolphins abandoned the impacted area by moving to another one nearby. The most meaningful evidence of long-term effects on survival or reproduction is the case of feeding programs of free-ranging bottlenose dolphins in Western Australia (Mann et al. 2000). In a 10-year study, differences in reproductive success were found between provisioned and non-provisioned females. First-year mortality of calves born to provisioned females was double that of calves born to non-provisioned females. Negative effects of human interactions on bottlenose dolphin calves may include their mothers changing behavior and engaging in different activities such as food partitioning and human interactions (Mann and Barnett 1999). Reduction in parental care was a suspected cause of the increased calf mortality (Mann and Kemps 2003). In most cases, this kind of information is unavailable or insufficient. Carrying out experiments becomes difficult due to spatial and temporal discontinuities along with many methodological obstacles when studying free-ranging dolphins (Bejder and Samuels 2003). Another way to resolve this challenge is to translate behavioral responses to energetic costs. Then, the additional cost due to interaction with tourism may be compared with the overall energy budget. However, estimates of energetic costs are difficult to make for free-living dolphins.

Whatever the final output, it is obvious that collecting baseline information and quantitative data about short-term reactions of dolphins to vessels is a necessary step in evaluating the potential long-term impact of dolphin watching. From a conservation perspective, it would be desirable to manage dolphin watching in a way that minimizes these short-term effects. From a sustainable and rational use perspective, dolphin populations should be exploited by tourism in a way that guarantees their persistence.

Dusky dolphin tours

Commercial tours that take paying passengers to see dusky dolphins are conducted off the coasts of South Africa, Argentina, and New Zealand. In South Africa, swimming with cetaceans is not permitted, and permitted tour companies are allowed to only view the dolphins. At present, there is no directed dusky dolphin tourism in South Africa, and there are just two tour companies with permits to view whales and dolphins that visit dusky dolphins regularly, but some unpermitted tourism does occur and violators of regulations regarding dolphins are at times prosecuted (M. Meÿer, August 2008, personal communication). In Argentina, dolphin tours are relatively new and largely unregulated. In New Zealand, dusky dolphin tours are well established, regulated by permits, and have been monitored consistently for over 10 years. For the remainder of this chapter, we focus on dusky dolphin interactions with tourism in Argentina and New Zealand, where it has received the most scientific attention.

Dusky dolphin tourism in Argentina

Dusky dolphin-watching tourism was developed in the late 1990s in Golfo Nuevo, northern Patagonia, Argentina (Coscarella et al. 2003). In Patagonia, Argentina, the World Heritage Protected Area, Península Valdés, represents an important location for national and international tourism year-round. The interest is mainly motivated by the presence of southern right whales and the possibility of whale-watching activities. At least 73 726 tourists watched whales from boats and 3000 trips were performed in 1997 (Rivarola et al. 2001), while at present more than 250 000 people come every year to watch whales. Dolphin watching based on dusky dolphins and Commerson's dolphins (*Cephalorhynchus commersonii*) started as an alternative attraction in 1997 (Coscarella et al. 2003).

The characteristics of dolphin-watching operations and their inter-annual changes and trends were analyzed by Coscarella et al. (2003). Dolphin-watching activities began at almost the same time in 1997 in Golfo Nuevo and Bahía Engaño, a location to the south where the main species encountered on tours is the Commerson's dolphin. Tourism is managed at the provincial administrative level in this region and the marine mammal laboratory (Laboratorio de Mamíferos Marinos Centro Nacional Patagónico CONICET) provided recommendations for dolphin tour regulations; however, these were not incorporated in the management scheme. Several years of study showed short-term changes in dolphin behavior (Coscarella et al. 2003), which led the provincial authorities to develop Codes of Conduct for dusky dolphins in Golfo Nuevo and for Commerson's dolphins in Bahía Engaño, but these tour activities are largely unregulated and guidelines unenforced.

In Golfo Nuevo, dolphin-watching boats operate from Puerto Madryn, and they use a wide area located on the western portion of this gulf (approximately 1000 km^2). An example of sighting trips performed in Golfo Nuevo is shown in Figure 11.4. Although dusky dolphins are the main species encountered on tours, other species including common bottlenose dolphins, common dolphins, and killer whales can also be found and approached (Table 11.1). There are no permits allowing specifically for dolphin watching and thus the activity is neither regulated nor controlled. Five tour vessels operating from Puerto Pirámides,

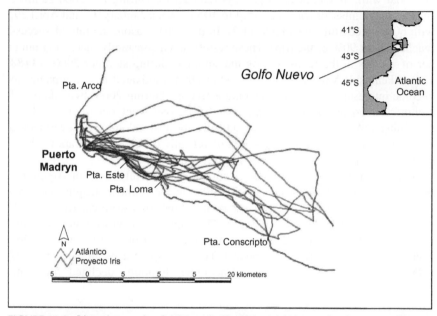

FIGURE 11.4   Lines show tracks of sighting trips for dusky dolphins operating from Puerto Madryn, Golfo Nuevo, Argentina.

TABLE 11.1   Species encountered on dolphin-watching trips in Golfo Nuevo, Argentina

	Puerto Pirámide (%)	Puerto Madryn (%)
Dusky dolphin (*Lagenorhynchus obscurus*)	5.14	34.78
Bottlenosed dolphin (*Tursiops truncatus*)	0.57	30.43
Killer whales (*Orcinus orca*)	1.71	4.35
Commerson's dolphin (*Cephalorhynchus commersonii*)	0.57	0.00
Trips with sightings	7.43	69.57
Trips without sightings	92.57	30.43

on the south side of Península Valdés, also visit dusky dolphins sporadically in Golfo Nuevo, and the dolphin sightings are mostly incidental during tours focused primarily on whale watching.

During the first dusky dolphin-watching season (January–March 1997) in northern Patagonia, a total of 80 trips were performed by only one commercial boat with 18 successful trips (less than 25%). During the 2000 summer season, the number of trips increased to 105 between January 11 and April 29, with 60% successful trips (Table 11.2). In the 2001 season, the rate of success reached about 90% of the trips. These results and a constantly increasing number of visitors in the region during the summer (during summer 2000, 43 188 visited Península Valdés versus 250 000 in 2005) motivated new companies to become interested in dolphin-watching activities. During 2004, two additional commercial boats operated; however in 2005, one went out of use and the remaining vessel only searched sporadically for dusky dolphins. Nevertheless, dusky dolphin-watching activities are not yet fully developed, and are likely to continue to expand in Argentina.

Perhaps because dusky dolphin tourism is relatively new in Argentina, there is considerable inter-annual variability in the number, length, and success of tours (Table 11.2). Between one and three trips were carried out each day (mean 1.54 trips/day, SD = 0.25). The trips lasted on average 2.8 hours (SD = 0.16 hours) and the time spent searching prior to the first sighting averaged 1.1 hours (SD = 0.22 hours). The boat spent a mean of 0.35 hours (SD = 0.43 hours) with dolphins. Encounter durations declined following

TABLE 11.2 Dolphin-watching activities in Golfo Nuevo, Argentina, 2000–2005

	2000	2001	2002	2003	2004	2005
Number of boats	1	1	2	2	3	2
Total number of trips	105	105	no data	no data	88	no data
Trips per day	1.65	1.93	1.48	1.36	1.22	1.62
Trip duration	02:40	02:43	02:58	02:56	03:04	02:39
Searching time	01:04	00:59	01:32	01:09	00:59	00:54
Sighting success (%)	60	89	52	47	83	70
Sighting duration	00:56	00:25	00:21	00:15	00:20	00:28

the initial 2000 season. The number of trips per day seems to be negatively correlated with the number of boats, since in 2004, the only year that three boats operated, the number of trips per day fell to 1.22 (Table 11.2).

Boat operators search for flocks of birds and, if dolphins are present, they slowly approach the group. Depending on the weather conditions and the activity of the dolphins, operators turn off the engine and allow some passengers to enter the water (this only occurred during 2000–2003). If dolphins show a high level of activity and aerial displays, operators induce dolphins to bow-ride by moving the boat faster. Although the first trip of each day is usually used to find the dolphins and then the same location is revisited during the next trip, there is no evidence to support the assumption that the same group is approached. Whether repeated visits of tour vessels to the same group have greater effects than visits to different groups remains to be investigated. Additionally, whether the same individuals are visited by tour vessels has yet to be examined.

During summer 2001 to 2005, surveys were carried out to detect possible bias of commercial boats towards specific groups or behaviors, as well as the immediate response of dolphins to different ways of approaching them (Dans et al. 2008). Both research and commercial vessels were used as observation platforms. Commercial operators were allowed to behave "normally" in searching tactic and during approach. Research vessel surveys consisted of random transects throughout the study area, until a group of dolphins was detected. Observations from both platforms were conducted under similar temporal, spatial and environmental conditions by experienced researchers.

A total of 192 groups were detected. Most of them were groups composed of adults and juveniles ($>80\%$), while groups with mothers and calves were less frequent (Figure 11.5, bottom). During research vessel surveys, a greater proportion of groups encountered were smaller in size than those encountered on commercial tours (Figure 11.5, top). Once a group of dolphins was detected, the boat approached to within 200 m and behavior was assessed. Then the research boat was maneuvered parallel to the dolphins, matching group speed and heading at a minimum distance of 100 m, while the commercial vessels generally drove straight toward the group, approaching to within 50 m or less. The predominant activity was re-assessed once the boat was closer to dolphins. This monitoring was conducted with 156 groups: 93 approached by commercial vessels and 63 encountered by the research vessel. The activity before the approach varied depending on whether observations were conducted from the research vessel or a tour vessel ($\chi^2 = 17.971$, df = 4; $P < 0.01$). Since they used seabird activity to find dolphins, it is not surprising that commercial vessels mainly found feeding groups (43%) and no resting groups (Figure 11.6). On the other hand, the research boat, systematically surveying random transects, found the same proportion of groups feeding and traveling (39 and 38%), and also found groups resting. Thus, tours had a bias toward encountering groups of feeding dolphins. The activity of dolphins once the vessel approached them also depended on the vessel, with dolphin traveling more often increasing in apparent response to approaches by commercial vessels ($\chi^2 = 14.168$, df = 4; $P < 0.01$).

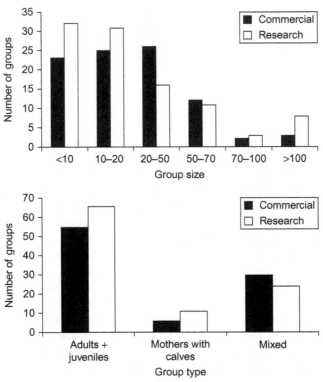

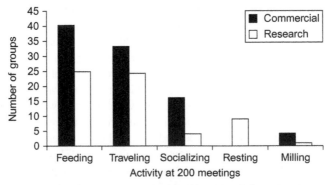

FIGURE 11.5   Distribution of size (top) and composition (bottom) of groups sighted during summer 2001–2005 in Golfo Nuevo, Argentina by research and commercial vessels are compared.

FIGURE 11.6   The number of dolphin groups engaged in each activity at the moment the boat detected and approached them at 200 m is compared for commercial tour vessels versus a research vessel in Golfo Nuevo, Argentina.

Dolphins showed different responses to vessel approach depending on their activity prior to the approach (Figure 11.7). Feeding and milling groups changed activity in response to half of the commercial boat approaches, while traveling, socializing, and resting groups showed fewer to no responses. When a group of

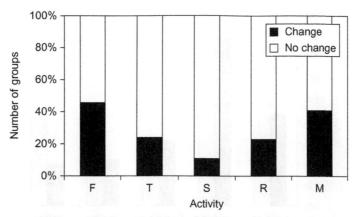

FIGURE 11.7 The proportion of time the activity of dolphins changed or did not change when approached by boats in Golfo Nuevo, Argentina (data from research and commercial boats combined) is compared based on the initial dolphin group activity (F = feed, T = travel, S = social, R = rest, M = mill).

traveling dolphins was approached by a commercial vessel, the dolphins continued traveling most of the time. The commercial vessel and the type of approach had an effect on the direction of change of activities (McNemar test $\chi^2 = 19.33$, df = 10; $P = 0.03$), while the likelihood of change from one activity to another remained equal in the case of the research vessel (McNemar test $\chi^2 = 15.6$, df = 10; $P = 0.11$; Dans et al. 2008).

When the research vessel was used as the observation platform, groups detected were approached to 100 m, and subsequently followed at this distance for as long as possible (30 minutes at least). Commercial vessels could visit the focal group at the same time. The research vessel approached dolphins from the side and with similar speed and direction as the dolphins, and appeared to minimize dolphin behavioral changes as a result, as also described by Lusseau (2003b). During the focal-group follow, the predominant activity (T = travel, M = mill, F = feed, R = rest) was recorded every 2 minutes (Lehner 1996). Transitions from one activity to another were calculated from control and treatment (when the commercial boat was also present) behavior sequences and modeled by Markov chains (Bakeman and Gottman 1997).

During the study period, 29 groups of dolphins were followed by the research vessel and visited by commercial vessels at the same time. Control and treatment chains were constructed from 706 and 308 sequences, respectively. Two transitions showed significant differences when commercial vessels were present: the transition M→F decreased and T→M increased (binomial test, $P < 0.05$). The model predicts that dolphins invest 41% of their time traveling and 22% feeding. In the presence of commercial vessels, time budget for feeding (Z-test, $P < 0.05$) and socializing (Z-test, $P < 0.01$) decreased while time budget for milling increased significantly (Z-test, $P < 0.001$, Figure 11.8a).

The time to return to feeding and socializing was longer in the presence of commercial tour vessels, while dolphins returned to resting and milling in less

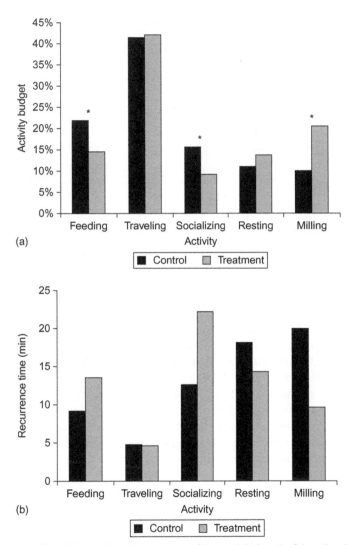

FIGURE 11.8 The activity budgets (a) percentage of time and (b) length of time (in minutes) for dolphin groups to return to an activity once the behavior has changed (recurrence time), predicted by matrix models are compared for control (no commercial tour vessel) and treatment (commercial tour vessel) activity transitions of dusky dolphins interacting with vessels in Golfo Nuevo, Argentina. Stars indicate statistically significant differences.

time (Figure 11.8b). Irrespective of original behavioral state, milling behavior was reached in a shorter period of time in the presence of commercial boats. Also, dolphins moved from feeding and socializing to traveling in less time in the presence of tour vessels than when they were absent.

Over the longer term, it is possible that the observed changes could modify activity patterns, ultimately time budgets, and therefore the population's energy intake–expenditure balance (Williams et al. 2006). Among the most important effects, feeding and socializing time decreased in the presence of tour vessels. A reduction in social interactions may have an effect on reproductive output. High levels of boat activity could decrease the time spent feeding enough to reduce energy intake.

In Golfo Nuevo, tourism activity and tour boats overlap in space and time with dolphin-feeding activity. Tour vessels could hamper foraging and prevent dusky dolphins from fully utilizing the resources within a portion of the habitat.

Several variables may be used in developing guidelines for boat operators to reduce effects, such as distance from dolphins and boat behavior during the encounters. Given observed differences between dolphin responses to tour vessels and the research vessel, vessel activity and distance are likely key components in minimizing effects on the dolphins' feeding time budget. Research vessels maintaining a greater distance and driving parallel to the animals had fewer effects than tour vessels that approached the dolphin groups more directly.

At present, there is little direct regulation of dolphin-watching activity in Argentina (Coscarella et al. 2003). Since the number of tourists and companies involved in dolphin-watching activity are not as high as whale watching, managers may have the perception that dolphin tour activity levels are not yet a problem. However, the constant growth of the tourism industry in the area of Península Valdés predicts that alternatives to whale watching will increase. Given the evidence that dusky dolphin foraging is disrupted by dolphin tours, there is the potential for the unregulated and expanding tourism industry in Argentina to result in biologically significant impacts to dusky dolphins. Therefore, this situation should be carefully monitored and managed to protect both the dolphins and the sustainability of dolphin tourism in Argentina.

Dusky dolphin tourism in New Zealand

Dusky dolphin tourism has been extensively developed in New Zealand since the late 1980s (reviewed by Würsig et al. 1997, 2007, see also Chapter 13).

Whale- and dolphin-watching tourism has undergone rapid growth in New Zealand over the past two decades, with the number of permitted dolphin tour operators increasing from none to over 75 since the late 1980s (reviewed by Orams 2004). A number of studies have shown behavioral responses of marine mammals to tourism in New Zealand. Tour vessels affected respiration rates, time at the surface between dives, and re-orientation rates of sperm whales (*Physeter macrocephalus*) off Kaikoura (Richter et al. 2003, 2006). New Zealand fur seals responded to tourist approaches from both land and sea, with habituation in areas where there were high levels of interaction (Boren et al. 2002). Studies of common bottlenose dolphins in the Bay of Islands during the 1990s showed behavioral responses to tour vessels (Constantine et al. 2004) and indicated long-term sensitization to disturbance (Constantine 2001). More recent research in Doubtful Sound and Milford Sound has indicated a number of effects of tour activities on bottlenose dolphins, including changes in dive behavior (Lusseau 2003a), displacement from areas by tour activities (Lusseau 2004), and changes in residency patterns (Lusseau 2005). Motor noise appears to be a key element in these interactions, and effects were less pronounced if vessels were driven carefully in accordance with New Zealand Department of Conservation (DOC) guidelines (Lusseau 2006). Changes in dolphin behavior in response to tour activities have also been noted for Hector's dolphins (*Cephalorhynchus hectori*; Bejder et al. 1999) and common dolphins (Constantine and Baker 1997, Stockin et al. 2008) in New Zealand.

There are several locations to view and swim with dusky dolphins in New Zealand. Dolphin tours that include dusky dolphin viewing are conducted in the Marlborough Sounds on the north shore of the South Island by two tour companies (in Queen Charlotte Sound and out of French Pass), on the central west coast of the South Island by two tour companies (out of Greymouth and Punakaiki, although these tours focus primarily on Hector's dolphins and common dolphins), in Fiordland on the lower west coast (although sightings of dusky dolphins are rare), and off Kaikoura on the east coast of the South Island by four companies.

Dusky dolphin tours began in Kaikoura as a side attraction to sperm whale watching in the late 1980s (Te Korowai O Te Tai O Marokura 2008). Tourists going out to watch sperm whales were taken to view dusky dolphins when they were conveniently nearby or sighted between sperm whale surfacings. Tours that focused completely on viewing and swimming with dusky dolphins began as a small operation in 1989; at first guests were offered these trips on an on-demand basis, until demand was such that scheduling trips became possible (Te Korowai O Te Tai O Marokura 2008). Since 1989, tour operations have grown and merged in Kaikoura (Simmons and Fairweather 1998, see also Chapter 13).

During the mid 1990s, dusky dolphin interactions with tours were frequent and increasing (Barr and Slooten 1999). From 1994 to 1995, the proportion of time (during daylight hours) large dusky dolphin groups off Kaikoura were observed interacting with vessels increased from 65% to 79%, and most (84%) of these interactions were with commercial tour vessels (Barr 1997). During 1997–2003, results from vessel-based research indicated a somewhat lower visitation rate, with vessels other than the research vessel present just 55% ± 4.8% of the time large groups were monitored (93% of these were tour vessels), and swimmers interacting with dolphins 12% ± 1.7% of the time (mean ± SE, $n = 169$ groups, Markowitz 2004). This discrepancy may be due to different sampling periods, including more effort later in the day and in winter when tours were less frequent in the latter study, rather than a change in the rate of dolphin–tour interactions, as tourism operations continued to increase during the late 1990s. A voluntary midday rest period, beginning in 1999, may also have reduced the percent of time dusky dolphins spend with tour vessels. More recent investigations found that more vessels interact with dusky dolphins on the weekend compared to weekdays, resulting in less "vessel-free" time for the dolphins on weekends (Duprey et al. 2008). This is most likely a result of the influx of non-commercial recreational vessels into Kaikoura on weekends (Duprey et al. 2008).

Following a review of dusky dolphin tourism that found greater than the mandated three vessels were present 8% of the time and vessels failing to operate within DOC approach guidelines 7% of the time (Barr and Slooten 1998), the DOC placed a 10-year moratorium on commercial permits for dusky dolphin tourism off Kaikoura. Researchers are currently engaged in a 2.5-year study, reviewing dusky dolphin tourism off Kaikoura prior to the end of the moratorium in 2009. At present, there are four commercial tour operators taking tourists to interact with dusky dolphins in the Kaikoura area, two vessel-based tour companies and two aerial tour companies (Figure 11.9). The two boat-based commercial tour companies (Figure 11.9a,b) hold permits to view and/or swim with wild dusky dolphins in the Kaikoura area. One of these boat-based companies has three permits, allowing up to a maximum of 50 trips per week. Each of the three licenses allows a maximum of 13 swimmers in the water at one time. These permits are restricted to viewing or swimming with dolphins and this business focuses on interactions with dusky dolphin groups in the Kaikoura area (Figure 11.9, top). Another boat-based company holds four permits allowing a maximum 112 trips per week to view whales and dolphins in the Kaikoura area. The main focus of these tours is sperm whale watching along the Kaikoura coast, but the whale watch vessels also visit dolphin groups between whale surfacings or at the beginning/end of a tour (Figure 11.9, center). The other two commercial tour operators in the Kaikoura area have permits for viewing whales and dolphins from the air, using a fixed wing airplane and helicopter (Figure 11.9, bottom).

Both shore-based studies (Barr 1997, Yin 1999) and vessel-based research (Markowitz 2004) found that dusky dolphin groups had slower swimming speeds and rested most heavily at midday. These observations, combined with increasing vessel traffic, led to the establishment of a "rest period," a voluntary code of conduct for vessels interacting with dusky dolphins in the Kaikoura area (Department of Conservation 1999). Following meetings with the DOC in 1999, commercial tour operators agreed to a 2-hour time period, from 11:30 to 13:30, when they would not approach dusky dolphin groups during the peak summer tour season (December–March; Department of Conservation 1999). The voluntary code of conduct was meant to complement mandatory regulations already present in New Zealand's Marine Mammal Protection Regulations. The seasonal rest period reduced traffic interacting with large groups of dusky dolphins at midday, but was unsuccessful at eliminating traffic altogether, and while the code of conduct was adhered to strictly by one boat-based commercial tour operator, it was not by the other commercial tour operator (Duprey 2007).

The commercial tours to view and swim with dusky dolphins in the Kaikoura area are remarkably consistent in their structure, showing little variation over the past 11 years (Table 11.3). Dusky dolphin swim tours are on average just over 2 hours long, including travel time as well as time for swimmers to gear-up, swim with dolphins, and view and photograph the dolphins post-swim (Duprey 2007).

FIGURE 11.9 Commercial tour operations in Kaikoura, New Zealand that take passengers to see dusky dolphins include (a) dolphin swim tours, (b) whale-watching tours that visit dolphins briefly, and (c) aerial tours from fixed wing planes and helicopters.

Over 90% of swims occur with large groups (>50 dolphins), and swims are never conducted with small mother–calf nurseries (Markowitz 2004). Calves were present in large groups during 71.4% of swims monitored in 1997–1999 (Markowitz 2004) and 72.4% of swims monitored in 2006 (Duprey 2007). A commercial tour may make as few as one swim drop or as many as seven swim drops, with a mean just over three swim drops per tour (Table 11.3). Up to 13 tourists are allowed to swim from each tour vessel at any one time, but on average slightly fewer (11–12) enter the water during any one swim drop (Markowitz 2004).

TABLE 11.3 Consistency of swim-with-dolphins tours off Kaikoura, New Zealand 1997–2007

Year	Source	Tour length (hours)	Swim drops per trip	Swim duration (min)
1997–1999	Markowitz 2004	2.1	3.4	8.3
2005	Duprey 2007	2.0	3.3	8.1
2007	Markowitz and Markowitz 2007	2.3	3.5	8.6

The distance traveled between swim drops varies considerably depending on the behavior of the dolphins. For example, in 2005 the distance between swim drops ranged from 43 m to 3404 m, with a mean distance of 2.2 km traveled from the first to the last swim of the tour (Duprey 2007). In the winter months, mean tour vessel speeds during interactions with dolphin groups are higher, and the mean duration of interactions is shorter than in summer (Markowitz and Markowitz 2007), likely due to the tendency of dolphin groups to be more mobile and wide-ranging in winter (Würsig et al. 1997, 2007).

On average, less than one-fourth of the tour is spent swimming with the dolphins (mean total swim duration = 26–30 minutes of a >2 hour tour, Table 11.3). Approach speeds prior to swim drops are generally slow (e.g., in 2007 mean approach speeds were <9 km/h or <4 knots; Markowitz and Markowitz 2007). During swims, the tour vessel is most often stationary (Markowitz 2004). Following the swim, tour vessels remain with the dolphins for roughly 30 minutes on average, as passengers change into dry clothes, take photographs and listen to an onboard interpreter discuss the biology and conservation issues involving dusky dolphins and cetaceans worldwide (Duprey et al. 2008).

Although biologically significant impacts of tourism on dusky dolphins in New Zealand have not been demonstrated, some behavioral responses of dolphins to tour activities off Kaikoura have been noted. These include decreased mean swimming speed in the presence of tour vessels (Yin 1999, Markowitz 2004), an increase in noisy leaping rate (Barr 1997), as well as changes in direction and behavioral state (Würsig et al. 1997). In Kaikoura, dusky dolphins feed predominantly at night, so tour vessel visits during the daytime rarely disrupt foraging. Noisy leaps apparently can function in large groups as signals directing group movement (Markowitz 2004). Therefore, it is possible that increased leaping in the presence of vessels indicates an increase in communicative behavior, as increased whistling has also been observed among bottlenose dolphins interacting with swimmers (Scarpaci et al. 2000). Leaping is a high energy activity, and could, therefore, represent an energetic "cost" of interactions with tour vessels.

Leaping rates of bottlenose dolphins have also been documented to increase in the presence of tour vessels (Lusseau 2006). Of course, we cannot rule out the possibility that such signals are directed toward the people rather than other dolphins.

Such short-term behavioral responses may or may not indicate long-term or population level effects (Bejder and Samuels 2003), and to date there has been only limited information suggesting the possibility of long-term effects of tourism on dusky dolphins at Kaikoura. Using a surveyor's theodolite to track dolphins from shore, Brown (2000) compared dusky dolphin movement patterns to those documented prior to tourism in the late 1980s by Cipriano (1992), and found a greater tendency for dolphins to move southward later in the day. Although this does not appear to have resulted in a shift in the overall distribution of dusky dolphins around Kaikoura, this topic is currently under further investigation (Markowitz and Markowitz 2007).

Studies to date fail to show biologically significant impacts of tours on dusky dolphins off Kaikoura. Large groups (those most often visited by tours) slowed down when more vessels were present (Yin 1999, Markowitz 2004) and turned toward vessels as often as turning away (Markowitz 2004). The most common response of dusky dolphins to vessels off Kaikoura was to ride the bow of approaching vessels (Würsig et al. 1997). Observations from a research vessel indicate that dusky dolphins in large groups can be observed engaging in bow-riding about half the time vessels are present (Markowitz et al. 2004).

Some characteristics of the dusky dolphin population off Kaikoura make these dolphins well-suited to interact with this sort of tourism activity on a regular basis. The large number of individual dusky dolphins passing through Kaikoura each year (estimated $>12\,000$) with regular seasonal turnover (estimated residence time $= 103 \pm 38.0$ days, mean ± 1 standard error) probably limits the potential for long-term sensitization or chronic stress (Markowitz 2004). As the dolphins are mainly nocturnal foragers off Kaikoura (see Chapter 5), tour activities rarely have the potential to disrupt dolphin foraging, and much of the daylight hours, other than the midday period when the dolphins rest most heavily, are spent engaged in social interactions. Whether dolphin activity patterns and social structure are significantly altered by interactions with tour vessels is currently under investigation.

As we continue our studies of dusky dolphin interactions with tours along the Kaikoura coast, we find new indications of how the tour industry might be improved. Further investigation of the behavior and movement patterns of the dolphins prior to, during, and following such approaches (before–during–after method, Bejder and Samuels 2003) will be needed to confirm whether, and to what extent, various approach types change dolphin behavior. Additional factors currently under investigation include the effect of various vessel and engine types, approach and departure speeds, and interaction duration, as these vary between dolphin swim tours and whale watch tours that visit the dolphins.

While the immediate, local socio-economic benefits of dusky dolphin tourism are evident, inconspicuous benefits related to the experience and educational value of dolphin tours may have far more profound and lasting consequences, by enhancing public awareness of conservation issues. However, as "minimal effects" do not necessarily mean "no effects," continued monitoring of this growing industry is important to ensure long-term sustainability and the continued health of wild dusky dolphin populations interacting with people.

Comparing the effects of tourism in Argentina and New Zealand

Our studies indicate differences in dusky dolphin responses to tourism in Argentina and New Zealand. While dusky dolphin tourism is a much larger industry in New Zealand, the effects of tour activity appear to be less than in Argentina. We believe these different responses are the result of both ecological and management differences between the two locations.

Dusky dolphins off Kaikoura feed mainly at night when the deep scattering layer is shallow enough for dolphins to feed (never below 130 m; Benoit-Bird et al. 2004), whereas dusky dolphins in Argentina engage in coordinated foraging during the day (Würsig and Würsig 1980). Consequently, dusky dolphin foraging is unlikely to be affected by dolphin tourism off Kaikoura, but has been shown to be disrupted by tour vessels in Patagonia. Dusky dolphins gather off Kaikoura in larger groups during the daytime than they do in Argentina. Therefore, there is the potential for relatively few dolphins to interact directly with tour vessels and swimmers at any one time compared with Argentina, where groups are smaller.

In addition to these ecological differences, there are striking differences in the management of dolphin tourism between Argentina and New Zealand. Tours in Argentina are not regulated through any sort of permitting process or official guidelines. In New Zealand, all tours are operated under permit, the number of permits is limited (just two boat-based tour companies in Kaikoura), and permits are accompanied by conditions and guidelines related to approach procedures and swim operations. Clearly, these regulations have not overly hampered the development of a lucrative dolphin tourism industry, but they have apparently contributed substantially to the sustainability of the industry. Compliance with both mandatory regulations and voluntary codes of conduct should be regularly reviewed, and regulations updated based on the latest scientific information regarding what human actions affect the dolphins. Generally, however, these policies appear to be effective at minimizing the effects of tourism on dolphins, but there remain potential problems such as the effects of seasonal increases in relatively unregulated pleasure boat traffic on dolphins.

We conclude, therefore, that while tourism based on dusky dolphins has great potential for both conservation and business, it should be carefully regulated, actively managed, and consistently monitored to ensure the continued health of both dusky dolphin populations and the dolphin tour industry. If this can be accomplished, the economic incentive to local communities in preserving robust, healthy dolphin populations can be very great indeed, and the potential global benefit of conservation education through dolphin tourism is even greater.

SUMMARY AND CONCLUSIONS

Human interactions with dusky dolphins include direct harvest, fisheries, changes to habitat, and tourism. Due to limited population information combined with high levels of direct and indirect mortality from dolphin harvest and fisheries by-catch, dusky dolphins are listed by the IUCN as a species lacking sufficient information to fully assess conservation status.

Dusky dolphins have been hunted off Peru for over three decades, where they are used as fish bait and food for human consumption. During the late 1980s and early 1990s, dusky dolphins were killed by the thousands each year in directed hunts using nets and harpoons. In the mid-to-late 1990s, new legislation restricted dolphin harvests, but also made it more difficult to assess the number of dolphins harvested as hunting practices became more clandestine. Direct harvest of dusky dolphins continues today, but it is estimated that the number of dolphins harvested has dropped to less than 1000 per year. However, these numbers may still be quite high for an already depleted population. Clearly, direct harvest has been the single greatest threat to dusky dolphins off Peru, and has occurred at levels far exceeding what is likely sustainable.

Incidental mortality of dusky dolphins caught in fishing gear has been documented off Argentina, occurring at a rate which may not be sustainable. During the mid 1980s, estimated annual mortality of dusky dolphins in midwater trawls for shrimp at night off Patagonia reached nearly 8% of the present estimated population size, primarily affecting reproductive females. Greater regulation of fishing gear during the mid 1990s led to a tenfold reduction in estimated incidental mortality from this fishery. Following depletion of hake stocks in the late 1990s, there was growth of the midwater anchovy fishery, resulting in a level of dusky dolphin mortality currently unknown due to lack of monitoring. Purse seine fisheries for anchovy and mackerel in the Buenos Aires Province intentionally set nets around schools of dusky dolphins, resulting in mortality of an unknown but possibly high total number of dolphins (e.g., incidental mortality at a single port was estimated at roughly 100 dolphins per year in the 1990s). Further regulation and monitoring will be necessary to fully assess and mitigate population impacts of fisheries on dusky dolphins in Argentina. Incidental takes of dusky dolphins in fisheries off the coasts of New Zealand and South Africa are not thought to be large, and probably have a negligible effect on dusky dolphin populations in these areas.

People also interact with dusky dolphins by introducing changes to their habitats. In New Zealand, extensive development of aquaculture overlaps in some areas with dusky dolphin foraging habitat. Studies show that foraging dusky dolphins do not use areas within existing marine farms as often as adjacent areas, indicating that the farms inhibit efficient foraging. Further, marine farming can have indirect effects related to changes in plankton levels, water chemistry, and benthic ecology. Although there is no evidence for population impacts (e.g., increased mortality, decreased reproduction) due to marine farming, there is clearly potential for habitat displacement of hundreds of dolphins. Based on research and recommendations of dusky dolphin biologists, local governments and national agencies have acted to protect some areas from large-scale aquaculture development. However, there remains great economic incentive to increase the number of marine farms. Consequently, several decisions are currently under appeal in New Zealand courts.

Environmental contaminants also have the potential to affect wild dolphin populations. As high-level predators, dolphins often have relatively high concentrations of contaminants due to bioaccumulation. Because of their distribution in southern temperate waters, dusky dolphins are exposed to relatively low levels of organochlorines but higher levels of pesticides. Post-mortem analyses have shown elevated levels of both pesticides and organochlorines in dusky dolphins, which may be linked to immunosuppression leading to a higher incidence of some diseases. To date, the effects of environmental contamination on dusky dolphins appear to be relatively low, but they could be greatly amplified through interactions with other anthropogenic impacts.

Dusky dolphin tourism has grown substantially over the past two decades. Tourism has the potential to provide conservation benefits by encouraging local communities to preserve healthy dolphin populations and by providing tourists with a greater appreciation for wildlife. Tourism can also affect the daily lives of dolphins, and in extreme cases even impact the health of wild populations. We compared dusky dolphin tourism in Argentina and New Zealand, in locations where the ecology of the dolphins and the level of dolphin tour regulation vary considerably. Our studies indicated that the ecology of the animals can greatly influence the way in which they respond to tour vessels, and that appropriate regulation can help reduce the potential for biologically significant tourism impacts. We conclude that careful monitoring and regulation of tourism are important to promote development of a truly "dolphin friendly" industry.

ACKNOWLEDGMENTS

Research in New Zealand was accomplished with the help of many hard-working colleagues and full-time field assistants, including Tracy McKeown, Joana Castro, Wendy Markowitz, Cathryn Owens, Heidi Petersen, Kim Andrews, April Harlin-Cognato, Alejandro Acevedo, Cindy McFadden, Terry Brock, Monika Merriman, Laura Boren, Belinda White, Stacie Carlyle, Jody Weir, Adrian Dahood, Sierra Deutsch, Amy Beier, Heidi Pearson, Rodney Honeycutt, and Anita Murray. Additional assistance was provided by many capable, enthusiastic Earthwatch volunteers. Jack Van Berkel at the University of Canterbury Edward Percival Field Station supplied lab space and logistical support. Field research was financially supported by the New Zealand Department of Conservation, the Earthwatch Institute, the National Geographic Society, the Marlborough District Council, Massey University, University of Otago, and Texas A&M University. Boat maintenance was supplied by John Hocking, Murray Green, and Steve Howie. A number of tour operators assisted with the research, including Dennis Buurman, Lynette Buurman and Ian Bradshaw of Dolphin Encounter, John MacPhail of Wings Over Whales, Kauahi Ngapora, Lisa Bond and other staff of Whalewatch Kaikoura, Maurice Manawatu of Māori Tours Kaikoura, Danny and Lynn Boulton of French Pass Sea Safaris, Les and Zoe Battersby of Dolphin Watch Marlborough, and Ian and Pam Montgomery of Okiwi Bay Lodge. Bernd and Melany Würsig provided guidance and support throughout the work. We wish to thank Centro Nacional Patagónico (National Research Council of Argentina) and the University of Patagonia for logistical and financial support given throughout this study. We also wish to thank fisheries officers, officers of the Fishery Government Agencies of Chubut and Santa Cruz Province, the National Coast Guard and the fishing companies Harengus S.A. and Alpesca S.A. Research in Argentina was accomplished with the help of many hard-working assistants and colleagues at Centro Nacional Patagónico (CENPAT), including Griselda V. Garaffo and Mariana Degrati. Mike Meÿer, of Marine and Coastal Management, South Africa, supplied information regarding human interactions with dusky dolphins along the South African coast.

Chapter 12

Human Interactions with Dusky Dolphins: A Management Perspective

Simon Childerhouse[a,1] and Andrew Baxter[b]

[a]*Department of Marine Science, University of Otago, Dunedin, New Zealand*
[b]*New Zealand Department of Conservation, Nelson, New Zealand*

Within a 6-hour drive of each other are two completely different New Zealand marine environments where dusky dolphins and humans regularly interact. Both of these sites are characterized by the utilization of the marine environment, but in quite different ways. One of these, Kaikoura, is arguably the marine mammal eco-tourism capital of New Zealand, with dusky dolphins, sperm whales, and fur seals the target of viewing and swimming operations by tens of thousands of tourists every year. The other is Admiralty Bay, a bay in the outer Marlborough Sounds that contributes to a multimillion dollar marine farming industry but which is also an important winter feeding site for dusky dolphins. Both sites make significant contributions to the New Zealand economy, but where do the dusky dolphins fit into the picture? Well, the answer to that is not straight-forward.

As with any commercial venture, economic imperatives will determine the success or failure of the eco-tourism and marine farming sectors but at what cost to the dolphins? While there are clear economic reasons to protect dolphins where they support an eco-tourism industry—one does not want to "kill the goose that lays the golden eggs"—there are no similar economic drivers in the marine farming industry. Even with eco-tourism, the lure of increased profits can potentially sway where the balance lies, particularly with regard to the level of behavioral change or interference that is deemed acceptable.

The New Zealand Government is enthusiastic about these two economic power houses, eco-tourism and marine farming, and encourages their growth and development. At the same time, the conservation and protection of dolphins is a national priority enshrined in law. The challenge for management is in finding a balance between these two objectives, one that allows development to occur while at the same time protecting the marine environment and the species in it, including dusky dolphins. Well-meaning humans can have very different perceptions of where the balance lies, and sadly, compromise among them is often decided in the courts.

[1]Current address: Australian Marine Mammal Centre, Australian Antarctic Division, Hobart, Australia

There is no doubt that without a strong management framework, dusky dolphins in New Zealand would be significantly worse off than they are today. So when you next visit Kaikoura and see dolphins leaping offshore, or even savor a Marlborough Sounds green-lipped mussel, remember that the balancing of utilization and conservation of the marine environment represents an ongoing challenge. For this balance to be successful, the process needs the willingness of all parties to be reasonable, to compromise where possible, and to acknowledge that there may be real biologically determined bottom lines that cannot be crossed. The data feeding into decision-making must also be guided by the best available science.

Simon Childerhouse, Dunedin, New Zealand, 18 May 2009.

INTRODUCTION

Dusky dolphins (*Lagenorhynchus obscurus*; Figures 12.1 and 12.2) have a discontinuous range across the Southern Ocean (Van Waerebeek and Würsig 2002), with populations around South America, from northern Peru south to Cape Horn and from southern Patagonia north to about 36°S, including the Falkland Islands; off south-western Africa from False Bay to Lobito Bay, Angola; and in New Zealand waters, including off Chatham and Campbell Islands. Populations also inhabit waters surrounding oceanic islands of the Tristan da Cunha archipelago in the mid-Atlantic, the Prince Edward Islands, and Amsterdam Island in the southern Indian Ocean (Van Waerebeek et al. 1995, see also Chapter 2). Dusky dolphins have been confirmed from southern Australian waters but are rare there (Gill et al. 2000). This species is regarded as coastal or "semi-pelagic," and is usually found over the continental shelf and slope (Jefferson et al. 1993, Aguayo-Lobo et al. 1998) but has also been reported in offshore, oceanic waters. There are few reliable estimates of abundance for this species and therefore its status is listed as "data deficient" by the International Union for Conservation of Nature (IUCN 2008).

Given this widespread southern hemisphere and primarily coastal distribution, dusky dolphins regularly interact with humans. These interactions can take many forms, such as large-scale directed killing for fish bait in Peru (Van Waerebeek et al. 1997), incidental drowning in trawl operations in Argentina (Crespo et al. 1997b), or loss of habitat due to expansion of aquaculture in New Zealand (Lloyd 2003), through to interactions often perceived as benign, such as tourism operations in New Zealand (Barr and Slooten 1999). These interactions are described in more detail in Chapter 11. The scope, magnitude, and impact of these and other dusky dolphin–human interactions vary considerably, as do the management approaches that have been used to address them.

This chapter describes the differing approaches that have been used to manage dusky dolphin–human interactions across the southern hemisphere, with a particular emphasis on management in New Zealand waters. The approach taken is to examine both national and international frameworks and summarize management actions taken. Two New Zealand case studies are also discussed in detail.

FIGURE 12.1 Dusky dolphins off Kaikoura, New Zealand (photo by Dennis Buurman).

FIGURE 12.2 A pod of dusky dolphins at Kaikoura, New Zealand (photo by Kim Westerskov).

MANAGEMENT FRAMEWORKS

National frameworks

New Zealand

Marine mammals in New Zealand waters are managed under a range of statutes, the foremost of which are the Marine Mammals Protection Act 1978 (MMPA) and the Marine Mammals Protection Regulations 1992 (MMPR). With a clear focus on marine mammal welfare, these two pieces of legislation provide solely for the protection, conservation, and management of marine mammals. Neither promotes tourism or other uses, whether commercial, recreational, cultural, or scientific, nor do they address related matters such as tourism quality control or people safety.

Both the MMPA and the Regulations are administered by the Department of Conservation (DOC), the principal government agency responsible for marine mammal welfare in New Zealand. As part of this role, the DOC responds to stranded or injured marine mammals, including the occasional dusky dolphin, often with public assistance.

Any activity that would intentionally disturb or harm a marine mammal requires a permit under the MMPA. The Act also allows for the establishment of marine mammal sanctuaries that, as the name implies, afford a higher level of protection for marine mammals within their boundaries. Sanctuaries have been used to protect threatened or endangered species such as Hector's dolphins (*Cephalorhynchus hectori*) and New Zealand sea lions (*Phocarctos hookeri*), though none have been created specifically for the protection of dusky dolphins or other non-threatened species. Population management plans can also be developed for managing and mitigating fishing-related mortality; however, no population management plans have ever been finalized.

The focus of the MMPR is on managing the effects of marine mammal viewing. Their particular aim is to prevent adverse effects on and interference with marine mammals. There are two principal mechanisms for achieving this.

First, the MMPR stipulates operating conditions for commercial operators (or anyone else) when in the vicinity of whales, dolphins, or seals. Many of these conditions relate to the operation of vessels around marine mammals, notably speeds, orientation, approach distances, and numbers of craft. Unlike for whales, the Regulations do not stipulate a minimum approach distance to dolphins; however, boats are required to approach dolphins from behind and to the side and vessels cannot circle dolphins, obstruct their path, or cut through pods (Figure 12.3). A maximum of three vessels may be within 300 m of dolphins and these must keep to a slow speed (i.e., "no-wake" or no faster than the slowest dolphin) unless departing, when speed may be gradually increased up to 10 knots to out-distance dolphins. Swimming with dolphins is allowed, though not with pods with very young calves.

Second, the Regulations establish a permit regime for anyone wanting to take people to view or come into contact with marine mammals commercially. Various assessment criteria are listed, including the need to avoid significant adverse effects on the normal behavior of marine mammals, and consideration of cumulative effects, educational value, and operators' experience and knowledge. Over and above the standard operating rules, commercial tourism effort can be controlled through conditions on permits and limits on the number of permits issued.

In addition to the DOC, regional councils (local government) and the Ministry of Fisheries also have key responsibilities in the management of marine mammals in New Zealand waters. Under the Resource Management

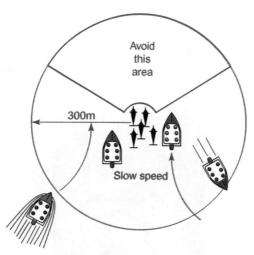

FIGURE 12.3   Basic dolphin-watching rules in New Zealand waters. Refer to text for further explanation (from Department of Conservation, New Zealand).

Act 1991 (RMA), regional councils (including some district councils) over-see the management of a variety of uses and activities that might directly or indirectly affect marine mammals. These include, for example, coastal naviga-tion, discharges, structures, and marine farming. The Ministry of Fisheries is the Government agency responsible under the Fisheries Act 1996 for the sus-tainable management of fishing in New Zealand, which includes managing the adverse effects of fishing on protected species such as marine mammals.

The Treaty of Waitangi and the relationship that Māori, the indigenous peo-ple of New Zealand, have with the natural world are recognized through vari-ous New Zealand statutes. *Kaitiakitanga* (guardianship) over natural resources is a central part of Māori culture, with many *iwi* (Māori tribes) regarding dusky dolphins and other marine mammals as *taonga* (treasures) (Box 12.1). Consultation with *iwi* over the management of natural resources is, therefore, a common and important theme across central and local government agencies.

While on the face of it all these legislative tools and processes provide a high degree of protection for dusky dolphins and other marine mammals in New Zealand waters, they do not confer absolute protection. Importantly, both the Fisheries Act and the RMA are *enabling* pieces of legislation, providing for use and development of natural resources in addition to their protection.

The purpose of the Fisheries Act is to "provide for the utilization of fish-eries resources while ensuring sustainability." While it also aims to avoid, remedy, or mitigate adverse effects of fishing on the aquatic environment, the Fisheries Act only seeks that "associated or dependent species should be main-tained above a level that ensures their long-term viability." Thus, a certain level

BOX 12.1

A Māori Perspective on Nature and the Sea

Maurice Manawatu, Māori Tours Kaikoura, Kaikoura, New Zealand

Maurice Manawatu speaking in front of god Maui, who fished up the North Island of New Zealand while standing on the canoe of the South Island. Maurice is the head of Māori Tours Kaikoura, a one-half day excursion to share Māori history, legends, appreciation, and natural healing (photo by Bernd Würsig).

> Toi tu te Marae o Tane
> Toi tu te Marae o Tangaroa
> Toi tu te Iwi
> [If the house of Tane (God of the Forest) survives,
> Likewise the house of Tangaroa (God of the Ocean),
> The people will live on]

To explain about Māori connections to the environment, we need to start at the beginning of time. Māori people are tribal and the creation story has varying versions depending on the tribe telling the story. Even so, the essence of the story is the same throughout New Zealand.

In the beginning, Mother Earth (Papatuanuku) was married to the Sky Father (Ranginui). Although our tribe says she was married first to the God of the Ocean (Tangaroa). Mother Earth and her husband the Sky Father were in a parental embrace, and their children were trapped between them; there was no daylight. The children discussed separating their parents to create light, and after much debate, they decided they would. Each child tried without success until Tane (the God of the Forest) managed to separate his Mother and Father (the Earth and Heaven).

With the separation successful, Tane set about trying to create humans. He and the other gods mated with different female spirits, and earthly plants, animals, and our marine life were created. Tane then went to Mother Earth for help creating humans, and she provided him with all the richness of the Earth that he and the other gods used to create the first human being; a woman named Hine Ahuone.

This story, handed down from generation to generation, tells us that we are related to our natural environment. The gods who created earthly plants, animals and marine life are the same gods who created us; we are one and the same.

The gods also created the spectacular physical environment we live in today, and the story of Aoraki is one that speaks of this. He was a god who came down from the heavens after the separation, to see for himself the work Tane and the other gods had completed. Before leaving the heavens, Aoraki said his Karakia (prayers) to give him safe passage to and from the heavenly realms. Unfortunately, he said the wrong Karakia to return, and as a result he and his canoe crew hit a reef and they all turned to stone. The stone of Aoraki is what we know as Mount Cook today. His crew are part of the Southern Alps, the mountain range dividing the east and west coast of the South Island of New Zealand. Stewart Island to the south is the anchor of that canoe.

The Aoraki story speaks of the South Island, and is part of the tradition of the Ngai Tahu, the South Island's main tribe. Each of the 18 subtribes that make up Ngai Tahu have their own stories that are more specific to their areas. One such story from my people, the Ngati Kuri, subtribe of Kaikoura, is about Matamata.

Matamata was a sperm whale who befriended one of my ancestors. Whenever the ancestor sailed the oceans, he was always accompanied by Matamata, and it is said that Matamata was his protector. When that ancestor passed away, Matamata disappeared, but the story tells us that whenever a direct descendant of my ancestor is in danger or about to pass away, Matamata will make an appearance again.

Our links to the marine mammals of Kaikoura are an important part of who we are today. As the Matamata story shows, they were our guardians in bygone days, and they remain part of our core. One of the modern success stories of our tribe is the establishment of Whale Watch Kaikoura; a world renowned company that takes visitors to share in the delights of seeing sperm and other whales, dusky and other dolphins, fur seals and a wide variety of seabirds in their natural habitat.

These creatures provide an economic base for our people to prosper. Yet in today's world, where the very existence of nature is often at risk, the roles need to go both ways. Now we must also take on the role of protector, and ensure that just as they are our guardians, we return that favor so that whales and dolphins and other creatures of the sea continue to exist.

Māori culture is built on traditions and oral history, and these stories are part of our identity and the connection we have forged with the environment in which we live. We believe that when we walk, our ancestors walk with us, through times of difficulty and success.

We are taught from birth that if we remove the core of ourselves and ruin the fundamental nature of the world we live in, then there will be nothing left for our children, their children and the generations to come.

No Reira Nga Kaumatua nga pakeke nga tamariki mihinui kia koutou katoa
(To the elders, the adults, and the children, I extend greetings to you all)

FIGURE 12.4   Dusky dolphin drowned in a set net, Golden Bay, New Zealand (photo by Louise de Jaeger).

of fishing-related mortality is accepted under this Act (Figure 12.4), and it is generally not an offense to accidentally or incidentally harm marine mammals while legitimately fishing, provided the incident is reported. Nonetheless, the 2008 ban on set netting and trawling from many coastal areas around the South Island and along the north-western coast of the North Island to protect the endangered Hector's dolphin illustrates both the significance of the Fisheries Act and the role of the Ministry of Fisheries in the management of marine mammals around New Zealand.

In a similar vein, the RMA manages use, development, and protection of resources in a way that "enables people and communities to provide for their social, economic, and cultural well being." Important sustainability caveats (i.e., meeting the needs of future generations, safeguarding life-supporting capacity, and avoiding, remedying, and mitigating adverse effects) and various "matters of national importance" (notably the preservation of natural character and the protection of significant habitats of indigenous fauna) are built into the RMA. The New Zealand Coastal Policy Statement (a national policy promulgated under the RMA) also provides guidance about the protection of significant habitats, communities, and other natural values. Yet without detracting from the importance of all these guiding principles, application of the RMA generally demands a weighing up of all factors—social, economic, and environmental—and decisions often result in a balancing of need versus protection.

Dusky dolphins are not listed as being threatened in New Zealand, and management of dusky dolphin–human interactions in New Zealand to date has largely been limited to managing the effects of marine mammal watching and, to a lesser extent, aquaculture. Both of these issues are discussed in more detail as case studies later in this chapter. There are no marine mammal sanctuaries or fisheries management rules aimed specifically at the protection of dusky dolphins, though the fisheries restrictions implemented in 2008 to protect Hector's dolphins could have positive spin-offs for those dusky dolphins living in the same general vicinity. Some captive dusky dolphins were held in captivity in the 1970s and 1980s, but none are today.

Australia

In Australia, the Environment Protection and Biodiversity Conservation (EPBC) Act 1999 is the primary piece of federal legislation that manages cetacean–human interactions. The EPBC Act seeks to ensure adequate marine mammal protection based on sound science, while also allowing appropriate opportunities for the development of sustainable marine industries and recreational practices. Conservation and management of threatened marine species involves determining the threats faced by marine species; preventing, mitigating, and/or managing those threats; and supporting the recovery of species until they can be removed from the EPBC list of threatened species. In addition to federal legislation, State and Territory Governments are responsible for conservation and protection of cetaceans in coastal waters, out to the 3 nautical mile limit. This includes responding to stranded and entangled cetaceans and managing whale- and dolphin-watching activities.

Threatened species are listed under the EPBC Act. Recovery Plans and Action Plans are developed for specific species and groups of species (e.g., Bannister et al. 1996, Department of the Environment and Heritage 2006) to identify and address threats and improve species conservation and protection. The EPBC Act establishes the Australian Whale Sanctuary, giving high level protection to whales, dolphins, and porpoises within Australian waters. The Sanctuary includes all Commonwealth waters from the 3 nautical mile state waters limit to the boundary of the Exclusive Economic Zone (i.e., out to 200 nautical miles) and further in some places. Within the Sanctuary it is an offense to kill, injure, or interfere with a cetacean. Severe penalties apply to anyone convicted of such offenses. All States and Territories also protect whales and dolphins within their waters.

Activities in the Australian Whale Sanctuary that may impact on whales, dolphins, or porpoises may require a permit. It is also an offense to "injure, take, trade, keep, move, harass, chase, herd, tag, mark or brand a whale, dolphin or porpoise" in the Australian Whale Sanctuary without a permit. All "takes" (including by-catch) of marine mammals are required to be reported to the Federal Government. There are existing mechanisms for reporting by-catch. Once this occurs, the nature and extent are considered, and if the government deems it necessary to reduce or eliminate by-catch, they work with the fishing industry to modify practices to minimize by-catch issues (e.g. By-catch Action Plan).

Permits to "take" a marine mammal may be issued only after all impacts of the activity have been taken into account. Permits cannot be issued to kill a whale, dolphin, or porpoise or to take one for live display. Any organization planning to undertake activities that may have a significant impact on whales, dolphins, and porpoises must seek approval from the Minister for the Environment, Heritage and the Arts. Permission is only granted if the activity meets rigorous assessment and approval criteria to ensure that no cetacean populations are harmed and that their habitat is protected. At present there are no accredited Management Plans under the EPBC Act for the by-catch of any marine mammals.

Dusky dolphins are uncommon in most Australian waters and there appear to be few dusky dolphin–human interactions. They are listed as migratory under the EPBC Act and the species does not have a Recovery Plan. There are no dusky dolphins held in captivity in Australia. There has been no specific management action taken to protect or conserve this species at Federal, State, or Territorial Governments in Australian waters.

South Africa

All dolphins in South Africa are protected under the Marine Living Resources Act (MLRA) 1998. Within this Act, there are several Regulations that promote the protection of dolphins, specifically:

> No person shall, except on the authority of a permit, engage in fishing, collecting, killing, attempting to kill, disturbing, harassing, keeping or controlling of, or be in possession of, any dolphin or porpoise or any part or product thereof at any time.

The definition for disturb or harass includes:

> deliberate driving (of) a fishing vessel or vessel through a school of dolphins or porpoises.

> No person shall (a) feed any wild dolphin or porpoise; or (b) advertise or engage in any fishing vessel or vessel trip, whether for gain or not, which is intended to provide for a swim-with-dolphins experience.

Best (2007) reports that dusky dolphins are occasionally caught as by-catch in several South African fisheries, including inshore set nets and purse seines (see also Chapter 14). While fishers who incidentally catch marine mammals are not automatically exempt from prosecution under the MLRA, in practice they are unlikely to be prosecuted if the by-catch was accidental and reported. There are no dusky dolphins in captivity in South Africa. There is no specific management or conservation of dusky dolphins in South Africa, other than their protection under the MLRA.

Peru

Van Waerebeek et al. (1997) summarized management action by the Peruvian Government in response to high levels of by-catch and directed hunting in the 1980s and 1990s of small cetaceans, including dusky dolphins. The Peruvian

Ministry of Fisheries (MIPE) first outlawed the exploitation of small cetaceans in 1990, but this rule was widely ignored and poorly enforced. An official report stated that "the decree has not totally eliminated the capture of small cetaceans in some ports" (Arias-Schreiber 1995). In 1994, MIPE issued a new Ministerial decree, in essence reiterating the 1990 ban but in more strict terms. Efforts to publicize and enforce this measure were more successful, and cetacean meat gradually disappeared from the public market (Van Waerebeek et al. 1997). In 1996, a common law went into effect which prohibited all captures and trade for seven odontocete species, including dusky dolphins. A subsequent amendment exempted live-capture fisheries for dolphinaria from the ban. The retention of small cetaceans, including dusky dolphins, caught either incidentally or deliberately in fisheries is still banned by Peruvian law (Majluf et al. 2002). While the number of dusky dolphins appearing in fish markets has declined significantly, by-catch is still an issue in some places at the time of publishing this book.

Chile

All small cetaceans have been officially protected in Chile since 1977, with dusky dolphins considered "Vulnerable" (Aguayo-Lobo et al. 1998). Under the amended *Ley de Caza* (Hunting Law) all cetaceans are considered a "manageable resource" and their exploitation is regulated by the *Ley General de Pesca y Acuicultura* (Fisheries and Aquaculture Law) enforced by the *Subsecretaria de Pesca* (Sub. Pesca., i.e., National Fisheries Service). The *Decreto Exento* (Decree) No. 225 (passed in November 1995 by the Sub. Pesca.) established a 30-year ban on killing of small and large cetaceans (and most other marine mammals) in Chile. Live take of marine mammals for exhibitions or research purposes is permitted if authorized by the Sub. Pesca. Chile has also set up at least one Marine Protected Area with a cetacean focus (Francisco Coloane in the Chilean fjords near Punta Arenas established in 2003) but none have been created specifically for dusky dolphins. There is no specific management of dusky dolphin–human interactions in Chile.

Argentina

There is no specific legislation or management of dusky dolphins in Argentina, although they are caught in several fisheries (Crespo et al. 1997b). Marine mammals are not seriously considered as part of the equation when fishery management is discussed, even though it has been demonstrated that by-catch levels have been unsustainable, and may still be (Crespo and Hall 2001, Dans et al. 2003a, 2003b). Dusky dolphin watching, which takes place in Patagonia (Coscarella et al. 2003), is nominally managed at the provincial level. However, in effect, dolphin watching is unregulated and lacks a management scheme, although recommendations about such a scheme have been made by local researchers (E. Crespo, personal communication).

International frameworks

Convention on International Trade in Endangered Species of Wild Fauna and Flora (CITES)

The Convention on International Trade in Endangered Species of Wild Fauna and Flora (CITES) was established to regulate international trade in species threatened with extinction. Its aim is to ensure that international trade in specimens of wild animals and plants does not threaten their survival. At present, the Convention has more than 170 signatories. Dusky dolphins are listed in Appendix II, which means that they are considered as a species that is not necessarily threatened with extinction, but trade in them must be controlled in order to avoid utilization incompatible with their survival. Therefore, any international trade must be regulated by signatory countries through the CITES Secretariat and Conference of Parties. At present, there is no known international trade of dusky dolphins.

Convention on Migratory Species (CMS)

The Convention on Migratory Species (CMS) is an international treaty aiming to conserve terrestrial and marine migratory species and their habitats on a global scale. Dusky dolphins are presently listed in Appendix II of the CMS. Dusky dolphins meet both criteria for listing under the CMS; they are a migratory species, and have an unfavorable conservation status or would benefit significantly from international cooperation. While there are presently no specific Agreements or Memoranda of Understanding that relate to dusky dolphins, a Regional Agreement for South American small cetaceans has been proposed that, if successfully completed, will cover the species.

International Whaling Commission (IWC)

The International Whaling Commission (IWC) was set up in 1946, under the International Convention for the Regulation of Whaling. The purpose of the Convention is to provide for the proper conservation of whale stocks and thus make possible the orderly development of the whaling industry. The IWC has always regulated the catches of the large whale species, but the smaller species of whales, dolphins, and porpoises (commonly known as "small cetaceans") are not covered in the Convention. Hence, the IWC has no ability to regulate direct and incidental catches of small cetaceans, although member governments are split on the exact interpretation. However, the IWC Scientific Committee frequently draws attention to potential conservation problems, and the Commission generally promotes cooperation between coastal and range states to conserve and manage small cetaceans. Although the Commission does not set regulations for small cetacean management, the IWC Scientific Committee addresses matters related to the conservation of small cetacean species at its annual meetings, and a review of dusky dolphins was last undertaken in 1996. There has been no formal consideration of the species since this time, but there have been many scientific research papers relating to dusky dolphins presented and discussed by the IWC Scientific Committee.

Convention on the Conservation of Antarctic Marine Living Resources (CCAMLR)

The Convention on the Conservation of Antarctic Marine Living Resources (CCAMLR) came into force in 1982, as part of the Antarctic Treaty System, in pursuance of the provisions of Article IX of the Treaty. It was established mainly in response to concerns that an increase in krill catches in the Southern Ocean could have a serious effect on populations of krill and other marine life; particularly on birds, seals, and fish, which depend on krill for food. Based on a review by Van Waerebeek et al. (2004), Leaper et al. (2008) developed a list of species that are potentially ecologically important south of the CCAMLR boundary (between 45°S and 60°S depending on longitude). This list did not include dusky dolphins because the southern limit of their distribution does not appear to overlap with the CCAMLR boundaries. Given this conclusion, it is unlikely that the CCAMLR will deal with issues related to dusky dolphins.

International Union for the Conservation of Nature (IUCN)

The IUCN is the world's oldest and largest global environmental network and produces the "Red List" of threatened species. The present Red List has dusky dolphins listed as "data deficient" due to a lack of abundance and trend information, and the fact that regional populations have not been thoroughly and evenly evaluated (IUCN 2008). The Cetacean Specialist Group of the IUCN plays an important role in identifying problems of conservation of the world's dolphins, whales, and porpoises, and brokering approaches to their solution. Their summary of the global situation is given in *Dolphins, Whales and Porpoises: 2002–2010 Conservation Action Plan for the World's Cetaceans* (Reeves et al. 2003). The IUCN Action Plan highlights issues of concern for dusky dolphins (e.g. incidental by-catch in fisheries, deliberate killing for bait), and provides potential solutions to cetacean conservation problems.

Other

There are several other potential international frameworks for the management of dusky dolphins. For example, the United Nations Environment Program (UNEP) and the Food and Agriculture Organization of the United Nations (FAO) co-developed a Global Plan of Action for Marine Mammals. This plan includes reference to, but little specific information about, dusky dolphins. The central goal of the plan was to bring governments together to agree and harmonize their policies for marine mammal conservation.

CASE STUDIES OF NEW ZEALAND MANAGEMENT APPROACHES

Management of tourism on dusky dolphins at Kaikoura

Kaikoura, on the east coast of the South Island of New Zealand (Figure 12.5), is arguably New Zealand's marine mammal watching capital (see also Chapter 13). There are several reasons for this distinction, foremost being the abundance and variety of marine mammals often encountered close to shore. The Kaikoura Canyon is a key factor in this pattern, with water depths of 800–1000 m within only a few kilometers offshore from Goose Bay (Figure 12.6). Kaikoura is located on a major tourist route and visitors have a high probability of success-ful encounters with marine mammals. Sperm whales (*Physeter macrocepha-lus*), dusky dolphins, and New Zealand fur seals (*Arctocephalus forsteri*) are

FIGURE 12.5 The Kaikoura coast, New Zealand.

FIGURE 12.6 The Kaikoura Canyon and neighboring bathymetric features (from National Institute of Water & Atmospheric Research, New Zealand).

the main species seen, though Hector's dolphins are also resident and several other species migrate along the coast annually.

While dusky dolphins occur close to shore during the day, sometimes feeding on fishes, most feeding occurs at night in deeper water offshore on mesopelagic fishes and squid (Benoit-Bird et al. 2004, see also Chapter 5). Activity during the day is often social, with resting, playing, and sexual behavior being frequently observed (Würsig et al. 1997, Barr and Slooten 1999, see also Chapter 8). Dusky dolphins alternate between various behavioral phases during the day, seemingly most receptive to human interaction during their periods of play (Barr 1997, Barr and Slooten 1999). At other times, notably when they are resting or feeding, they are more reluctant to interact with boats or people in the water. Nursery pods (i.e., separate groups with mothers and young calves) are particularly nervous of approaches by boats.

Two boat operators offer trips to view dusky dolphins along the Kaikoura coast. Based from a small harbor at South Bay (Figures 12.5 and 12.7), Dolphin Encounter and Whale Watch Kaikoura are permitted to operate as far north as the Clarence River, but in practice there is relatively little dolphin watching north of Kaikoura Peninsula. Most activity is to the south, centered on the region between the Peninsula and the Conway River, especially in the vicinity of the Kaikoura Canyon (Figures 12.5 and 12.6). Dolphin watching sometimes occurs even further south when dolphins (and whales) are not found closer to South Bay.

Dolphin Encounter specializes in vessel-based dolphin viewing and swimming (Figures 12.8 and 12.9), targeting dusky dolphins (see also Chapter 13). Operating up to three vessels at a time and a maximum of eight trips per day, Dolphin Encounter offers trips throughout the year, though most activity occurs from November to May when sea temperatures are the warmest and there

FIGURE 12.7   South Bay boat harbor, Kaikoura, New Zealand (photo by Andrew Baxter).

FIGURE 12.8   Commercial dolphin viewing at Kaikoura, New Zealand (photo by Dennis Buurman).

are more tourists. A lower level of winter viewing also occurs. Trips last for approximately 3 hours, with dusky dolphins seen in pods ranging from small groups through to hundreds and sometimes in excess of 1000 individuals.

Although sperm whales are the focus of Whale Watch Kaikoura's trips, dusky dolphins are also a feature, normally on the homeward journey. Whale Watch Kaikoura operates four vessels (Figure 12.10), with each boat allowed up to four 3-hour trips in a day.

The vessels used by the whale and dolphin watching fleet at Kaikoura have changed significantly over the years. During the initial developmental stages of

FIGURE 12.9   Swimming with dusky dolphins at Kaikoura, New Zealand (photo by Dennis Buurman).

FIGURE 12.10   Viewing dusky dolphins on the homeward journey of a whale-watch trip, Kaikoura, New Zealand (photo Whale Watch Kaikoura Ltd).

the industry, the vessels were typically small, fast, and highly maneuverable, ranging from 6 to 13 m in length. The exposed and changeable nature of the Kaikoura coast's sea conditions, coupled with the need to remove the vessels from the water each night (on trailers), dictated the use of such vessels. However, vessel size increased progressively during the 1990s, especially following the development of a sheltered harbor facility at South Bay (Figure 12.7). The whale and dolphin watching fleet now comprises much larger vessels: 18 m for Whale Watch Kaikoura (Figure 12.10) and ranging from 8 to 13 m for Dolphin Encounter (Figure 12.8). There has also tended to be a shift from mono-hull vessels to catamarans. Both water-jet propulsion and propeller-driven craft are used, though all of Whale Watch Kaikoura's vessels now have jet propulsion.

FIGURE 12.11 An aircraft used for whale and dolphin watching, Kaikoura, New Zealand (photo Whale Watch Air Ltd).

Three operations (four permits) offer scenic flights off the Kaikoura coast to view marine mammals, concentrating mostly on sperm whales but also viewing dusky dolphins. Flights provide a totally different view of the animals and are an alternative when the sea conditions are too rough for the vessels to operate, or for those with limited time or who are less inclined to brave the open sea. Fixed wing aircraft (two permits; Figure 12.11) or helicopters (two permits) are presently in operation. These permits are not limited to a set number of daily trips, though each permit may only operate one aircraft at a time.

The DOC monitors commercial operators' compliance with the MMPR and permit conditions, mainly by randomly placing staff incognito aboard permitted vessels. This "secret shopper" approach has proven to be the most cost effective and workable method for compliance monitoring at Kaikoura, being much more discrete and requiring fewer staff and less operational resources than a patrol vessel. Any public complaints are also investigated, either directly or with follow-up surveillance. Although there is a DOC boat stationed at Kaikoura, the exposed nature of the coast can limit its use and patrol work is mainly carried out over the summer when sea conditions are the most favorable and recreational boating activity is greatest.

Educating recreational boat users about appropriate behavior around dusky dolphins or other marine mammals at Kaikoura has been, and remains, a significant challenge. The Kaikoura coast is a very popular fishing, diving, and boating location for locals as well as visitors from further afield, notably neighboring Marlborough and Canterbury. On a good summer's day, there can be over 100 recreational vessels along the coast, and Kaikoura's reputation for marine mammal watching has resulted in a growing number of people in small boats seeking encounters with marine mammals. There are no licensing

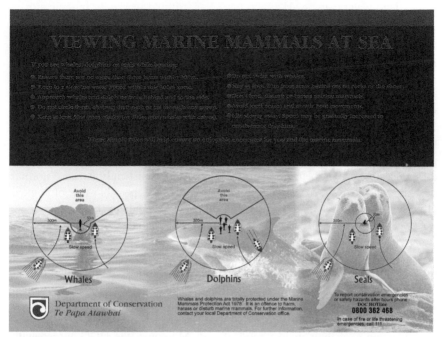

FIGURE 12.12 Marine mammal watching sign used at Kaikoura (from Department of Conservation, New Zealand).

requirements for recreational boats in New Zealand, and most boat owners do not belong to clubs, thus making them a difficult target audience for educational campaigns. Despite the presence of strategically placed signs outlining appropriate behavior around marine mammals (Figure 12.12) and a similar pamphlet also being available, observations by DOC staff suggest that the level of understanding by recreational boat users about appropriate boating behavior in the vicinity of marine mammals is generally low.

Interest by others wanting to set up new dolphin-watching operations at Kaikoura grew over the 1990s. Following a similar approach to that adopted for whale watching, DOC did not issue any new dolphin-watching permits or allow existing permits to expand until research on the impacts of tourism on dusky dolphins was carried out. This research was undertaken by researchers from Otago University, with field work carried out between 1993 and 1995 (Barr 1997). The major findings, conclusions, and management recommendations from this study are summarized in Barr and Slooten (1999).

The study found very high boat–dusky dolphin interaction rates, with dolphins accompanied by boats for a large part of the time during daylight hours (65% of observation time in the first year increasing to 78% in the second year). Several behavioral differences were noted, including: more leaps and slaps in the presence of boats from late morning; pods accompanied by boats were more compact during mid to late afternoon; and there were significantly more leaps and directional changes when a mix of different boat types (tourism vessels, recreational boats, and commercial fishing boats) was present after mid morning. Dolphins appeared to be more sensitive to disturbance after late morning, and the most obvious changes in dolphin behavior tended to occur when the operating regulations were not followed.

In most cases, the statistical power for detecting differences between when boats were and were not present was low, partly because dolphins were unaccompanied by boats for only a small proportion of the time. In some cases, even large differences in dolphin behavior were not detectable statistically because normal dolphin behavior was so variable. In other words, it is possible that biological effects were occurring, but the study was unable to detect them statistically.

Barr and Slooten (1999) made several recommendations based on their research findings, notably reducing boat activity from late morning to early afternoon and placing a cap on the amount of dolphin-watching effort. As a direct consequence of this study, Kaikoura dolphin-watching boat operators agreed to implement a voluntary time-off period between 11:30 am and 1:30 pm each day over the peak summer period (from December 1 to March 31), thus providing dusky dolphins with a 2-hour period of less intensive dolphin viewing over their main period of rest. Moreover, the DOC agreed with the researchers' recommendations that there should be no further increase in the amount of dolphin-watching effort at Kaikoura, declaring a 10-year moratorium on new dolphin-watching permits at Kaikoura. Some minor changes to swimmer numbers were also implemented, largely for administrative purposes.

The moratorium on new dolphin-watch permits expired in November 2009. In anticipation, the DOC commissioned a 3-year research program into the effects of current and potential future tourism activity on dusky dolphins at Kaikoura. This investigation was completed in mid 2009 but it's findings were not available at the time of writing this chapter. Its key research objectives were: quantifying the type, level, and operational extent of existing dusky dolphin-watching activity; determining and assessing the effect of this activity on the welfare of dusky dolphins, including the efficacy of the voluntary time-off period; and assessing the possible effects of any future expansion in commercial dusky dolphin viewing and swimming at Kaikoura.

Management of the impacts of mussel farming on dusky dolphins in the Marlborough Sounds

The changing face of mussel farming

Founded on the pioneering spirit of a few entrepreneurial individuals over the 1970s (Dawber 2004), mussel farming has developed into a significant industry within New Zealand. Mussel farming produced 97 000 tonnes of green-lipped

FIGURE 12.13 Long-line culture of green-lipped mussels, New Zealand (photo by Kim Westerskov).

mussels (*Perna canaliculus*; Figure 12.13) in 2006, equating to an estimated NZ$224 million in sales (New Zealand Aquaculture Council 2007). In the same year, mussels contributed around 63% of total aquaculture sales in New Zealand, the balance coming predominantly from the farming of chinook salmon (*Oncorhynchus tshawytscha*) and Pacific oysters (*Crassostrea gigas*).

The aquaculture industry has a goal of expanding total marine farming production in New Zealand to NZ$1 billion per annum by 2025 (New Zealand Aquaculture Council 2006). The New Zealand Government has also signaled its commitment to and support for the aquaculture industry (Ministry of Economic Development 2007, Ministry of Fisheries 2008).

Although mussels were grown initially on lines suspended from large rafts, the advent of long-line culture in the mid–late 1970s rapidly changed the face of the industry (Dawber 2004). This development, together with other improvements in culture and harvesting technology (Figure 12.14), led to a fairly rapid growth of mussel farming, especially through the early 1980s. Close to two-thirds of New Zealand's mussel farming occurs in the Marlborough District in the north of the South Island, and virtually all of this is in the Marlborough Sounds.

Despite the effects of two moratoria on new applications (1996–1999 and 2002–2004) and varying economic conditions, there was a steady overall increase in the amount of permitted marine farming space in Marlborough over the three decades from 1977 to 2007, expanding from around 370 ha in 1980 to 3000 ha in 2007 (Figure 12.15). Spatially, mussel farming dominates the marine farming industry in Marlborough.

Historically, mussel farming in the Marlborough Sounds comprised relatively small individual farms (generally around 3 ha) separated by a gap of 50 m or more and located in sheltered bays rather than along more exposed coastlines (Figure 12.16).

FIGURE 12.14   A mussel harvester at work in Admiralty Bay, New Zealand (photo Department of Conservation).

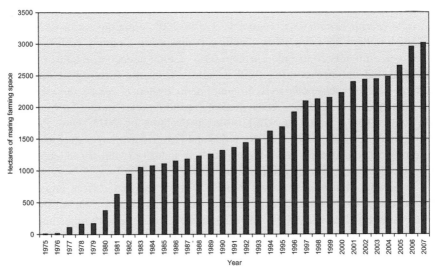

FIGURE 12.15   Hectares of marine farming space permitted in Marlborough, New Zealand (from New Zealand Marine Farming Association).

Farms were developed in a ribbon-like fashion along the coastline from 50 m to a maximum of 200 m offshore from the low-water mark, in depths ranging from around 10 to 40 m (Figures 12.17 and 12.18). This general pattern, which more or less continued until the late 1990s, was driven initially by navigational considerations and a desire by regulatory authorities to keep marine farming close to shore and "tucked away" from the middle of bays. Mussel farming was concentrated in Pelorus Sound, Croisilles Harbour, Port Underwood, and some

FIGURE 12.16   Mussel farm in Admiralty Bay, New Zealand (photo Department of Conservation).

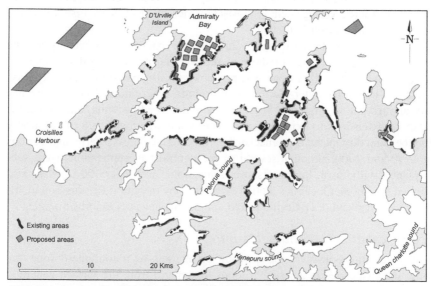

FIGURE 12.17   Existing mussel farms and new applications (1999–2000) in Pelorus Sound and the outer Marlborough Sounds, New Zealand (adapted from Lloyd 2003).

large bays in the outer Sounds, while Queen Charlotte Sound was kept relatively free of farms in recognition of this area's high recreational use.

However, following the lifting of the first moratorium in July 1999, applications for mussel farms beyond the traditional 200 m "limit" began to be received, heralding a new era in the development of marine farming in Marlborough. This shift in emphasis was the result of nearshore space zoned

FIGURE 12.18   Mussel farms in Admiralty Bay, New Zealand (photo by Robin Vaughn).

for marine farming becoming fully utilized, mussel farmers' ongoing search for productive water space and the industry's realization that the Marlborough Sounds Resource Management Plan was not such an obstacle to offshore marine farming as was previously thought.

Mussel farmers began to apply for offshore extensions to existing marine farms, as well as for large offshore or mid-bay farms, many 40 ha or more in area (Figure 12.17). This trend was perhaps best illustrated in Admiralty Bay; a 2750 ha bay in the outer Marlborough Sounds, where mussel farmers eventually applied for 14 large mid-bay farms (most exceeding 40 ha) and numerous offshore extensions to existing farms. The result was that much of Admiralty Bay became dominated by a checker-board pattern of mussel farm applications in 1999 and 2000, although several of the central bay applications were subsequently withdrawn. Large offshore farms, some well over 100 ha, were also applied for in other localities, notably Beatrix Bay in central Pelorus Sound, off the south-west end of D'Urville Island and off the heads of Pelorus Sound.

Ecological effects of mussel farming

Long-line culture involves the growing of mussels on a continuous rope, the long-line suspended in a series of loops (droppers) from the surface generally to a maximum depth of around 15 m (Figure 12.19). The droppers are typically inter-linked between a pair of surface backbone lines fixed in turn to a string of large buoys. The backbones are traditionally about 110 m in length for standard-shaped farms, with anchor warps extending beyond to large concrete anchor blocks or in some cases screw anchors. Each long-line is positioned about 20 m from its neighbor, though wider separation distances are sometimes used in less productive areas or more exposed locations. The result is parallel rows of buoys on the surface (Figures 12.16 and 12.18) supporting a three-dimensional array of mussel lines hanging in the water column (Figures 12.13, 12.19, and 12.20).

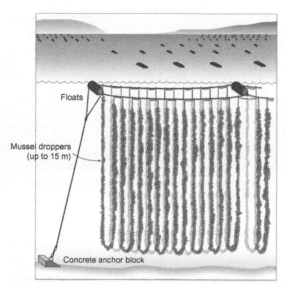

FIGURE 12.19 The typical configuration of a mussel long-line, New Zealand (adapted from Lloyd 2003).

FIGURE 12.20 Mussel long-lines suspended underwater, New Zealand (photo by Kim Westerskov).

FIGURE 12.21   Fluttering shearwaters offshore of a mussel farm in Admiralty Bay, New Zealand (photo Department of Conservation).

The ecological effects of mussel farming on seabed and pelagic communities did not feature very highly in the assessment of applications prior to the 1990s. Following the enactment of the RMA, greater attention began to be placed on the ecological effects of mussel farming. Impacts on seabed communities of mussel faeces, pseudofaeces, "shell-drop," and harvesting debris were initially the main focus (e.g., Department of Conservation 1995). Later, plankton depletion and carrying capacity issues began to feature prominently in mussel farming assessments, particularly with the shift in attention to larger offshore farms and as marine farming began to dominate certain bays within the Marlborough Sounds.

In addition to carrying capacity and wider food chain effects, applications for offshore mussel farms brought to light a range of issues that largely had not been considered before, including effects on the habitat of seabirds and marine mammals. The risks to marine wildlife from marine farming litter had long been recognized, and the potential of large cetaceans becoming entangled in mussel farm lines was highlighted when a Bryde's whale (*Balaenoptera edeni*) was washed up in 1996 on Great Barrier Island, Auckland, with a mussel spat-catching line trailing from its mouth and body (Lloyd 2003). People were also broadly aware of the possible effects on wildlife of noise and general disturbance from maintenance and harvesting vessels (Figure 12.14), though there was little or no information on the scale and significance of these issues. However, it was not until 2000 and 2001 that the direct impacts of mussel farms on the habitat of seabirds (Figures 12.21 and 12.22) and dolphins (Figures 12.23 and 12.24) first surfaced as issues.

This does not mean that such habitat effects had not been occurring previously, but rather that these issues had effectively "flown under the radar" until offshore marine farming came on the scene. Butler (2003) discussed the possible impacts of mussel farming on the endemic king shag (*Leucocarbo carunculatus*), and Lloyd (2003) provided an overview of potential effects on marine mammals and seabirds.

FIGURE 12.22 Mussel farm buoys provide a convenient roost for some seabirds, such as this spotted shag (*Stictocarbo punctatus*) (photo Department of Conservation).

FIGURE 12.23 Leaping dusky dolphin in Admiralty Bay, New Zealand. Note the mussel farm in the distance (photo Department of Conservation).

Admiralty Bay

Over the late 1990s, a team of researchers from Texas A&M University began to study dusky dolphins in Admiralty Bay, mostly as an extension to their dusky dolphin research at Kaikoura, amply described in this book.

FIGURE 12.24   Dusky dolphins in Admiralty Bay, New Zealand. A mussel harvester works alongside a marine farm in the distance (photo Department of Conservation).

When the various offshore marine farms were originally applied for in Admiralty Bay over 1999–2000, the importance of the wider Admiralty Bay area for dusky dolphins was not well understood. The value of Admiralty Bay was not documented by Duffy and Brown (1994), nor was this area identified as significant dolphin habitat by Davidson et al. (1995) or in the Marlborough Sounds Resource Management Plan. Dusky dolphins were known to be present in this general area, but it was not until the personnel of Texas A&M University began their research in Admiralty Bay that the significance of the bay for this species started to become clear (Markowitz et al. 2004).

The Marlborough District Council considered the various offshore applications in Admiralty Bay, approving some but declining others. Virtually all of these applications were subsequently appealed to the Environment Court—New Zealand's court which considers appeals over resource consent decisions— by either the applicants or other parties which had originally objected to the applications.

Texas A&M University's earlier research coincided with the "gold rush" of mussel farming applications in Admiralty Bay, and in July 2000 Tim Markowitz and April Harlin (now Harlin-Cognato), both then Ph.D. students at Texas A&M, wrote to the Department of Conservation, Marlborough District Council and others expressing their concern about the potential impact of such development on dusky dolphins visiting Admiralty Bay. Both the Department and the Council subsequently contributed funding and logistical support over the next several years for further studies on the distribution, abundance, and behavior of dusky dolphins in Admiralty Bay relative to marine farming in the bay.

Several key findings from Texas A&M University's research in Admiralty Bay proved crucial in the Environment Court hearings on the various marine farming applications: the unique characteristics of Admiralty Bay as dusky dolphin habitat, notably its shallow and relatively confined nature; habitual return of dusky dolphins to Admiralty Bay over winter and spring; relative abundance and wide distribution of dusky dolphins in the bay; particular bait-ball foraging behavior employed by dusky dolphins in Admiralty Bay; disruption and virtual displacement of bait-ball foraging by mussel long-lines; and overall significance of Admiralty Bay as dusky dolphin habitat. There was no suggestion that dusky dolphins might become entangled in the mussel farm lines or that their travel through the farms would be totally impeded.

The first of the New Zealand Environment Court hearings occurred in 2004 for two 42 ha mussel farms positioned in the middle of Admiralty Bay. As with most Environment Court hearings, no single issue was central to the case; rather, a range of interrelated matters were considered including economic and social benefits, navigation, public access, recreational use, wildlife values, natural character, and landscape. Nevertheless, the effects of the marine farms on dusky dolphins and particularly their habitat featured prominently in the Court's decision. Section 6(c) of the RMA requires as a matter of national importance "the protection of ... significant habitats of indigenous fauna" and, importantly, the court held that Admiralty Bay was significant habitat of dusky dolphins. Moreover, the court ruled that Section 6(c) required the habitat "to be protected for its own sake," irrespective of other (e.g., population level) effects on the dolphins. The Environment Court's decision was to decline both these large mid-bay applications.

At the time of writing this chapter, applications for one offshore farm and several offshore extensions were being considered by the Environment Court, though no final decisions had been released. Again, impacts on dusky dolphins and their habitat featured prominently in the court hearings.

CONCLUDING REMARKS

The generally coastal distribution of dusky dolphins means that they frequently inhabit areas where human activities occur. Inevitably, many of these interactions result in impacts on the dolphins. These interactions range from issues such as mortality from directed killing or incidental fisheries by-catch, to habitat loss due to aquaculture, or disturbance from tourism operations (see Chapter 11). Given these impacts, there is a need to manage human activities to ensure that the competing demands of both dolphin and human use of the marine environment are balanced, and that dolphins are protected from detrimental impacts.

In most parts of their range, dusky dolphins are poorly understood, and the impacts upon them are even less so. This lack of knowledge is perhaps the single biggest obstacle to the management of populations. When there is a lack of data on a dolphin population coupled with an absence of information about specific impacts on that population, there is no foundation from which regulatory authorities can make informed and appropriate management decisions. However, even in situations where there is useful scientific information, competing demands from human users and the pressure they can apply to regulatory authorities can have a significant bearing on decisions.

Given the lack of information about dusky dolphins and anthropogenic impacts upon them, there has been little specific management of dusky dolphin populations across the southern hemisphere, except in relation to tourism and aquaculture in New Zealand. However, many areas where dusky dolphins are found have some form of generic legislation (e.g., MMPA or the equivalent) that provides some level of protection to small cetaceans, although there are none that offer any specific protection to dusky dolphins. In general, this legislation promotes the conservation and protection of small cetaceans at a high level, rather than providing specific objectives and measurable standards. Such generic legislation can be useful but in reality allows governments to say they are providing protection and management, when in fact there may be little or no tangible action.

Another common feature of management is a low level or even complete lack of compliance and enforcement of regulations across many parts of the southern hemisphere. This is a worldwide problem common to multiple facets of marine protection, and relates mainly to the complexity and expense of monitoring compliance and undertaking enforcement. This gap, more often than not, leads to a lack of adherence to any regulations by human users, and therefore many detrimental activities continue despite apparent regulatory control.

The great diversity of potential and actual impacts on dusky dolphins from human activities across the southern hemisphere means that a range of management approaches are likely to be required. Management action is further complicated by the wide-ranging and migratory behavior of dusky dolphins, that highlights the need for multi-national management strategies and approaches. There are some international fora that offer an avenue for joint management approaches (e.g., CMS), and some progress has been made in this direction, but it will require the active engagement and participation of countries to progress these. Although such agreements are for the most part non-binding, they establish a common expectation and understanding among governments, and can provide momentum for subsequent multilateral treaties. Furthermore, while these agreements frequently lack regulatory enforcement capability, they can provide a focus for individual governments to take further action.

There have been different approaches taken by governments and other agencies attempting to manage dusky dolphin and human interactions. Legislative approaches are perhaps the most common, and these provide a framework for regulation and management. Within these frameworks, differing approaches have been used to protect dusky dolphins, including permitting systems to limit the number and type of tourism operations and trips, restrictions on aquaculture development in significant habitat, controls on fishing methods that catch dolphins, and the outright banning of live capture for display. Alternatives to regulation may include methods such as voluntary codes of conduct. These voluntary measures can be useful, but their effectiveness has been questioned (e.g., Scarpaci et al. 2003, Wiley et al. 2008). Voluntary codes have been used by tourism operators at Kaikoura in New Zealand and at Peninsula Valdés in Argentina, but there has been no assessment of their efficacy.

Managing anthropogenic impacts to improve conservation and protection of dusky dolphins will require a fundamental shift in several key areas: (1) there needs to be a significant increase in the understanding of dusky dolphin populations and of the threats that face them, based on robust and comprehensive science. High-quality science is crucial to inform decision makers; (2) legislative and regulatory mechanisms need to be in place to ensure that conservation and management of dusky dolphins is able to be undertaken; (3) most current regulatory mechanisms that could help protect dusky dolphins put the burden of proof on the regulatory authority. An alternative approach which should be considered is to put the burden of proof on human users of the marine environment who would be required to demonstrate that they do not have an impact on dolphins before their proposed activity is permitted. The scientific rigor of their research should be thoroughly evaluated before it is accepted. Furthermore, costs of demonstrating an impact or lack of impact should be borne by the user; (4) there must be adequate education, compliance and enforcement of management decisions to ensure that they are followed or there is little point in making them; (5) given the complexity and difficulties in determining true cause and effect relationships, the precautionary principle must be applied in decision making; and (6) given the highly mobile and migratory nature of dusky dolphins, there need to be binding multi-national regional agreements to manage issues across international boundaries and in international waters.

ACKNOWLEDGMENTS

We thank the following people for providing valuable information and advice: Peter Best, Graeme Coates, Enrique Crespo, Sonja Heinrich, Miguel Iñiguez, Tim Markowitz, Mike Meÿer, Carlos Olavarria, Dave Paton, Emer Rogan, Koen van Waerebeek, and Bernd Würsig. Steve Dawson and Sam DuFresne provided useful comments on an earlier version of this chapter, as did several anonymous reviewers.

Chapter 13

Dolphin Swimming and Watching: One Tourism Operator's Perspective

Dennis Buurman

Encounter Kaikoura, Kaikoura, New Zealand

Kaikoura . . . a name that has become synonymous with marine mammals and seabirds. Since 1988 sea-based operations in Kaikoura have carried hundreds of thousands of visitors to the world of the whale, the dolphin, the albatross, and the fur seal. Imagine, if you will, a coastal village located on a peninsula that juts out into the Pacific Ocean. In the background, not far from the township, protruding in their rugged splendour, rise the Seaward Kaikoura Mountains, their tops more often than not crested with snow. This in itself creates a sort of amphitheatre, a backdrop, against which the massive tail flukes of a sperm whale descend, the explosive antics of a dusky dolphin leap, or the expansive wings of a soaring albatross span.

Dennis Buurman, Kaikoura, New Zealand, December 2008

INTRODUCTION

Situated on the east coast of New Zealand's scenic South Island, Kaikoura has moved from the wilderness into a position of both national and international importance. Gone are the days when the only claim to fame for this small coastal village was its name, *kai*, the Māori name for food and *koura*, the name for crayfish (rock lobster), a delicacy for which the waters around Kaikoura are renowned. Kaikoura can now confidently assume the mantle of being among the top sites in the world to observe marine mammals and seabirds.

So why Kaikoura? What sets this small coastal village with a population of only 4000 apart from the many locations in the world where marine wildlife can be readily viewed? The main reasons are the easy access to these marine entertainers due to their close proximity to land and a pre-existing infrastructure of the main highway, shops, motels, eateries, and such.

Many places around the world can boast of having marine mammals in their waters, but finding them can be frustrating and uncertain. Kaikoura is blessed with a deep ocean canyon system, the Kaikoura Canyon, which approaches very close to shore, providing a smorgasbord of marine life that supports a food chain from the smallest phytoplankton to the largest of the toothed whales, the sperm whale (*Physeter macrocephalus*). Whales, dusky dolphins (*Lagenorhynchus obscurus*), New Zealand fur seals (*Arctocephalus forsteri*), and seabirds from tiny prions to huge albatrosses can be viewed year-round, with operators having very short distances to travel, meaning multiple trips can be made daily. Prime examples of this are the tour operators Dolphin Encounter, who can operate up to seven tours (3-hour dolphin-watching and swimming excursions) per day, and Whale Watch Kaikoura, who during the long summer days, run up to eight tours per day, with a trip duration of 2.5 hours.

Visitor numbers to Kaikoura have climbed staggeringly since the inception of marine tourism in the late 1980s. While the marine life is the star attraction, an array of supporting activities on both land and sea has developed as well as the infrastructure and service industries required to cater for the estimated 1 million plus visitors that stay or pass through Kaikoura annually.

However, whales and dolphins are the undisputed icons of the seas off Kaikoura. Mainly juvenile male sperm whales occur year-around over and near the Kaikoura Canyon system—often with blows visible from land. The attractive dusky dolphins are undoubtedly the acrobatic specialists of the dolphins, and are marvellous ambassadors for their dolphin kind. In Kaikoura, they entertain both those who come to watch them and those confident enough to don a wetsuit and enter the chilly waters and interact with them.

So how did the swimming with dolphins story begin in Kaikoura, or indeed, in New Zealand? To answer this, we go back to the summer of 1989–1990.

THE DOLPHIN ENCOUNTER STORY

In 1989, my brother Rik Buurman and his friend Ian Bradshaw, two Kaikoura rock lobster fishermen, began scheming ways of providing alternative attractions for tourists who had been drawn by the lure of sighting sperm whales, a successful tourism activity started only two years earlier. The young men strategized as they worked their craypots, eyeing the boat loads of tourists that sped by in high-powered 6 m rigid hulled inflatables. Daily they would add to the plans formulated on the previous days.

Their initial idea was to provide a boat-based scuba diving and fishing activity. They had also heard from one of the whale-watching companies, how on occasions when their tours had encountered large pods of dusky dolphins, pleas by passengers onboard the boats to get in the water with the dolphins, had resulted in some spectacular feedback by those adventurers, some in swimsuits, some in the "altogether," who interacted with the dusky dolphins. So the idea of tagging on an option to swim with dolphins seemed a logical choice.

Discussions with the New Zealand Department of Conservation, the government body responsible for administering marine mammal permits, resulted in an attitude of: "give it a go and see if it works out."

So, with limited resources but plenty of enthusiasm, they launched their new venture mid-way through summer 1989–1990. The initial name for their operation, "Dolphin Mary Charters," was a reflection on the short life of Ian Bradshaw's much loved sister, Mary, tragically lost in a car crash a year before. With tight funds and little credit, they purchased a 6.5 m catamaran with two new 115 hp outboard engines, and the expedition was under way (Figure 13.1).

Rik and Ian set about taking the news of their operation to the visitor. With photos in hand, nightly they visited the small number of backpacking establishments in Kaikoura, looking for even the slightest glimmer of interest shown by a generally disinterested audience as they announced the purpose of their visit. As most backpackers had their hearts set on seeing whales, their advances were mostly greeted by silence. However with the eye of a wily auctioneer, they became adept at picking out any who showed the remotest interest and, given the chance, would accost them and begin flicking through photographs that proved you could dive, catch fish, and swim with dolphins in Kaikoura.

Once the attention of a backpacker was won, the rest was not so difficult, as within a short time, others overhearing the conversation would gather round and next morning this success was confirmed by "bums on seats" as the "Dolphin Mary" left the slipway at South Bay in its quest to dive, catch fish, and swim with dolphins.

Soon the backpacker gossip chain began to achieve the type of marketing results normally reserved for those with unlimited budgets, and in their second season of operations, Rik and Ian dropped the diving and fishing to concentrate on the dolphin swimming. Now their nightly forays into the growing number

FIGURE 13.1 New Zealand's first dedicated dolphin swimming vessel, "Dolphin Mary," surveyed to carry nine passengers plus crew (photo by Barbara Todd).

of backpacking establishments began to pay dividends as word spread about swimming with dusky dolphins in the waters off Kaikoura. An attraction rivaling the watching of sperm whales was born.

A formal application to the Department of Conservation for a marine mammal permit was made by the partnership in 1990, for three trips per day carrying up to 8 (and in 1992, another permit for 15) passengers for swimming purposes per trip.

In 1991, Rik and Ian extended to me and my wife Lynette the offer to join the venture as partners, in exchange for added business acumen and finances. I had been operating a successful crayfish business for 14 years, and had the stability and capital to help. I sold my crayfish quota, boat, and gear, and the dolphin operation moved into its next phase. A new vessel surveyed for 25 passengers, 15 swimmers, and 10 spectators, was commissioned for delivery in November 1991, store, office, and wetsuit changing rooms were set up in the town center, a bus purchased, and staff of two guides and an office assistant were hired to help operate the business. Rik and Ian were the skippers, I was general manager, boat launcher, and early-morning dolphin spotter, and Lynette kept the books (Figures 13.2 and 13.3).

In summer 1990/91, approximately 1300 visitors swam with the dolphins. Over the 1991/92 summer, about 5000 visitors swam with or viewed dolphins, so the operation had grown tremendously in one year. The name was changed to Dolphin Encounter to better identify the experience for tourists.

In the early years, we mistakenly thought that dusky dolphins only inhabited the inshore waters around Kaikoura during the summer months of October to April. It was presumed by us that toward the end of autumn, the dusky dolphins moved north, either to the top of the South Island or even further afield, along the east coast of the North Island, returning in October when the females bore their calves in the relatively protected waters of the inshore Goose Bay area. Hence tour operations were advertised to coincide with this suspected cycle. In the first

FIGURE 13.2   Our rather colorful headquarters in 1992 (photo by Dennis Buurman).

three seasons (1989/90 to 1991/92), once the dusky dolphins were thought to have "moved off," in late March early April, the tour operation involving the dolphins went into recess. Indeed, in the winter of 1992, the first year with an established operational base, the office was closed and did not open again until just before the heralded resumption of dusky dolphin activity in October.

But then a lone bottlenose dolphin (*Tursiops truncatus*) showed up and Dolphin Encounter began to take paying passengers to view and swim with this sociable dolphin, who was termed "Maui" (after the Polynesian god). Once the duskies returned very close to shore in October, our focus returned to them, but on many occasions, as the 1992/93 season got underway, we had combined tours where we swam with the duskies and on returning to port, came across "Maui" and as a bonus, had a swim with her as well (Figures 13.4 and 13.5).

FIGURE 13.3 Our new vessel delivered late in 1991 with a group of swimmers on board (photo by Dennis Buurman).

FIGURE 13.4 "Maui" the bottlenose dolphin in her favorite pursuits. Here she is seen chasing a whale-watching vessel off the coast at Goose Bay (photo by Dennis Buurman).

FIGURE 13.5 Rubbing "Maui" with seaweed was something she seemed to enjoy (photo by Dennis Buurman).

As we gained more knowledge about dusky dolphins in winter, often in excess of 1000 dolphins in a pod, we realized that winter dolphins were further offshore than summer ones, and that indeed many if not all winter dolphins were of a different type from further afield (and from where is not known to date). This was corroborated by the researchers who have written extensively about this partial migration, elsewhere and in this book. In 1994, the first serious attempt was made at operating a year-round operation by venturing offshore in search of pods of dusky dolphins during the winter months of June to September.

Our methods of finding the dusky dolphins took on new meaning over the winter months. In summer it was normally a case of our vessels traveling to the south from the marina at South Bay and locating the dolphins within the Goose Bay, Haumuri Bluffs area. The dusky dolphins move into the inshore waters as the day dawns, after feeding over the edge (further offshore) of the shelf during the night. During the day, they spend their time close inshore, generally meandering back and forth, in no immediately apparent pattern, and then moving offshore in the late afternoon (Figure 13.6).

In winter, the method of detecting dolphins is far more complex and is a lot more involved. While sperm whales in Kaikoura can be located relatively easily by the use of hydrophones, dusky dolphins do not emit a sound of similar amplitude or frequency, which means that detecting them is reliant on our visual prowess, good eyesight and a pair of binoculars.

Daylight hours in winter dictate that our trip schedule is normally reduced to two trips per day this time of the year. Our first trip of the day in winter is the 8:30 am tour, as opposed to 5:30 am in summer. The day begins by having one of our crew taking a powerful (80 × 20) pair of binoculars onto the Kaikoura Peninsula to search for dolphins. This vantage point allows us to scan both north and south looking for a dark patch on the water which indicates a pod. Other signs to look for are a splash as a dolphin leaps, which could indicate the presence of a group. While it sounds simple in theory, in practice, finding the

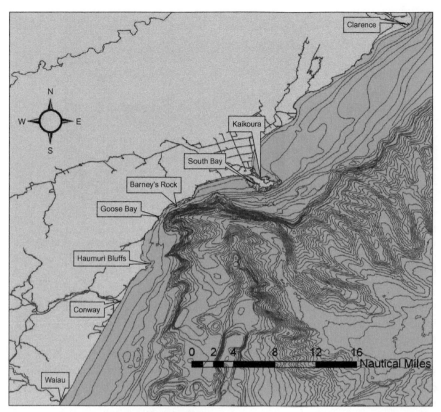

FIGURE 13.6   Chart showing operational area.

dolphins in winter can be difficult and arduous. Visitors being fitted with wetsuits and briefed at our headquarters are quite oblivious to the efforts going on to locate dolphins so that the trip can proceed successfully. As a last resort, if dolphins cannot be located in this manner, an airplane is chartered to search the vast area for dolphins. As well, local fishermen help the tourism operation by passing on coordinates of dolphin pods they have encountered (Figure 13.7).

During the late 1990s and early 2000s, New Zealand rated as one of the most sought-after tourist destinations in the world, particularly with the back-packer traveler. Basking in this popularity, tourism became New Zealand's number one foreign exchange earner. At this time visitor numbers to Kaikoura were also at their peak. For those in the tourist trade, the tourist season got underway in October and went right through to April before there was any significant decline. Our dolphin trips during these months were fully booked for the allowable swimming positions and it was not uncommon to have a standby list of 4 or 5 pages of eager visitors looking for the opportunity of interacting with the dusky dolphins in their natural environment. This resulted in us having to employ more staff on the boats (skippers and guides) and on our reservation

FIGURE 13.7   A pod of dusky dolphins moving north off the Kaikoura Peninsula, combined
with a large swell, provided a stunning photo opportunity for the photographer on this trip in the
midst of winter. The snow-covered Seaward Kaikoura Range makes a fantastic backdrop (photo by
Dennis Buurman).

team, as well as a dedicated boat and machinery maintenance team. During the
high summer season, our staffing numbers grew to around 20, dropping to half
that number over the quieter winter months.

We had also gone through the process of upgrading our vessels. New vessels meant we could carry more passengers and provide more comfort for our
clients, but swimmer limits on each permit did not change. Although our swimmer numbers were restricted, there were no restrictions on those only wishing
to view the dolphins, so larger, more comfortable vessels enabled us to carry
more spectators, at a maximum of 30 (Figure 13.8).

In the 1999/2000 season, Rik Buurman decided to move onto other ventures and his share-holding was bought out by the other partners. In 2000, the
business expanded by purchasing a third permit (and boat) which had a limited
allocation of eight trips per week, with no more than two trips on any one day.
So to summarize, we now had three permits, two allowing three trips per day
and one allowing for eight trips per week. This led to a further increase in staffing levels, with numbers expanding to 25 in the high season.

In 2004, the Dolphin Encounter operation was moved to larger facilities,
and was also re-named Encounter Kaikoura to reflect the fact that not only dolphin watching, but also bird watching and other products were being offered
(not discussed in detail here). The number of passengers we were carrying to
swim with or view the dusky dolphins had now grown to between 24 000 and
27 000 annually (Figure 13.9).

Throughout the duration of what is now a long involvement with tourism and
marine mammals, the only constant is the fact that, to be frank, there is no constant.
The dusky dolphins have led us through an epic journey and while we have been
able to develop an understanding of their idiosyncrasies, the challenges of operating

FIGURE 13.8 Our vessel *Lissodelphis* built in 1998 added a new dimension to our operation allowing for the ability to carry more spectators to view the dusky dolphins (photo by Dennis Buurman).

FIGURE 13.9 Our complex situated on the Esplanade in Kaikoura provides for a more spacious atmosphere as well as allowing for additional activities to the core business (photo by Dennis Buurman).

within the realms of nature, with all its vagaries, has been character-building material.

After so many years of interaction with dusky dolphins, we have built up a knowledge and understanding of these intrepid dolphins as well as a sizable database on their movements within our area of operation. Typically, the dolphins are located inshore in the summer months and further offshore in winter. However, as with many wild species, one can never predict, or be certain, that they will be constrained within the prescribed pattern or behaviors we set them by our observations. One thing we have learned from our long association with dusky dolphins is that when you think you have them "figured out," they do something completely different to confuse you. Having said that, over the years we have not seen any significant alteration to their overall behavior.

While we are the only operator to swim with dusky dolphins in Kaikoura, there are four other vessels (Whale Watch Kaikoura) as well as two fixed wing and two helicopter operators with permits that include the ability to view dusky dolphins. So understandably there is a lot of activity with dusky dolphins, particularly during the long summer days when visitations to the dolphins can be reasonably constant throughout the day. From December 1 to March 31 each year, sea-based operators observe a voluntary "time off period" between 11:30 am to 1:30 pm each day, when they avoid contact with dusky dolphins.

Our database of the locations of dusky dolphins in the Kaikoura region provides the longest continuous record of dusky dolphin movements and distribution in Kaikoura. From the GPS positions recorded by our skippers on each trip, it is indicated that dolphins have not noticeably shifted away as a result of tourism interactions.

Trips typically follow a set pattern. Once our clients who have opted to swim have been suited up and comprehensively briefed at base, they are transported by bus to South Bay where the dolphin boats are waiting. Once on board, and after a further safety briefing, the skippers set off in search of dolphins, normally to the south. The skippers and naturalists on board interpret what is seen, and what is hoped to be seen, throughout the cruise, and we believe that this setting of the stage of dusky dolphins in their natural environment is critically important to educating our excited tourists while we have their attention!

It is interesting to see the reactions of visitors once the dolphin pod has been found and is being approached. For many, this is the first time they have seen dolphins in the wild, let alone hundreds of them all tightly grouped together, effortlessly displaying their acrobatic prowess with the ease and accomplishment of a finely tuned Olympic athlete. Depending on their behavior at the time, the dolphins will often approach the boat, giving passengers their first close-up glimpse of the dolphins and the distinctive gray-striped body coloration from which they derive their name. It is not uncommon for our crew on the boats to witness profound emotional scenes, such as the exhibiting of tears, as momentarily passengers are overcome by the sheer joy of this first encounter. Although we have been involved with this activity for a considerable length of time, one never tires of the satisfaction that is associated with bringing enrichment to those that choose to take the dusky dolphin experience. And this enrichment, we know, will in many cases lead to a greater appreciation of nature, and in some cases to greater involvement of helping nature. If our tourists, even a small percentage, can be so motivated, then we are truly pleased and proud.

Dealing with wild animals, be they on land or sea, always brings with it a caution, a realization that nothing is predetermined. With the dolphins there are so many factors which come into play, particularly when you add weather, sea conditions, and water visibility into the vagaries of dolphin behavior. One never knows from one trip to the next what nature will throw at you. But duskies are special and seem to "want to" interact with humans much of the time. It is quite astounding, when you analyze out the level of interaction we have with these wild dolphins, that we can enjoy the measure of success we do.

We judge the success of our swims by the level of interaction between dolphins and swimmers. When dolphins go out of their way to interact with the swimmers by swimming in and around them, this is the highest degree of success. We encourage our swimmers to involve the dolphins by performing such actions as making themselves as "dolphin like" (impossible to most) as possible, diving down, maintaining eye contact, and making noises through their snorkels. The latter also adds the extra dimension of providing entertainment for those watching onboard the boats, as the assorted sounds emitted by the aquatic melee resonate across the waters.

We always emphasize the fact that the success of our tours, particularly the swim, is dependent on the dolphins. If they are in the mood, interaction can be incredible; however, if they are preoccupied with their own activities, the best our swimmers can hope for is a glimpse of them as they continue on their way. Sometimes, the dolphins are so intent on their own actions that the swim has to be terminated—for to continue swim attempts would be to harass the mammals.

Many a tale could be told on the occasions a tour was successfully concluded with only one or two dolphins staying with a group of 13 swimmers for a swim duration of 30–40 minutes. There have even been occasions when this number of dolphins have entertained three boatloads of swimmers at one time! Situations like this are simply humbling when one attempts to comprehend the sense of privilege felt as a result of such an interlude (Figure 13.10).

FIGURE 13.10 For those coming to New Zealand, the interaction with dusky dolphins often proves to be the highlight of all the experiences on offer through the country (photo by Dennis Buurman).

FIGURE 13.11 A dead dusky dolphin after an orca attack on a pod of duskies. The photo was taken on a disposable camera by one of our crew on the boat (photo Dolphin Encounter).

Over 17 seasons there are many stories to tell, the sight of killer whales (*Orcinus orca*, or "orca" for short) always brings a thrill both to the tour operators and visitors alike. There have been a number of dramatic incidents involving orca taking dusky dolphins (Figure 13.11). While this is part of nature and the wild environment, the sight of duskies being tracked and hunted down by a pod of orca is as full of suspense and trepidation as the most intriguing thrillers on the big screen. Our boats have witnessed dolphins being taken in full view of everyone onboard. It goes without saying that emotions among the passengers and crew are very mixed after such an encounter.

We also live with the reality that due to the unpredictability of the weather and the marine mammals, there are many occasions when those who have their heart set on playing out the dream of interacting with Kaikoura's dusky dolphins leave without fulfilling this dream. Others go to extreme lengths to ensure they will not be disappointed, either staying on, returning at a later date, or even rescheduling flights not to miss out on the chance.

The memories of a large part of our lifetime are endless; the stories of events, people, our visitors, our staff, and the wildlife would fill volumes. In looking back, from small beginnings to becoming one of New Zealand's iconic attractions, the success of Dolphin Encounter has not merely come about by chance. It has developed by a certain degree of innovation accompanied by good business practice and an understanding of the need to provide top-quality service and hospitality. The dedication and professionalism of management and staff have proved fundamental to the ongoing success that we have experienced.

Our operation is a partnership where the partners brought a range of talents that fitted neatly into providing the necessary skills to successfully operate and develop our dolphin experience. From the outset, we developed a philosophy based on our statement of intent which we established early in our operating years. Part of this states "We must instill in our clients a knowledge and understanding of

FIGURE 13.12   The magic of dusky dolphins and their infectious exuberance is what makes operating our marine mammal tours in Kaikoura an absolute privilege (photo by Dennis Buurman).

the mammals and their environment, so that our clients depart serious in purpose and endeavour to preserve what this world has left and work at retrieving what has been lost."

But the culminating element to our story of success is the magic of the dusky dolphins and the impact they have on those that take the opportunity to share in this experience. Elementary to this is that we have steadfastly focused on giving those who take our tour, a natural experience, not one that is fabricated, but one where any interaction between human and mammal is on the terms of the mammal (Figure 13.12).

The profile of our business is such that it is open to public scrutiny. We are in a position of being accountable to the Department of Conservation and to the general public. Because of the way in which our business is perceived, we are constantly approached for such things as work experience, internships, financial support for community groups, and sponsorship opportunities. We are committed to the local community and to assisting where possible.

We believe that we have embarked upon and are committed to environmental best practice in a measurable and sustainable way, with beneficial outcomes. Our commitment is to continue to advance education, conservation, and preservation as an integral part of our day-to-day operation. More than ever, we realize the impact of continued global abuse of our environment, and if major environmental disasters are to be averted, we must do whatever is possible to work towards ensuring that sustainability, as well as improvement of our natural ecosystems.

Chapter 14

Neglected But Not Forgotten—Southern Africa's Dusky Dolphins

Peter B. Best[a] and Michael A. Meÿer[b]

[a]Mammal Research Institute, University of Pretoria, c/o Iziko South African Museum, Cape Town, South Africa

[b]Marine and Coastal Management, Department of Environmental Affairs and Tourism, Cape Town, South Africa

In December 1990, a young dusky dolphin was brought to the South African Museum. It had been found dead on the beach opposite Robben Island, Table Bay, by a jogger late the previous evening. Externally it looked quite healthy, but as the standard dissection unfolded, we encountered something quite unusual. Although there were no gross external signs of trauma, the rear of the cranium was completely shattered. Despite the absence of any bullet, the most likely explanation seemed to be that this young dolphin had been shot, even though the beach was quite deserted at the time and there were no reports of an attempt to euthanize a dolphin. The full stomach (a most unusual occurrence in a normal stranding) was also an indication that the animal may have been shot at sea and not on the beach.

Two months later, a second dusky dolphin was found dead some 20 km further north on the same stretch of coast; although it was freshly dead, blood was oozing from the mouth. Dissection revealed almost identical features to the first animal—a shattered back to the skull but with no obvious external signs of damage (and no bullet). Given the unlikelihood of two such similar occurrences happening by chance, we started to look for a pattern. Discrete enquiries revealed that fishing vessels from a particular harbor had been operating in the area on both occasions, and we began to suspect that the victims may have been shot while bow-riding, possibly as a retaliatory act for inadvertently interfering in the fishing operation.

It was another 8 years before a third victim arrived at the Museum, another dusky dolphin seen floating dead at sea before washing ashore on the west coast of the Cape Peninsula, with blood oozing from its mouth. This time the execution was not so professional: besides a shattering of the left condylar region of the skull, there were two overlapping entry wounds in the axillary region of the left flipper, and copious amounts of blood in the body cavity. This time we also had a smoking gun—observers on the

*shore at the stranding site reported that two boats had earlier that day driven through a
school of feeding dolphins and seals and thrown a net for bait.*

*Although "seal" is a four-letter (obscene) word to fishermen at the Cape, it is clear
that their wrath can extend to other perceived competitors in the marine environment,
including dolphins.*

Peter Best, Cape Town, South Africa, November 2008

INTRODUCTION

In 1828 John Edward Gray at the British Museum of Natural History described
a new species of dolphin from the Cape of Good Hope as *Delphinus (Grampus)
obscurus*, later referred to as *Prodelphinus obscurus* (Flower 1885) and eventually included in the genus *Lagenorhynchus* by True (1889) (see also Chapter 1).
Despite being the type locality of the species, the region has been singularly
wanting in providing further information on this little dolphin.

While it is true that there has been no systematic research project on dusky
dolphins in southern Africa, information has been collected opportunistically as
a by-product of a number of other research projects, and this chapter attempts
to summarize what knowledge has accumulated in the process.

APPEARANCE

Size and morphometrics—sexual dimorphism

The largest individuals among 51 male and 53 female dusky dolphins examined
in South Africa were 1.90 m and 1.91 m long, respectively, indicating there is
no major difference in size between the sexes. To examine whether the same
tendencies exist for sexual dimorphism in certain body measurements as were
apparent in Peruvian animals (Van Waerebeek 1993a), comparisons have been
confined to sexually mature individuals, of which there were 9 males and 16
females potentially available (see below). Apart from differences in the position
of the anus (and consequently the size of the anal girth) in the two sexes, which
were not tested for as they are common to most cetaceans, the anterior length
of the flipper, its maximum width, the length of the dorsal fin base, the width of
the tail flukes and their depth, were all examined (Table 14.1).

For statistical analysis, the anterior length of the flipper and its maximum
width were summed into a combined flipper index, as were the span of the tail
flukes and their depth (as measured from the notch to the leading edge) into a
fluke index. Mature males proved to have larger flipper and fluke indices than
mature females (Wilcoxon Mann–Whitney $z = 2.425$, one-tailed $P = 0.0077$,
and $z = 3.031$, one-tailed $P = 0.0012$, respectively). No significant difference
could be detected in the length of the dorsal fin base between sexes (Wilcoxon
Mann–Whitney U-test $z = 0.923$, one-tailed $P = 0.178$), possibly because

TABLE 14.1 Comparison of flipper and fluke sizes between sexes in dusky dolphins from South Africa

	Flipper index			Fluke index	
% total length	Female	Male	% total length	Female	Male
39.5–39.99	1		51–51.99	2	
40–40.49		1	52–52.99	2	
40.5–40.99	2		53–53.99	1	
41–41.49	3		54–54.99	3	1
41.5–41.99	2		55–55.99	1	
42–42.49	2	1	56–56.99	2	
42.5–42.99	1		57–57.99	4	
43–43.49		1	58–58.99		2
43.5–43.99		1	59–59.99		
44–44.49	2		60–60.99		2
44.5–44.99	1	3	61–61.99		1
45–45.49			62–62.99		
45.5–45.99		2	63–63.99		1
46–46.49					
46.5–46.99					
47–47.49		1			
Total	14	10		15	7

this is a rather subjective measurement. It was not possible to investigate other sexually dimorphic differences in dorsal fin shape proposed by Van Waerebeek (1993a), as appropriate measurements were not included in the standard protocol, and fin tracings were unavailable.

Dentition

Counts of visible teeth in sexually mature individuals ranged from 26–32 in the upper jaw and 24–30 in the lower jaw (Table 14.2). Mean values in each jaw are 1–2 teeth less than their equivalent values in dusky dolphins from Peru (Van Waerebeek 1993a), but whether this is a real regional difference or simply the result of a different interpretation of what constitutes an erupted tooth is unclear.

TABLE 14.2 Counts of visible teeth in sexually mature dusky dolphins from South Africa

	Males	Females
Sample size	8	14
Left upper	26–31, mean 28.5	27–31, mean 28.8
Right upper	26–32, mean 28.5	26–31, mean 28.4
Left lower	25–30, mean 27	25–30, mean 27.2
Right lower	25–30, mean 27	24–29, mean 26.6

Length:weight relationship

Data for 42 dusky dolphins captured at sea (or found dead with evidence pointing to a death at sea) had a length/weight regression

$$W = 14.23 \ L^{2.848}$$

with an r^2 value of 0.970, where W = weight in kilograms and L = total length in meters (Figure 14.1).

Another 36 dusky dolphins, most of which originated from strandings, were lighter for their size (by 11–12%) than those caught at sea, presumably because they were debilitated in some way, with a length/weight regression of

$$W = 12.62 \ L^{2.849}$$

A visual comparison of the weights of non-juvenile animals (more than 1.5 m long) captured at sea in summer (October 15–April 14, $n = 12$) and winter (April 15–October 14, $n = 27$) suggests that there is no marked seasonal change in body weight (Figure 14.2).

DISTRIBUTION AND MOVEMENTS

Dusky dolphins occur along the west coast of southern Africa, probably largely within the influence of the cold Benguela Current system. The northernmost record is a photograph of two animals off Lobito Bay, Angola, at about 12°22'S, published in the magazine *SA Yachting* and purported to be taken by a member of the South African Navy (Kramer 1961). Between 1958 and 1959, three visits to the port of Lobito were made by a total of eight minesweepers and seaward defense boats belonging to the South African Navy, while in 1957 and 1959,

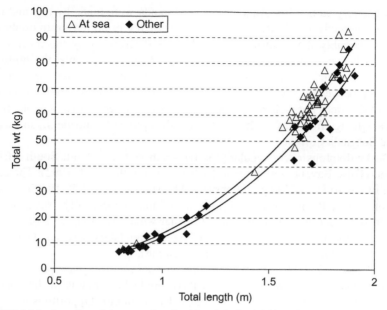

FIGURE 14.1 Length–weight regressions for (above) dusky dolphins taken at sea and (below) other dusky dolphins (mostly stranded) from South Africa.

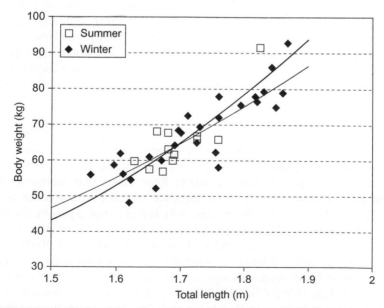

FIGURE 14.2 Comparison of length–weight regressions of non-juvenile dusky dolphins taken at sea off South Africa in summer and winter.

other courtesy calls to the Angolan port of Luanda were made by one and two frigates, respectively (Wessels 2002). There is therefore no reason to doubt the locality. Possibly the species' northern limit is determined by the position of the front between the warm Angola and cold Benguela currents, the northern boundary of which ranges seasonally between 12.6 and 15.5°S and the southern boundary between 15 and 17.2°S (Meeuwis and Lutjeharms 1990). Although Weir (2007) does not include the species in her faunal records for Angola, most of her sighting effort was north of 11°S, and thus probably outside the influence of the Benguela Current.

Dusky dolphins occur regularly as far south as Cape Point, South Africa, and into False Bay, but rarely range farther east (Findlay et al. 1992). The easternmost record is a sighting on February 22, 2001, at 34.622°S, 19.356°E, just east of Danger Point, and is well documented photographically. Nevertheless, it is the only such record in 10 years of regular whale-watching activities in the region, confirming its singularity (W. Chivell, Dyer Island Cruises, Gansbaai, South Africa, July 2008, personal communication).

Findlay et al. (1992) plotted the known sightings of the species along the west coast, from which it seemed there might be a hiatus in distribution between 27 and 30°S, around the Orange River mouth. Opportunistic sightings of dolphins from diamond-mining vessels in this area have also been almost exclusively of Heaviside's dolphins (Patti Wickens, De Beers Marine Corporation, Cape Town, South Africa, July 2008, personal communication), although the validity of species identification might be an issue. This region corresponds to the position of the Luderitz Upwelling Cell and Orange River Cone (LUCORC), in which the combination of a surface hydrodynamic and a subsurface thermal barrier could limit transport of ichthyoplankton from the southern to the northern Benguela system. This "barrier" would apply principally to pelagic fish such as anchovy (and to a lesser extent sardine and horse mackerel) but not to mesopelagic species such as light and lanternfishes (Lett et al. 2007). Satellite telemetry data (Eric Olsen and Mette Mauritzen, Institute for Marine Research, Bergen, Norway, August 2008, personal communication) and at-sea observations have also indicated that although the LUCORC area contains several breeding colonies of the Cape fur seal (*Arctocephalus pusillus*), it is not a recognized feeding area for the species. Hence the hiatus in dusky dolphin sightings around the Orange River mouth may represent a real discontinuity in distribution as a consequence of suboptimal feeding conditions. Nevertheless, whether this discontinuity is real or not, a genetic comparison of individuals from Namibia and South Africa using microsatellite markers failed to reveal any significant population structure (Cassens et al. 2005).

Sightings of dusky dolphins in southern African waters occur mainly in water less than 500m deep, although schools have been encountered in water over 2000m deep (Findlay et al. 1992). The species occurs regularly in the nearshore zone, including just outside the back line, from where live specimens were occasionally caught by beach seine for oceanaria.

EXTERNAL COMMENSALS AND PARASITES

Cyamids were found on 2 of 50 dusky dolphins examined in South Africa. One animal carried 47 *Scutocyamus antipodensis* on the body surface and appendages and 1 in the region of the mouth: it was a 1.66 m male taken at sea with no signs of debilitation and of normal weight (68 kg). A second dolphin had a single cyamid on its dorsal fin (Best 2007). Otherwise no external parasites were found. Five dolphins taken in a trawl off northern Namibia, however, all bore between 5 and 20 stalked barnacles (*Xenobalanus globicipitis*) each, attached mainly to the trailing edges of the fluke (24), dorsal fin (21) and flippers (5) (Leif Nøttestad, Institute of Marine Research, Bergen, Norway, July 2008, personal communication).

In Peruvian waters, *Xenobalanus* occurred on 38.9% of 707 individuals, while no cyamids were reported from the same sample of individuals (Van Waerebeek et al. 1993).

FOOD AND FEEDING

Food remains recovered from the stomachs of 36 dusky dolphins from South Africa included 42 978 identified prey items, representing 24 different taxa (Table 14.3). These were almost all identified from trace items such as otoliths (for fish) and beaks (for squid), and differential digestion rates may well mean that the trace items did not present an unbiased sample of prey species; in particular, the durability of squid beaks probably resulted in the cephalopod component being over-represented. Seventeen of the identified items were fish and seven cephalopods, and the predominance of fish in the diet is also shown by the low proportion of cephalopods in the overall diet (0.6% by number and 8.1% by weight) and the fact that only one cephalopod (*Loligo*) was found in more than 14% of stomachs, compared with seven fishes (hake *Merluccius*, pelagic goby *Sufflogobius*, horse mackerel *Trachurus*, anchovy *Engraulis*, pilchard *Sardinops*, hatchet fish *Maurolicus*, and a lanternfish *Lampanyctus*). The three most important prey species by mass were *Trachurus* (34.7%), *Merluccius* (22.9%), and *Lampanyctus* (12.8%), with arrow squid *Todarodes* (5.3%) and *Sardinops* (5.2%) a somewhat distant fourth and fifth, respectively. The stomach contents of five dusky dolphins taken incidentally in a midwater trawl in the northern Benguela system were reported as containing 98% *Trachurus* (Jens-Otto Krakstad, Institute of Marine Research, Bergen, Norway, July 2008, personal communication).

Comparing the relative incidence per stomach of fish from an epipelagic (*Trachurus, Engraulis, Sardinops*) and mesopelagic (*Maurolicus, Lampanyctus*, a lanternfish *Diaphus*, butter snoek *Lepidopus* and Myctophidae) habitat, there were only two of 31 stomachs from South Africa in which fish from one or another ecotype did not form more than 90%. This reciprocal relationship presumably reflects the utilization of two different foraging strategies. Separate

TABLE 14.3 Summary table for identified stomach contents of 36 dusky dolphins from South Africa, with mass data derived from measurements of otoliths or beaks

	Frequency eaten	Number eaten	Mean weight (g)	Mass eaten (g)	% by number		% by mass	
					Summer	Winter	Summer	Winter
Fish								
Merluccius	0.500	709	57.65	40872.92	0.63	2.45	7.84	28.58
Macrouridae	0.056	17	14.00[a]	238	0.00	0.07	0.00	0.18
Sufflogobius	0.222	484	6.19	2993.682	0.00	2.01	0.00	2.31
Trachurus	0.583	8371	7.40	61919.18	12.91	24.64	23.73	38.86
Engraulis	0.306	1124	4.99	5605.335	2.22	2.93	4.83	2.52
Sardinops	0.194	147	63.44	9325.1	0.60	0.14	11.80	2.77
Scomber	0.028	2	69.55	139.1	0.01	0.00	0.29	0.00
Liza	0.083	39	76.72	2991.916	0.00	0.16	0.00	2.31
Thyrsites	0.028	2	390.60	781.2	0.00	0.01	1.61	0.00
Etrumeus	0.056	6	10.96	65.76	0.03	0.00	0.14	0.00
Maurolicus	0.278	11203	0.62	6982.684	32.43	21.07	8.99	2.02
Myctophidae	0.222	1441	1.23[b]	1772.43	0.82	5.34	1.00	0.99
Diaphus	0.028	238	1.23[b]	292.74	1.26	0.00	0.60	0.00

Lepidopus	0.083	374	18.29	6841.176	0.02	1.54	0.11	5.23
Lampanyctus	0.167	18576	1.23	22813.43	48.70	38.92	28.79	6.82
Pterothrissus	0.028	3	50.00[c]	150	0.00	0.01	0.00	0.12
Neocyttus	0.028	1	50.00[c]	50	0.00	0.00	0.00	0.04
Cephalopods								
Loligo	0.222	23	98.79	2272.1	0.06	0.05	1.81	1.07
Sepia	0.028	2	0.70	1.4	0.00	0.01	0.00	0.00
Ommastrephidae	0.056	16	71.25[d]	1140	0.00	0.07	0.00	0.88
Sthenoteuthis	0.111	56	17.51	980.4	0.29	0.00	1.99	0.01
Todarodes	0.139	87	108.75	9461	0.02	0.34	6.48	4.87
Todaropsis	0.028	8	30.76	246.1	0.00	0.03	0.00	0.19
Lycoteuthis	0.028	49	6[e]	294	0.00	0.20	0.00	0.23
Total		42978		178229.7				

Summer = 15 October–14 April, winter = 15 April–14 October.

[a]Applying mean otolith diameter of 5.4 mm to otolith size/body mass data for two macrourid species in Labropoulou and Papaconstantinou (2000) and taking average.

[b]Taken as equivalent to the value for Lampanyctus.

[c]Arbitrary weight using size of species relative to other fish species eaten.

[d]Taken as the mean weight of Sthenoteuthis, Todarodes and Todaropsis eaten.

[e]Comparing mean rostral length of 2.2 mm to statement by Imber (1975) for rostral length of 4.2 mm.

utilization of surface shoaling fish and those associated with the deep scattering layer has also been described in New Zealand. In shallow waters, dusky dolphins tended to feed during the day, either in a coordinated fashion, herding schooling fish close to the surface (Markowitz et al. 2004), or in small subgroups at low prey densities. In deeper waters, dusky dolphins fed at night when the deep scattering layer migrated up to within 130 m of the surface (Benoit-Bird et al., 2004).

Although mullet (*Liza richardsoni*) were not a major dietary component overall, when they did feature, they tended to form a substantial proportion of the food eaten (34.6% and 43.9% by mass in two of the three stomachs concerned). This indicates that some individuals come close inshore to forage in extreme nearshore waters on occasion.

Reconstituted weights of prey items found in each stomach have been created from the dimensions of trace items (otoliths and beaks), using published or unpublished regressions of body mass against trace item dimensions. Those from stranded animals ranged from 7.4 to 25 362 g, with a mean of 4072 g ($n = 10$), while those from animals taken at sea ranged from 73.6 to 15 394 g, with a mean of 5300 g ($n = 26$). In a two-tailed test, these means are almost significantly different (Mann–Whitney Wilcoxon test, $P = 0.054$), but biologically one might expect the stranded animals to contain less food. By comparison, a captive female dusky dolphin consumed between 1870 kg and 3100 kg of food per year, or an average of 5.1–8.5 kg per day (Kastelein et al. 2000). This suggests that the reconstituted food mass (with the possible exception of the cephalopod component, because of the longer retention time of squid beaks) may represent intake of food over approximately one 24-hour period.

A comparison of stomach contents between animals from summer (October 15–April 14; $n = 16$) and winter (April 15–October 14; $n = 20$) suggests some seasonal differences, with greater intake of *Sardinops*, *Maurolicus*, and *Lampanyctus* in summer and *Merluccius* and *Trachurus* in winter. Excluding 10 stranded individuals (seven in summer and three in winter), the size of the average reconstructed stomach mass per dolphin is also less in summer than in winter (4059 vs. 5957 g), but not significantly so (Mann–Whitney Wilcoxon test $z = -1.617$, one-tailed $P = 0.053$). A captive female showed a similar tendency for a higher food intake in autumn and winter compared to spring and summer (Kastelein et al. 2000), so despite the lack of statistical significance, there may be a tendency for dusky dolphin feeding intensity in the wild to vary seasonally.

The stomach contents of two dolphins caught from the beach in a seine net early in the morning contained predominantly Myctophidae (*Lampanyctus* and *Diaphus*) or hatchet fish (*Maurolicus*), comprising 51.7% and 91.2%, respectively, of reconstituted food mass. The presence of these mesopelagic fish in stomachs suggests that the animals had been feeding offshore during the night and had subsequently moved inshore (Sekiguchi 1994). A similar diurnal onshore–offshore movement apparently associated with feeding on organisms associated with the deep scattering layer has been described for dusky dolphins in New Zealand waters (Cipriano et al. 1989, see also Chapter 5).

The subsurface foraging behavior of dusky dolphins in the northern Benguela Current has been observed acoustically using a scientific echosounder and multi-beam sonar. While feeding on a school of *Trachurus*, the dolphins descended to a maximum depth of 162 m, with mean swimming speeds increasing from about 1 m/s at a depth of 100 m to 3.3 m/s when diving down to 150 m. They attacked the fish from below while swimming at about 120 m depth, forcing the fish school towards the surface; some of the attackers worked in pairs. The response of the fish was to form vacuoles in the school, increase packing density or make avoidance dives of over 60 m to reach deeper and darker waters (Matteo Bernasconi, Institute for Marine Research, Bergen, Norway, August 2008, personal communication).

REPRODUCTION AND GROWTH

Thirteen live or freshly dead neonates (with a raw unhealed umbilicus) ranged from 81.5 to 91 cm in length (average 86.6 cm), and occurred between January 9 and March 8, with a mean date of January 29. Another 16 dead and decomposing neonates ranged in length from 72 to 88 cm (average 81.8 cm), and occurred between January 15 and April 5, with a mean date of February 12. The smaller average sizes and later dates of these records are very probably consequences of their poor body condition and uncertain dates of stranding, respectively, and so should not be taken as being inconsistent with the data from the live or freshly dead individuals. Although there are no near-term fetuses available, our best estimate of the size at birth would therefore be 86–87 cm, and the season of births defined as January–March, with a peak at the end of January/early February.

There are data available for six fetuses, one from April, four from July, and one from August, and 10 calves, classified as such from their size and healing or healed umbilicus. The latter ranged from 92.1 to 120.3 cm, and occurred between February 14 and April 19. In attempting to quantify the rate of fetal and neonatal growth, the sizes of fetuses, neonates, and calves have been combined and plotted against their respective dates of occurrence (Figure 14.3), and a Bartlett's best-fit regression used, given that both variables (size of dolphin and date of death) are in effect random samples (Simpson et al. 1960). This produces the equation

$$Y = 0.321x - 12.24,$$

where y = date of death (from April 1 in year 1) and x = total length in cm.

Dividing the size at birth (86.6 cm) by the estimated fetal/neonatal growth rate (0.321 cm/day, 95% CI 0.267, 0.358) indicates a period of linear fetal growth of 270 days, or 8.9 months. Substituting the mean weight at birth of 8354 g ($n = 10$) in the allometric equation of Calder (1982),

$$t_0 = 7.25M^{0.19},$$

where t_0 is the initial period of non-linear embryonic growth and M is the mean mass at birth (in grams), we obtain a value of 40.3 days. On this basis, total gestation would be $270 + 40 = 310$ days or just over 10 months.

The smallest mature female from the subregion was 1.68 m ($n = 16$), and the largest immature 1.76 m long ($n = 20$), suggesting that puberty is reached at an intermediate length. Grouping by 5-cm intervals implies that 50% maturity in females is reached at 1.7–1.75 m in length (Table 14.4).

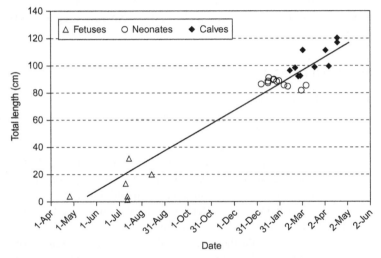

FIGURE 14.3 Growth in fetal and neonatal dusky dolphins off South Africa.

TABLE 14.4 Maturity status of female dusky dolphins from South Africa

Total length (m)	Number examined	Number mature	% mature
<1.6	4	0	0
1.6–1.649	4	0	0
1.65–1.699	9	1	11.1
1.7–1.749	4	2	50
1.75–1.799	6	4	66.7
1.8–1.849	5	5	100
1.85+	4	4	100
Total	36	16	

Of the 15 mature females (excluding one in captivity), two were ovulating, five were pregnant, three were lactating, and five were resting. None was simultaneously lactating and pregnant or ovulating, and the frequency of reproductive classes (albeit in a very small sample) would seem to indicate a 2- or 3-year rather than an annual calving frequency. The two ovulating animals were so classified because no embryo could be found despite there being an apparently active corpus luteum in an ovary. They died in March and May, respectively, which would be consistent with a breeding season in March/April, a 10- to 11-month pregnancy and birth at the end of January/early February. However, there is always the possibility that the ovulating animals were in fact pregnant with an early conceptus that went undetected, or that they were unsuccessful ovulations (the tendency for *Lagenorhynchus* ovaries to contain a large number of corpora albicantia that are probably not representative of successful pregnancies was highlighted by Harrison et al. 1972). The three lactating females occurred between February and April, too early to provide any clue to when lactation might end.

Information on testis size (weight and/or length) is available for 37 males. The relationship between testis length and weight was expressed by the function

$$W = 4.469e^{0.177L},$$

where L is the length of larger testis (in centimeters) and W is the combined weight of testes excluding epididymes (in grams). The overall fit was good ($r^2 = 0.899$), and the relationship has been used to estimate testis weights in the three cases where only lengths were available (Figure 14.4). Dolphins up

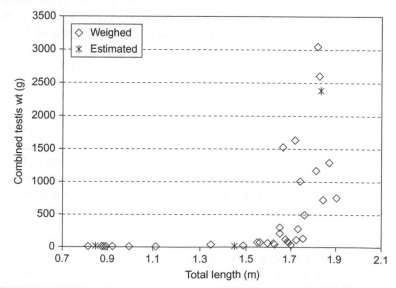

FIGURE 14.4 Relationship between combined testes weight and body length in dusky dolphins from South Africa.

to 1.63 m in body length had small testes, weighing less than 80 g combined ($n = 18$). Dolphins over 1.76 m in length had large testes (combined weights of 500 g or more, $n = 8$). Using criteria of a combined testis weight of 610 g and testis length of 24 cm at maturity (Van Waerebeek and Read 1994), the smallest mature male from the subregion was 1.66 m ($n = 10$), the largest immature was 1.76 m ($n = 27$) in length, and the size of the male at sexual maturity was therefore probably within this range.

With only 10 mature males, and a seasonal coverage from October to March, there were insufficient data to test for any seasonal variation in testis size.

BEHAVIOR

In southern African waters, dusky dolphins have been described as occurring in groups of 2–800 with a mean of 35.3 ($n = 100$); there is no significant difference between the size of schools in summer or winter (Findlay et al. 1992). The few sightings made in deeper water are of larger groupings, ranging from 50 to 300 with a mean of 212.5 (Best 2007), although this figure may be influenced by the lower probability of spotting small schools in the sea conditions often prevailing offshore.

Since 1995, a total of 521 sightings of dusky dolphins has been made incidental to other cetacean research work in the Western Cape (Table 14.5). All this effort has been expended in nearshore waters less than 100 m (and usually less than 50 m) deep, so is unrepresentative of the total habitat of the species. Seasonal coverage is also incomplete, with only 26 sightings between May and September, due to reduced effort over the winter months. From January onwards, the percentage of schools inshore containing calves is much higher than in the months of October to December, with a peak incidence of 38% in February. This presumably represents an influx of neonates into the population, and as such coincides with the season of births as predicted from fetal and neonatal growth. The proportion of calves estimated in a school also increases over the same time, from 17 to 18% in October and November to almost 40% in March, with over half the schools at that time judged to consist entirely of cow–calf pairs. This indicates strong segregation of adult females by reproductive condition. Although sample sizes are small, the mean number of animals in schools containing calves was much larger in October and November (46–79) than in subsequent months, dropping to 6–9 in December and January and then increasing to 17–24 from February to April. Taken in conjunction with the proportion of calves per school, these data can be interpreted as the departure of individual near-term females in a perinatal condition from the larger schools in December and January, and aggregating with other females accompanied by young calves as the calving season develops.

Overall, 93.1% of nursery groups between January and April contained fewer than 40 individuals, with a mean of 17.5, but this was highly influenced

Table 14.5   Incidence and composition of schools of dusky dolphins containing calves, South Africa, 1995–2007

Month	Total sighted	Number with calves	% with calves	Nursery schools	
				Mean % calves present	Mean school size
January	56	11	19.6	32.9	9.0
February	71	27	38.0	37.1	20.7
March	110	32	29.1	39.4	17.8
April	108	31	28.7	34.7	24.0
May	3	0	0		
June	1	0	0		
July	0	0	0		
August	7	0	0		
September	15	0	0		
October	67	4	6.0	17.5	46.8
November	47	3	6.4	18.9	79.3
December	36	3	8.3	41.7	6.7
Total	521	111	21.3		

by a few very large sightings: the mode in 101 sightings was at 8 dolphins (Figure 14.5). At this time of year, nursery groups ($n = 57$) were encountered in sea surface temperatures of very similar ranges (10–18°C, 9–18°C) and means (13.8°C, 13.7°C) to 237 other groups. They were also found in water depths ranging from 2.5 to 20.7 m (mean 7.4 m), with other groups in water depths of 2.1 to 56 m (mean 9.1 m). This difference is statistically significant (Mann–Whitney test, $z = -204$, two-tailed $P = 0.0414$), implying that in late summer and spring, nursery groups tend to occur in shallower water than other groups (Figure 14.6), even within nearshore waters.

Because of the lack of survey coverage over winter, the subsequent fate of the nursery schools is unknown. By October, when the calves would be about eight months old, these schools have either disintegrated as a result of neonatal mortality or weaning and incorporation of the remainder into the population as a whole, or the calves have grown big enough to be effectively indistinguishable from other subadults.

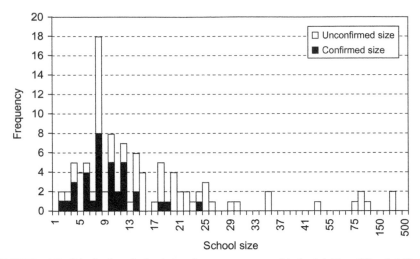

FIGURE 14.5 Distribution of school sizes for nursery groups of dusky dolphins off South Africa, January–April (confirmed size refers to those groups for which the size was established with some confidence, and larger groups naturally tend to be unconfirmed).

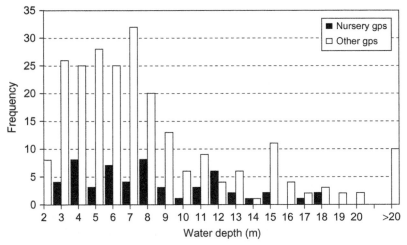

FIGURE 14.6 Distribution of nursery and other groups of dusky dolphins with inshore water depths off South Africa, January–April.

CONSERVATION THREATS

By-catch

Dusky dolphins on the southern African coast are involved as by-catch in several fisheries. Between 1988 and 2002, at least 16 dusky dolphins are known to have died as a result of becoming entrapped in mostly research-related midwater trawls, both off South Africa and in Namibia. Estimates of the cetacean by-catch in this fishery off South Africa are that it is "insignificant," with the reported catch of only one ("bottlenosed") dolphin in 2663 trawls between August 2002 and December 2005, and an extrapolated total catch of seven cetaceans for that period. Although the cetacean catch may have been under-reported owing to the position of the observers on the vessel, independent observations confirmed a very low cetacean catch rate (Petersen and McDonell 2006). The situation in the northern Benguela Current, however, where effort in this fishery has been much higher, is unknown: although observer reports apparently fail to indicate any cetacean by-catch (Andrea Hindley, MRI Whale Unit, University of Pretoria, June 2008, personal communication), this is considered unlikely to reflect the real situation. Eight dusky dolphins are known to have died in 22 research trawl surveys off Namibia post 2000, with about 40 trawls per survey, or a rate of one dolphin per 110 trawls: the dolphins are reported to normally stay clear of the trawl (Jens-Otto Krakstad, Institute for Marine Research, Bergen, Norway, February 2009, personal communication).

Four dusky dolphins are known, and a further two suspected from the circumstances of their "stranding," to have died as a result of entanglement in set nets along the west coast of South Africa between 1986 and 1998. These are seine nets set close inshore to catch mullet, mostly in summer, and the mortalities have occurred between January and May, with one in October. It is a condition of the permits issued for this fishery that the nets should not be left unattended, but this provision does not seem to be strictly adhered to or enforced. There is no observer program for this fishery, and therefore no indication of by-catch levels.

Between 1980 and 2004, a further four dusky dolphins are known to have died as a result of entrapment in purse seines, with a fifth probable victim found floating dead in an area used the previous night by purse seiners. Between 1979 and 1983, personnel of the Sea Fisheries Research Institute observed 209 throws by purse-seiners operating out of Walvis Bay (now Namibia). Eleven dolphins (species unknown) were caught, of which two failed to escape, giving a capture rate of 5.26 dolphins and a mortality rate of 0.96 dolphin per 100 throws (Rob Cooper, Marine and Coastal Management, Cape Town, July 2008, personal communication). In 1983, there was a total of 7013 throws off Namibia, which, if the above capture and mortality rates are applicable, would mean that some 369 dolphins (of all species) might have been captured, and 67 died during that year. Between May 24 and September 7, 1987, technicians from Marine and Coastal Management observed 62 throws by purse-seiners between St. Helena Bay and Hout Bay, South Africa. No dolphins were caught, but on two occasions 30 and 10 common dolphins (*Delphinus* sp.) escaped from the net before it was closed. Although there is some more recent official observer coverage of this fishery, there are no published data on cetacean by-catch. However, as an indication of the potential effect, the total number of throws between 2003 and 2007 in the South African fishery north of Cape Point, within the range of dusky dolphins, varied from 3640 to 9721, with an average of 6719 (Jan Van der Westhuizen, Marine and Coastal Management, Cape Town, July 2008, personal communication).

Seine nets deployed from the beach have been responsible for the death of at least one and possibly two dusky dolphins since 1988—this excludes attempts at live capture for oceanaria using beach seines. This fishery is very localized and sporadic in nature, takes place in daylight, and is targeted at specific shoals, so a high cetacean by-catch should not be anticipated, and possibilities for release are good. It does not have an observer scheme.

Deliberate takes

Other incidental takes have been recorded, including (besides the three animals shot and mentioned in the opening vignette) two dusky dolphins harpooned from fishing boats in 1984 and 1990. Such deliberate takes have been illegal in terms of South African legislation since 1973 (Best and Ross 1977). One dolphin was also taken accidentally when it became entangled in a monofilament hand line deployed from a research boat.

Other deliberate removals include 37 dusky dolphins that were taken in a live capture fishery for oceanaria between 1961 and 1981, of which 13 were immediately released (Best and Ross 1984). The remainder died either at capture or later while in captivity.

Pollutants

Persistent organochlorine pollutant levels were measured in the blubber of 12 dusky dolphins from southern Africa between 1977 and 1987. High tDDT/PCB ratios for 1977 and 1982 were not observed subsequently, and the low overall mean DDE/tDDT ratio of 0.76 (together with an even lower figure for nine Heaviside's dolphins) suggested that there was no significant input of DDT in coastal waters along the west coast of South Africa. This might be expected given the generally arid conditions of the coastal region and the scarcity of arable land (de Kock et al. 1994).

Besides better documentation of the numbers of animals involved in any by-catch, the most important information needed is a reliable estimate of population size and (ideally) trend, so that the possible impact of these factors can be evaluated.

DISCUSSION

It is interesting to compare aspects of the biology of the dusky dolphin population off southern Africa with those for the same species off Peru, as these two populations inhabit the two major coastal upwelling systems in the southern hemisphere, the Humboldt Current off Peru, and the Benguela Current off southern Africa, but genetically are highly differentiated (Cassens et al. 2005).

Physically, the dusky dolphins from southern Africa are slightly smaller than their counterparts from Peru (Van Waerebeek 1993b), with mature females measuring 1.68–1.91, average 1.8 m, compared with 1.68–2.05, average 1.89 m, and mature males measuring 1.66–1.9, average 1.8 m, compared with 1.76–209.5, average 1.87 m. This difference is also reflected in the smaller neonatal size in southern African dolphins (86.7 cm vs. 91.4 cm), although the methods of estimation are not identical (Van Waerebeek and Read 1994). The slightly reduced dentition of southern African dolphins may be correlated with their smaller skull size (Van Waerebeek 1993a).

Dusky dolphins from southern African waters appear to be more catholic in their feeding habits than those off Peru. No single prey species occurred in more than 59% of stomachs, while there were seven species that occurred in more than 20% of stomachs: off Peru, one species was found in 97.8% of stomachs and there were only two species that occurred in more than 20% (McKinnon 1994). This difference in prey diversity was reflected in the numbers and mass of each prey species eaten. Off Peru, one species (anchoveta, *Engraulis ringens*) comprised 92.5% by number and 83.8% by weight, with the next most important species by number being *Trachurus symmetricus* (2.5%) and *Loligo gahi* (2.2%), and by weight *Merluccius gayi* (5.1%) and *T. symmetricus* (3.5%). In southern African waters, there were three species that dominated numerically (*Lampanyctus* 43.2%, *Maurolicus* 26.1%, and *Trachurus* 19.5%), and three species by weight (*Trachurus* 34.7%, *Merluccius* 22.9%, and *Lampanyctus* 12.8%), with another two scoring higher than 5% by weight (*Todarodes* and *Sardinops*). Although the data from southern Africa were collected over a longer time span (roughly 14 years) than Peru (2 years) and so may have been less subject to the effects of inter-annual variability, there seems little doubt that anchoveta was by far the most important prey species for all reproductive classes of dolphins off Peru during the period of sampling (McKinnon 1994).

Despite a rather small sample from southern Africa, there also seem to be differences in the reproductive behavior of dusky dolphins from the two regions. The proportion of mature females that were resting (i.e., not pregnant or lactating) was 7/15 or 46.7% in South African dusky dolphins, but only 22/144 or 15.3% in dolphins from Peru (Van Waerebeek and Read 1994); these proportions are significantly different (chi-square = 30.3, $P < 0.001$). This suggests that females off southern Africa may be less fecund overall. The calving season in southern African waters is also some four months later than that off Peru, occurring in late summer rather than spring. This finding is essentially based on the incidence of neonates, and is unaffected by the different estimates of the duration of pregnancy for the two regions. The latter may be to some extent the consequence of a very small sample of fetuses from southern Africa: theoretical estimates of the fetal growth rate from the relationship with birth length proposed by Kasuya (1977) for small odontocetes would indicate 0.289 cm/day, and the total length of gestation estimated from the relationship with birth length proposed by Perrin et al. (1977) would be 11.3 months. Both these estimates are intermediate between the respective values calculated for southern African and Peruvian animals.

The different timing of the breeding seasons between the two populations may be linked to seasonal differences in biological productivity between the two regions, as shown by remote sensing of chlorophyll concentrations (Thomas et al. 2004). In the Peruvian region of the Humboldt system, upwelling is strongest in winter and weakest in summer, whereas in the southern Benguela it tends to be strongest in spring and summer (Carr and Kearns 2003, Fréon et al. 2006). These seasonal differences in productivity presumably feed through ultimately to differing seasonal availabilities of prey species in the two regions.

In a comparison of the seabirds of the Humboldt and Benguela systems, Crawford et al. (2006) postulated that for four species pairs that fed largely on anchovy and sardine in each system, the species from the Humboldt system exhibited reproductive characteristics favoring a more rapid increase than the equivalent species from the Benguela system. Crawford et al. (2006) proposed that this was an adaptation to the more frequent occurrence of environmental perturbations such as El Niño in the Humboldt than the Benguela system that could adversely influence prey availability. A similar hypothesis might explain some of the differences in the biology of the dusky dolphins off Peru and South Africa. In the Humboldt system, the dolphins can take advantage of the most productive region for small pelagic fish on earth, but with the diet dominated by a single species, they are vulnerable to fluctuations in its availability, particularly the multi-year periods of scarcity during regime shifts (Schwartzlose et al. 1999). In response, the Peruvian dolphins may have adapted by maintaining a larger body size (so providing a greater potential for energy storage) and exhibiting a potential for more frequent reproduction than their conspecifics off South Africa (enabling them to take advantage of periods of high prey availability, as postulated by Van Waerebeek and Reed 1994). The possible impact of environmental anomalies on the nutrition of dusky dolphins from Peru is indicated by the deposition of a layer of anomalous dentine correlated with the occurrence of the 1982/1983 El Niño. Such layers were found in the teeth of 15 of 18 mature females examined, but not recorded in the teeth of other small odontocetes from the region such as Burmeister's porpoise (*Phocoena spinipinnis*), bottlenose, spotted (*Stenella attenuata*), and common (*Delphinus* sp.) dolphins (Manzanilla 1989).

But why does the Peruvian population not switch prey species when anchoveta are scarce? At the time of the stomach sampling in Peru (1985 and 1986), catches of anchoveta from both the north/central Peru and south Peru/north Chile stocks were close to their minimum, and catches of sardine close to their maximum (Schwartzlose et al. 1999). If the commercial catches truly reflected the relative availability of these two species, then the dominance of *Engraulis* and scarcity of *Sardinops* in the dusky dolphin diet at that time would imply strong species selectivity by the predator and a reluctance to switch prey. Unlike the Benguela and the California Current systems, the anchoveta seems to have remained the dominant component of the Peru Current system for millennia, and (until the recent unprecedented removals through fishing) has not suffered periodic displacement by longer lived species such as sardine (Bakun 1996). Dusky dolphins off Peru may therefore have adapted to forage almost exclusively on anchoveta and be "unprepared" to switch to other species at relatively short notice.

In conclusion, the differences seen between the biology of dusky dolphins off Peru and in the southern Benguela may have evolved as adaptations to foraging in environments with very different variabilities and seasonal availability of prey.

ACKNOWLEDGMENTS

We are very appreciative of all those colleagues who have assisted us in the field with observations of and in collecting material from dusky dolphins: these include Meredith Thornton, Jaco Barendse, Johan de Goede, Simon Elwen, Deon Kotze, Seshnee Maduray, Keshnee Pillay, Desray Reeb, Dandelene Reynolds, Keiko Sekiguchi, and Stephan Swanson. A host of Earthwatch and other volunteers also provided invaluable support in several cetacean projects in which dusky dolphins were sighted. The following people kindly allowed us to cite their unpublished work or work in press: Erik Axelsen, Matteo Bernasconi, Bjørn Bjerland, Rob Cooper, Jens-Otto Krakstad, Mette Mauritzen, Erik Olsen, and Samantha Petersen. Jan Van der Westhuizen (Marine and Coastal Management) kindly provided data on effort levels in the South African purse seine fishery. Norbert Klages (Arcus Gibb Consulting) was able to provide missing data from stomach content analyses. Commander Mac Bisset (South African Navy, retired) assisted in tracking the movements of SA Navy vessels. Coleen Moloney advised on comparisons between the Humboldt and Benguela ecosystems. PB acknowledges the support of the National Research Foundation and Mazda Wildlife Fund.

Chapter 15

Patterns of Sympatry in *Lagenorhynchus* and *Cephalorhynchus:* Dolphins in Different Habitats

Sonja Heinrich,[a] Simon Elwen,[b] and Stefan Bräger[c]

[a]*Sea Mammal Research Unit, University of St. Andrews, Scotland, UK*

[b]*Mammal Research Institute, Department of Zoology & Entomology, University of Pretoria, South Africa*

[c]*German Oceanographic Museum (DMM), Stralsund, Germany*

March 28, 2006, Southern South America: *Several distant splashes to starboard catch my eye. Leaping dolphins! Our vessel has just entered the sheltered waters of the Beagle Channel and we are enjoying the welcome calm after a rocky ride across the (in)famous Drake's Passage that separates the southern tip of South America from the Antarctic Peninsula. We had just had an adventurous cruise in the Antarctic and are now north-bound for milder climates to travel along the Pacific coast of South America. The splashes draw closer. Our vessel attracts the dolphins' attention. A quick count suggests that the group consists of at least 15 fast-moving individuals. Within a minute, six of the dolphins jostle for the best position at our vessel's bow wave to catch a free ride. I had not been able to get a good look at them as they raced over, but assumed them to be Peale's dolphins which are regularly found in these southern channels, though they usually occur in smaller groups. Leaning over the side I can now see dark gray bodies with a light blaze across the flanks moving in and out of the ship's wave beneath me. None of them have the characteristic dark face of Peale's dolphins. These friendly acrobats are very clearly dusky dolphins, the slightly smaller relative to Peale's in the somewhat ambiguous genus* Lagenorhynchus. *Finding dusky dolphins here at the entrance of the Beagle Channel near Isla Nueva (55°S) is especially exciting as this constitutes pretty much the southern-most extreme of their known range. After several minutes of obvious fun, the six bow-riders turn away from the vessel and continue their high-speed energetic travel with the rest of the group.*

I do not see these boisterous dolphins again until about 2000 km further north along the Chilean coast. In the interim, we cruise through the remote, tannin-stained and rain-swept southern Chilean fjords where small groups of Peale's dolphins make up most of our cetacean sightings. At a few select locations, in tidal rips or in front of a glacier, we are treated to glimpses of yet another local species, the elusive Chilean dolphin. I am familiar with these little gray goblins from several years of research in an archipelago at the northern extent of the fjords where our team had been investigating how Chilean and Peale's dolphins differ in their ecology and coexist in these coastal waters. Seeing the small Mickey-Mouse ear shaped dorsal fins of Chilean dolphins reminds me of the rather similar and closely related Hector's dolphins in far away New Zealand, where I also met my first dusky dolphins years before. It seems curious that out of this group, only dusky dolphins range throughout the entire southern hemisphere, yet co-occur with closely related but geographically restricted species off each continent.

As our vessel reaches the open coast of central Chile, Peale's and Chilean dolphins fade from sight, and groups of 10–20 dusky dolphins come back into view. And as we enter the upwelled, nutrient-rich waters of southern Peru we are in for a dusky treat. Large splashes at the horizon again provide the cue. This time, however, we are approaching a feeding aggregation and an impressively large group of dusky dolphins with a handful of common dolphins thrown into the mix. We watch in fascination as the dolphins work in subgroups of several dozens to corral fish (most likely anchovy) to the surface where hundreds of sea gulls, cormorants, and boobies join to take their share of the fishy feast. Our vessel slowly moves ahead as more and more groups of dusky dolphins come into sight, pushing the total number encountered to well over 1400 dolphins within a radius of several kilometers. Such large aggregations of dusky dolphins are well documented for other coastal upwelling areas, yet they don't seem to be the status quo, and such group- ings are unheard of for related species such as Peale's, Chilean, or Hector's dolphins. What factors influence the different ecological strategies and mediate sympatry of these similarly sized and related species? We have much to know yet to answer these questions.

Sonja Heinrich, St. Andrews, Scotland, September 2008

DEFINING SYMPATRY

The co-occurrence of two or more species in the same geographic area— "sympatry"—is common in the marine environment where important resources such as prey are often ephemeral, clumped and patchily distributed. Sympatry can range from overlap in broad-scale distribution with animals not directly interacting with each other, to the formation of multi-species groups with coor- dinated activities. Similar species co-occurring in the same area are thought to compete for resources unless they occupy different physical locations or eco- logical niches. Studies on the distribution and habitat use patterns of dolphins have revealed a range of sympatric strategies based on behavioral differences, habitat, and resource partitioning (Baird et al. 1992, Gowans and Whitehead 1995, Ford et al. 1998, Herzing et al. 2003, Bearzi 2005a, 2005b, Parra 2006).

Dusky dolphins (*Lagenorhynchus obscurus*) occupy disparate habitats in the southern hemisphere and exhibit diverse foraging and social behaviors. In different parts of their range, dusky dolphins are broadly sympatric with at least 12 other dolphin species (delphinids) and have been seen associated with Risso's (*Grampus griseus*), southern right whale (*Lissodelphis peronii*), common

(*Delphinus* spp.), bottlenose (*Tursiops truncatus*), Hector's (*Cephalorhynchus hectori*), Heaviside's (*C. heavisidii*) dolphins and killer whales (*Orcinus orca*; Würsig and Würsig 1980, Würsig et al. 1997). Even putative hybrids have been described of dusky dolphins with common (Reyes 1996, Würsig et al. 1997) and southern right whale dolphins (Yazdi 2002).

Here, we focus on patterns of sympatry between species of two closely related southern hemisphere genera: *Lagenorhynchus*, to which dusky dolphins belong, and *Cephalorhynchus*. Although different members of these genera have largely overlapping ranges, they differ markedly in their ecology, ranging behavior, and social structure, thus making them interesting study species for a comparative approach. Despite a growing interest in coexisting dolphin communities, few studies have gone beyond describing their occurrence or direct interactions. Questions such as "To what extent do sympatric dolphins differ in their resource use across a range of habitats?" and "How do the same species interact in different habitats?" remain unanswered. Thus, little is known about the ecological mechanisms that allow sympatric dolphins to coexist. In the following, we review existing information for sympatric *Lagenorhynchus–Cephalorhynchus* species and describe differences and similarities across a range of populations, habitats, and continents, attempting to elucidate at least some of the patterns in their sympatry.

SOUTHERN HEMISPHERE DOLPHINS

Distribution and species relations

Dusky dolphins share the cool temperate coastal and shelf waters of the southern hemisphere with other small-bodied dolphins of the same genus and the relatively closely related genus *Cephalorhynchus*. Members of the *Lagenorhynchus* genus are slightly larger (205–220 cm) and heavier (100–115 kg) than the *Cephalorhynchus* dolphins (145–175 cm; 45–74 kg; Dawson 2002, Goodall 2002, Van Waerebeek and Würsig 2002). The two genera exhibit similar patterns of morphological and genetic divergence (Harlin-Cognato and Honeycutt 2006) and a discontinuous geographic distribution with widely separated subspecies or congeners occurring off both coasts of South America, southwestern Africa, several temperate and sub-Antarctic islands, and New Zealand (Dawson 2002, Goodall 2002, Van Waerebeek and Würsig 2002; Figure 15.1).

Several studies have suggested that the currently recognized genus *Lagenorhynchus* is polyphyletic (LeDuc et al. 1999, May-Collado and Agnarsson 2006, Harlin-Cognato et al. 2007), and have placed the three southern hemisphere species, dusky, Peale's (*L. australis*), and hourglass (*L. cruciger*) dolphins into two groups within the subfamily Lissodelphininae, separate from the *Lagenorhynchus* species in the North Atlantic (LeDuc et al. 1999, Harlin-Cognato et al. 2007). The Peale's dolphin and hourglass dolphin are thought to be monophyletic (LeDuc et al. 1999, Harlin-Cognato et al. 2007),

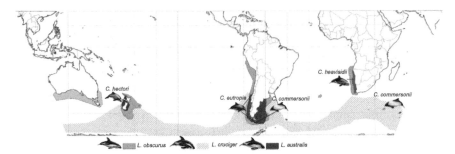

FIGURE 15.1 Approximate distribution of *Lagenorhynchus* and *Cephalorhynchus* dolphins in the southern hemisphere. Species ranges are indicated by shading, except for *C. eutropia* and *C. commersonii* (only outer limits denoted by lines, overlap with *L. australis*) (dolphin drawings courtesy of M. Würtz).

but have strikingly different distribution patterns and habitat requirements (Figure 15.1). Dusky dolphins have a circumpolar yet discontinuous distribution with three distinct subspecies suggested based on geographic differences in size, morphology, and coloration (Rice 1998). *L. obscurus fitzroyi* is thought to range from Peru to Argentina and the Falkland Islands and is broadly sympatric with Peale's dolphins in southern South America (Goodall et al. 1997). *L. o. obscurus* occurs from South Africa's Cape Agulhas to southern Angola, and *L. obscurus* subsp. inhabits the waters around New Zealand. Infrequent sightings and strandings indicate that dusky dolphins also occur (at least occasionally) off southern Australia (Gill et al. 2000). A further description of these trends is provided in Chapters 1 and 2.

The genus *Cephalorhynchus* is monophyletic and also belongs to the subfamily Lissodelphininae (Pichler et al. 2001, Harlin-Cognato and Honeycutt 2006). The four *Cephalorhynchus* species exhibit considerably more restricted ranges than the sympatric *Lagenorhynchus* species (Figure 15.1, Table 15.1). Only the Commerson's dolphin (*C. commersonii*) is found off more than one land mass (Dawson 2002), with subspecies off the southern east coast of South America (*C. commersonii*; Chile, Argentina, Falkland Islands) and around the Îles Kerguelen (*C. commersonii* subsp.; Robineau et al. 2007). In a small part of southern Chile (Strait of Magellan and Tierra del Fuego), the Commerson's dolphin is directly sympatric with the congeneric Chilean dolphin (*C. eutropia*). The latter is endemic to Chile and is patchily distributed from Cape Horn (55°S) north through the fjords and along the open coast to roughly 33°S (Goodall et al. 1988b). The Heaviside's dolphin ranges from the Cape of Good Hope off the west coast of South Africa to southern Angola (Best and Abernethy 1994). The Hector's dolphin is endemic to New Zealand, and has three main populations off the east, west, and south coasts of the South Island (Pichler et al. 1998). The highly endangered subspecies Maui's dolphin (*C. hectori maui*) is only found on the west coast of the North Island, and does not generally overlap with dusky dolphins.

TABLE 15.1 Ecological characteristics of southern hemisphere *Lagenorhynchus* and *Cephalorhynchus* dolphins

Dolphin species	Distribution	Habitat characteristics	Diet	Typical group size	Known or potential threats to populations[a]
Dusky, *L. obscurus*	Discontinuous across temperate southern ocean	Coastal and continental shelf, associates with cold water currents	Small schooling fish (e.g., sardines, anchovies, hake) and squid, deep scattering layer	2–500, gregarious, can form large aggregations of >1500	Direct catch for bait and consumption (Peru), incidental catch in fishing gear, tourism (New Zealand, Argentina), oil exploration (Argentina, Falkland Is.), habitat exclusion and degradation due to aquaculture (New Zealand)
Peale's, *L. australis*	Southern South America and Falkland Is.	Shallow coastal waters and occasionally continental shelf	Demersal and bottom fish, octopus, squid	2–10, occasionally form larger aggregations of <100	Incidental catch in fishing gear, some tourism, oil exploration and processing, habitat exclusion and degradation due to extensive aquaculture (Chile), past direct take for bait (Chile)
Hourglass, *L. cruciger*	Circumpolar in sub-Antarctic and Antarctic waters	Offshore, occasional coastal off South America	Fish, squid, crustaceans	1–16, occasionally form larger aggregations of <150	Some incidental catch in fishing gear, but not considered of concern
Heaviside's, *C. heavisidii*	South-western Africa	Coastal and shallow shelf waters	Demersal and pelagic fish (e.g., hake) and squid	1–10, occasionally form larger aggregations of <100	Incidental catch in fishing gear, tourism
Hector's, *C. hectori*	Endemic to New Zealand	Shallow coastal waters	Small surface schooling and benthic fish	2–8, occasionally form aggregations of up to 60	Incidental catch in fishing gear, tourism, some habitat exclusion and degradation due to aquaculture
Chilean, *C. eutropia*	Endemic to Chile	Shallow coastal waters	Small fish, octopus, squid, crustaceans?	2–10, occasionally form aggregations of >50 along open coast	Incidental catch in fishing gear, habitat exclusion and degradation due to extensive aquaculture, past direct take for bait
Commerson's, *C. commersonii*	South-east South Atlantic, some southern Chile, Kerguelen Is.	Shallow coastal waters	Small surface schooling and benthic fish, shrimp, squid	1–3, occasionally form large aggregations of >100	Incidental catch in fishing gear (Argentina), tourism, oil exploration and processing (Argentina, Chile), past direct take for bait (Chile)

[a]*Throughout the entire range unless country specified in parentheses.*

Broad-scale ecological patterns

The dolphin species discussed above (except one) are restricted to cool temperate coastal and shelf waters. Only hourglass dolphins have a predominantly oceanic and circumpolar distribution, occurring on both sides of the Antarctic Convergence and in cold currents associated with the West Wind Drift (Goodall 1997). Peale's dolphins have the smallest range of all the *Lagenorhynchus* species, as they are found only in the coastal and shallow shelf waters of southern South America. The neritic Peale's dolphin and the oceanic hourglass dolphins tend to occur in small groups with only occasional short-lived aggregations of several tens of dolphins (Table 15.1; Goodall 1997, Goodall et al. 1997). Dusky dolphins show a marked association with continental shelves and the cool waters of the Humboldt/Peruvian (eastern South Pacific), Falklands (western South Atlantic) and Benguela (eastern South Atlantic) currents as well as the cold water currents around New Zealand (Brownell and Cipriano 1999). They can be both neritic with usually small groups of less than 20 dolphins, and pelagic with groups of hundreds to thousands of individuals over the continental shelves of New Zealand, South Africa, and Peru (Würsig et al. 1997). These very large aggregations appear to be environmentally mediated and are likely to be a result of several groups joining when feeding on surface schooling prey. The three best-studied dusky dolphin populations show at least three different grouping patterns (Gowans et al. 2008): off the deep-water canyons of Kaikoura (New Zealand) they form a large, seasonally resident society; in the wide relatively shallow bays of Península Valdés (Argentina) dusky dolphins occur in strong fission–fusion societies from small to large groups and travel widely over an expansive continental slope; and in Admiralty Bay (New Zealand) they live in small, seasonally resident groups.

In contrast to the considerable variation in social structure and habitat use of *Lagenorhynchus* dolphins, the ecology and life history of *Cephalorhynchus* dolphins are more consistent. All species are neritic, usually occur in small groups of less than 10 dolphins, have low levels of association between individuals and exhibit relatively limited movements (Table 15.1; Dawson 2002).

Distribution patterns and sociality can be influenced by predation pressure. Deep-water sharks, such as sevengill (*Notorynchus cepedianus*), great white (*Carcharodon carcharias*), and shortfin mako (*Isurus oxyrinchus*), pose threats to dolphin species and/or populations that use deeper water (e.g., dusky dolphins, Heaviside's dolphins; Best 2007), but are unlikely to affect coast-hugging shallow water species (e.g., Chilean dolphins). Killer whales are known to take dusky dolphins and other species off Argentina (Würsig and Würsig 1980) and New Zealand (Constantine et al. 1998), but attacks on *Cephalorhynchus* dolphins have yet to be documented. Thus, predation pressure is not equal across species, populations, or even individuals, and is temporally variable depending on an individual's habitat use pattern and life history stage (e.g., calves are more vulnerable than adults; see also Chapter 7).

SYMPATRIC *LAGENORHYNCHUS* AND *CEPHALORHYNCHUS* DOLPHINS IN DIFFERENT HABITATS

A note on methods and our comparative approach

We have attempted to cover as many species and study areas as possible to capture the broad range of habitat characteristics and behavioral variation within and between the selected sympatric species. We have drawn heavily on published material, gray literature, unpublished information kindly provided by several colleagues, and findings from our own research. It quickly became apparent that a detailed comparison was hampered by a lack of systematically collected data for many species. Even when information was reported for one species, it was often not available for the sympatric species within the same area. Reported observations were also spatially and temporally skewed towards a handful of well-studied populations in relatively accessible nearshore habitat, usually limited to less than 5 km from shore, with data primarily collected during the climatically favorable season (i.e., usually in summer). Direct comparisons were further limited by heterogeneity in data collection and reporting. Thus, the painted picture of sympatric ecology should be viewed as an incomplete jigsaw puzzle, and generates hypotheses rather than conclusions. We hope that our review will provide some food for thought and stimulate interest in comparative studies of sympatric ecology, preferably conducted with standardized methodology and analyzed in a rigorous quantitative framework.

Sympatric relationships in New Zealand

Dusky dolphins are widely distributed in the cold temperate waters around the South Island of New Zealand (Constantine 1996, Würsig et al. 1997). Hector's dolphins are broadly sympatric with dusky dolphins, but have a more nearshore distribution and are absent from the deep waters of Fiordland. Despite this broad sympatry, direct interactions between the two species are rare. Co-occurrence of single-species groups and mixed-species groups have been noted in three areas: off Kaikoura on the east coast of the South Island, in the Marlborough Sounds, and off the west coast of the South Island (Bräger and Schneider 1998, Markowitz 2004).

The deep Hikurangi Trench comes close to shore (<1 km) just south of the Kaikoura Peninsula and is home to a well-studied population of over 12 000 dusky dolphins of which about 2000 can be found in the Kaikoura area at any one time (Markowitz 2004, Würsig et al. 2007). Dusky dolphins are found closer to shore during summer compared to winter. They also exhibit diurnal movements, inshore during the morning to rest and offshore at night to feed on prey associated with the rising deep-scattering layer (Würsig et al. 1997, Benoit-Bird et al. 2004). Group sizes vary with behavior and group composition from four to several thousand individuals, with nursing groups (those containing predominantly mother–calf pairs) averaging about 20 dolphins (Markowitz 2004, Weir 2007). Fewer than 300 Hector's dolphins inhabit the shallow coastal waters to the north and south of the Kaikoura Canyon area (between the Hapuku and Conway Rivers) year-round (Dawson et al. 2004). They are usually found in small groups (average 7 dolphins, Figure 15.2) in shallow nearshore waters (average depth 21 m, Figure 15.2). Interactions between the two species were rarely observed during two independent but temporally overlapping studies, one targeting dusky dolphins (Markowitz 2004) and the other Hector's dolphins (Bräger 1998). The dusky dolphin study found direct interactions mainly during summer (76%, $n = 17$) when dusky dolphins occurred closer to shore. Most interactions took place in shallow water (less than 30 m) and usually involved small groups of dusky dolphins (5–19) and Hector's dolphins (2–8). Dusky dolphin calves were present during 88% of these interactions. Mixed nursery groups with calves of both species were observed in four cases.

In contrast, the Hector's dolphin study did not record any interactions between the two species during 80 photo-identification surveys conducted between the Hapuku and Conway Rivers between March 1994 and April 1997. During that period, 45 individual Hector's dolphins were photo-identified and re-sighted repeatedly. The majority of individuals (69%) were only seen near the Hapuku River mouth and around Kaikoura Peninsula, a shallow-water area rarely visited by dusky dolphins. Another 27% of the individuals were only identified in the southern part of the study area (15 km south of Kaikoura Peninsula), and only two individuals (4%) were photo-identified on both sides of the deep-water canyon. Thus, both species differed markedly in their movement ranges and habitat use (see also Table 15.2). Hector's dolphins meet their habitat requirements in the nearshore waters, whereas dusky dolphins seem to only frequent these areas for resting, socializing, and shelter from predatory attacks (Constantine et al. 1998, Markowitz 2004, Würsig et al. 2007). The nature of the social interactions and mixed groups do not seem to incur competitive costs for limited resources (e.g., prey) as dusky dolphins do not forage close to shore in Hector's dolphin habitat.

Sympatric interactions on the west coast of New Zealand seem to follow a similar pattern, though are less well understood. The west coast of New Zealand is geographically diverse, ranging from the deep narrow fjords of Fiordland to the open coast of Westland which has an extensive shallow water belt (the 30 m contour extending to more than 2.5 km from shore). The distribution of both species along the coast reflects the variation of coast types, as well as seasonal variation in the environment (Bräger and Schneider 1998). The Westland coast is home year-round to the largest population of Hector's dolphins in New Zealand (Slooten et al. 2004). During 97 systematic boat-based sighting surveys conducted in the nearshore waters, Hector's dolphins were by far the most frequently encountered species ($n = 633$ sightings) and occurred in small groups of, on average, 7–9 animals in shallow water (3–15 m, Figure 15.2). Small groups of 2–100 (average 23) dusky dolphins were found during 14% of these surveys and

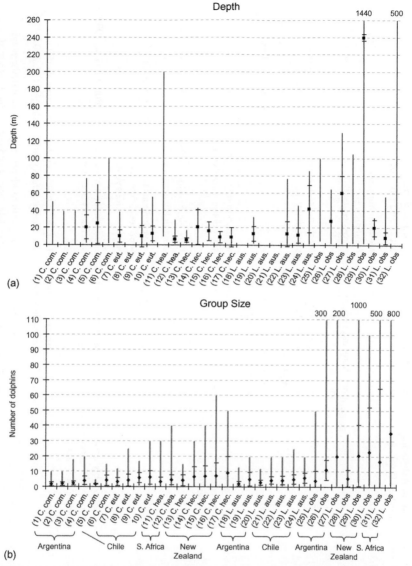

FIGURE 15.2 Variation in (a) bottom depth range and (b) group size for seven species at 21 different study locations in four countries. Mean with ±1SD error bar and range (line). Number on top indicates maximum value where these exceeded the scale. Information collected during systematic observations or dedicated surveys. Number in front of species name refers to data source: (1) Coscarella 2005; (2) Miguel Iñiguez, Fundación Cethus, Buenos Aires, Argentina, June 2008, personal communication to SH; (3) Miguel Iñiguez, Fundación Cethus, Buenos Aires, Argentina, June 2008, personal communication to SH; (4) Reyes 2008; (5) Lescrauwaet et al. 2000; (6) Robineau 1985; (7) Francisco Viddi, Centro Ballena Azul, Chile, July 2008, personal communication to SH; (8) Gibbons et al., 2002; (9) Viddi 2003; (10) Heinrich, unpublished data, Heinrich 2006; (11) Findlay et al. 1992; (12) Elwen 2008; (13) Bräger 1998; (14) Bräger 1998; (15) Bräger 1998; (16) Bräger 1998; (17) Bräger 1998; (18) de Haro and Iñíguez 1997; (19) Reyes 2008; (20) Gibbons et al., 2002; (21) Lescrauwaet 1997; (22) Heinrich, unpublished data, Heinrich 2006; (23) Francisco Viddi, Centro Ballena Azul, Chile, July 2008, personal communication to SH; (24) Viddi 2003; (25) Crespo et al. 1997; (26) Würsig and Würsig 1980; (27) Garaffo et al. 2007; (28) Vaughn et al. 2007; (29) Markowitz 2004, Weir 2007; (30) Bräger, unpublished data, Bräger and Schneider 1998; (31) Elwen 2008; (32) Findlay et al. 1992.

TABLE 15.2 Ranging and movement patterns of populations of *Lagenorhynchus* and *Cephalorhynchus* in adjacent habitats

Dolphin species	Study area, Country	Short-term ranges (km)	Maximum distance observed (km)	Diurnal/seasonal movements	Residency in study areas and site fidelity	Source
Dusky, *L. Obscurus*	40°–42°S, New Zealand	10–150	c.1000	Diurnal: inshore–offshore (non-winter); seasonal: alongshore and inshore–offshore	Medium-term (seasonal) residency, site fidelity across years	Würsig et al. 1991, Cipriano 1992, Markowitz 2004
Dusky, *L. o. fitzroyi*	42°–43°S, Argentina	20–100	c.800	Diurnal and seasonal: inshore–offshore and alongshore	Medium-term residency, site fidelity across years	Würsig and Würsig 1980, Würsig and Bastida 1986
Peale's, *L. australis*	53°S, Chile	c.70	c.300	Diurnal: alongshore; seasonal: inshore–offshore suspected	?	Lescrauwaet 1997
Peale's, *L. australis*	42°–43°S, Chile	10–40	c.45 (possibly under-estimated as suspected to move outside study area)	Seasonal: possible direction unknown	Medium-term residency, site fidelity across years	Heinrich 2006
Chilean, *C. eutropia*	42°–43°S, Chile	10–30	c.60	No	Long-term residency, high site fidelity across years	Heinrich 2006
Commerson's, *C. commersonii*	43°S, Argentina	?	c.250	Seasonal: possible alongshore	Mixed: Short- to medium-term residency, site fidelity across years	Coscarella 2005
Commerson's, *C. commersonii*	43°S, Îles Kerguelen	10–20	c.25 (possibly under-estimated as suspected to move outside study area)	Seasonal: inshore–offshore	Medium-term residency, site fidelity across years	de Bufrénil et al. 1989
Hector's, *C. hectori*	41°–44°S, New Zealand	8–53	c.106	Seasonal: inshore–offshore; diurnal: inshore–offshore?	Long-term residency, high site fidelity across years	Stone et al. 1995, Bräger et al. 2002
Heaviside's, *C. heavisidii*	32°–34°S, South Africa	40–80	c.150	Diurnal: inshore–offshore; alongshore movements unlikely	Long-term residency, high site fidelity across (at least 3) years	Best and Abernethy 1994, Elwen et al. 2006

Data from dedicated studies using photo-identification or radio-tracking/telemetry techniques (see sources for details).

tended to use deeper waters (7–32 m) than the Hector's dolphins (Figure 15.2), predominantly in summer. Their occurrence on the west coast is thought to be related to the warmer surface waters that occur during summer months (Gaskin 1968a) and likely affect the distribution pattern of the dolphins' preferred prey (Bräger and Schneider 1998). Five mixed-species groups were seen, consisting of 4–100 dusky dolphins and 2–7 Hector's dolphins, in waters less than 10 m deep. Dusky dolphin calves were present in three of these mixed groups.

In summary, interactions between Hector's and dusky dolphins in New Zealand appear to be rare and likely reflect differences in habitat use, diet, and foraging strategies. Hector's dolphins occur in small groups of 2–8 individuals (Table 15.1, Figure 15.2) and show a marked preference for turbid, shallow waters (Bräger et al. 2003) where they tend to hunt single, mostly benthic, prey (Slooten and Dawson 1988). Dusky dolphins, on the other hand, regularly occur in much larger groups and exhibit considerable plasticity in feeding behavior, with most tactics requiring cooperation between group members, thereby resulting in larger group sizes (Benoit-Bird et al. 2004, Vaughn et al. 2007).

When interspecific interactions do occur, they appear to be non-aggressive and incur no competitive cost, with both species temporarily using the same habitat. Interactions might also be beneficial to one or both species if, for example, larger mixed groups result in increased protection for one or the other species from predators, through greater awareness or decreased individual risk, particularly for nursing individuals or groups. Several attacks of killer whales on dusky dolphins off Kaikoura (and Argentina) have been documented, with dusky dolphins moving into very shallow water close to shore in response to the presence of killer whales (Würsig et al. 1997, Constantine et al. 1998, see Chapter 7).

Sympatric relationships in Africa

Dusky and Heaviside's dolphins are sympatric throughout the majority of the Benguela current system off the south-west coast of Africa. Dusky dolphins range from Cape Point, South Africa (about 35°S) to Lobito Bay, Angola (about 12°S), and extend from the shoreline to at least 500 m (and likely even 2000 m) depth (Findlay et al. 1992). Heaviside's dolphins have a slightly more restricted range and do not extend as far north, south or offshore, being found predominantly in waters less than 100 m deep (Findlay et al. 1992). The only exception to their sympatric distribution is a possible hiatus in the occurrence of dusky dolphins near the South African–Namibian border (Findlay et al. 1992).

The Benguela ecosystem is a cool eastern boundary current region that flows northward along the coast of southwestern Africa, from the Cape of Good Hope to southern Angola, and represents one of the most productive ocean areas in the world (Brown et al. 1991). The continental shelf off southwestern Africa is relatively deep and broad, ranging from about 35 to 70 km from the coast (maximum 150 km). The coastline is very exposed, being generally straight with few bays and it is dry, with few rivers.

Most information on the behavior and ecology of dusky and Heaviside's dolphins stems from three studies: a collation of large-scale surveys, incidental sightings, and stranding records (Findlay et al. 1992); comparative stomach content analysis of caught, by-caught, and stranded animals (Sekiguchi et al. 1992, Sekiguchi 1994); and a nearshore mark–recapture and satellite tagging study of Heaviside's dolphins (Elwen 2008). Dedicated photo-identification surveys ($n = 109$), primarily aimed at Heaviside's dolphins, were conducted in the nearshore waters of the Western Cape from Cape Town to Lamberts Bay. Dusky dolphins were encountered less frequently (180 groups, 3451 dolphins), but occurred in significantly larger groups than Heaviside's dolphins (825 groups, 5066 individuals) and in significantly deeper water (Figure 15.2). During 56 encounters (31% of all dusky dolphin sightings, 7% of Heaviside's sightings), both species were observed in close proximity (<50 m) and occasionally in mixed groups, including situations of single members of both species being sighted in groups of the other. No obvious interactions between the two species were visible at the surface except when both species came to bowride the research vessel. The larger and faster dusky dolphins tended to exclude Heaviside's dolphins from the bow wave.

Heaviside's dolphins take a variety of prey, including goby (*Sufflogobius bibarbatus*), horse mackerel (*Trachurus capensis*), and cephalopods, but nearly 50% of their diet consists of juvenile shallow water hake (*Merluccius capensis*; Sekiguchi et al. 1992). They exhibit well-defined diurnal inshore–offshore movements which are thought to be related to nocturnal foraging on juvenile shallow water hake that rise to the surface in the dark. Heaviside's dolphins are also found in higher densities in regions where long-term trends have shown juvenile hake to occur in predictably high abundance (Elwen 2008). Thus, the habitat use and movement patterns of Heaviside's dolphins appear to be closely linked to those of their major prey species.

Dusky dolphins in the Benguela Current appear to be more generalist predators, taking a wider range of species, including hake, lanternfish (*Lampanyctodes hectoris*), lightfish (*Maurolicus meulleri*), cephalopods (*Todarodes angolensis*), and particularly horse mackerel (Sekiguchi et al. 1992). They seem to alternate foraging behavior between feeding on surface schooling prey, prey associated with the deep scattering layer, or nearshore fish such as southern mullet (*Liza richardsonii*; Best 2007). A general diurnal inshore–offshore movement pattern similar to that in Heaviside's dolphins has not been noted. Dedicated land-based observations at St. Helena Bay suggest that dusky dolphin movements might track the location of a strong local upwelling cell. During environmental conditions (e.g., south-westerly winds) when the upwelling cell was suspected to form further offshore, dusky dolphins were notably absent from inshore waters that they otherwise used regularly (Elwen 2008). During 3 years of boat-based surveys along 400 km of coast, large groups of dusky dolphins (50–500) were only observed close to areas of known strong upwelling cells (Elwen 2008). Thus, dusky dolphin movements are likely to reflect the temporally varying distribution of mobile schooling prey that in turn associates with high primary productivity in strong upwelling areas.

There is considerable dietary overlap between dusky and Heaviside's dolphins. Where the same prey species are taken, the smaller Heaviside's dolphins tend to eat slightly larger prey items than dusky dolphins. This pattern seems opposite to that usually found in sympatric species or sexes where larger individuals tend to take larger and more diverse prey (Gittleman 1985, McDonald 2002). Prey size differences may be related to where prey is caught by each species as, for instance, hake generally move deeper as they grow (Iilende et al. 2001).

The lack of apparent inshore feeding by Heaviside's dolphins may act to reduce direct competition and facilitate the co-occurrence with dusky dolphins in nearshore waters. The nature and occurrence of interactions between these two species offshore remains unknown. Thus, observations on species interactions based solely on nearshore waters should be interpreted with caution. In general, sympatry of dusky dolphins and Heaviside's dolphins in the Benguela Current appears to be mediated by a combination of differences in diet, behavior, and habitat use. Heaviside's dolphins are found closer to shore, where they mainly seem to rest and socialize. They also appear to have a more specialized diet compared to dusky dolphins, and tend to take different sizes of the same prey species. (See Chapter 14 for further details of duskies off Africa.)

Sympatric relationships in South America

South Atlantic

Dusky dolphins are the most common dolphin species off the Patagonian coast of Argentina (Schiavini et al. 1999), and have been sighted from close to shore out to the 200 m depth contour (and the 200 nautical mile Argentine Economic Zone) across the relatively shallow and wide continental shelf (Crespo et al. 1997a). They are sympatric throughout most of this neritic range, with congeneric Peale's dolphins (Goodall et al. 1997) and Commerson's dolphins (Goodall et al. 1988a, Schiavini et al. 1999), but direct interactions between dusky dolphins and other species seem to be rare (i.e., are not well documented in the published literature). Dusky dolphins have been seen in the vicinity of, but not mixing with, Commerson's dolphins within 2 km of shore (Heinrich, personal observation). The lack of observed interactions can be attributed to differences in fine-scale habitat use, with dusky dolphins generally occurring farther offshore and in deeper water (Figure 15.2; Würsig and Würsig 1980, Crespo et al. 1997a, Goodall et al. 1997). Dusky dolphins also seem to be more common from 41°S–47°S, whereas Peale's dolphins are more frequently sighted south of 47°S (Crespo et al. 1997a, Goodall et al. 1997), and Commerson's dolphins south of 43°S (Goodall 1994). However, few systematic large-scale surveys with trained observers have been conducted, with most effort biased towards coastal areas (e.g., Crespo et al. 1997a, Schiavini et al.

1999). In addition, Peale's and dusky dolphins can be easily confused at sea due to their similar coloration, and there may be some incidents of misidentifications in the literature.

Based on localized studies at several coastal locations, Commerson's and Peale's dolphins seem to regularly occur in the same areas (Miguel Iñiguez, Fundación Cethus, Buenos Aires, Argentina, June 2008, personal communication to SH; de Haro and Iñíguez 1997, Reyes 2006, 2008). They are known to form mixed groups in which both species synchronize their behavior (Goodall et al. 1988a, 1997), and appear to be foraging together (de Haro and Iñíguez 1997). These two species are similar in their grouping behavior, with groups consisting of 1–5 dolphins that occasionally coalesce into larger aggregations (Figure 15.2 and references therein, Goodall et al. 1997, Reyes 2006). Groups of Commerson's dolphins were largest during cooperative feeding with on average 6 animals (SD = 3.2), and consistently smaller with on average 2 dolphins (SD = 1.5) during all other behaviors, including non-cooperative feeding (Coscarella 2005). Although individual groups (defined as individuals within a few body lengths of each other) tended to be small (<10 dolphins), large and relatively loose aggregations exceeding 200 dolphins in more than 80 subgroups have been observed during dedicated land-based observations (Coscarella 2005). Peale's dolphins also occur in small groups (4–5 dolphins) when feeding in kelp forests, and gather into larger groups (more than 15 animals) during cooperative feeding bouts further from shore (Lescrauwaet 1997, Schiavini et al. 1997, Viddi and Lescrauwaet 2005). This pattern has also been particularly well documented in dusky dolphins around Península Valdés (Würsig and Würsig 1980). Dusky dolphins are usually found in large groups (6–300 individuals; Figure 15.2, Goodall et al. 1997, Reyes 2006) with groups of more than 20 dolphins engaging in cooperative feeding more often and for longer (Würsig and Würsig 1980). Such schooling behavior usually occurs when prey is abundant but patchily distributed and is thought to increase foraging efficiency (Würsig 1986).

Dusky dolphins seem to prey mainly on pelagic and schooling species such as anchovies (*Engraulis anchoita*) or pelagic stages of demersal species such as squid (*Ilex argentinus, Loligo gahi*) and Argentine hake (*Merlucius hubbsi*; Koen-Alonso et al. 1998). Peale's dolphins are known for their strong association with nearshore kelp forests (*Macrocystis pyrifera*; Goodall et al. 1997, Lescrauwaet 1997), where they feed on a variety of demersal and bottom species (Schiavini et al. 1997). They might be more coastal generalist predators in areas farther offshore or where kelp forests are less abundant, taking both demersal and pelagic prey (Iñíguez and de Haro 1994). Commerson's dolphins in southern Argentina are also considered coastal generalist predators feeding mainly on or near the bottom (Bastida et al. 1988). The results from these diet studies match well with the behavioral and grouping patterns described above, and suggest niche separation between dusky dolphins and Peale's/Commerson's dolphins. However, several caveats and limitations need be mentioned here. Much of the information presented above has been compiled from a variety of

sources spanning decades of varying research efforts, using different method-
ologies, conducted at different locations, at different times of year and often
had limited sample sizes (e.g., diet studies). Thus, emerging patterns should be
interpreted with caution and should be tested with carefully designed comparative studies (for examples see Box 15.1, Bearzi 2005b, Parra 2006).

BOX 15.1

Comparative Ecology—Lessons from Chilean and Peale's Dolphins off Chiloé

Sonja Heinrich, Simon Elwen, and Stefan Bräger

Distribution and habitat data collected during seven summers (and one winter)
of systematic boat-based sighting surveys ($n = 382$) revealed a distinct pattern of
small-scale habitat partitioning among Chilean and Peale's dolphins in the Chiloé
Archipelago, southern Chile. Chilean dolphins were consistently found only in
several selected bays and channels. Predictive habitat modeling (using logistic
regression) showed that their preferred habitat characteristics consisted of relatively turbid and shallow waters (<20m), close to shore (<500m; Heinrich 2006).
Peale's dolphins also preferred shallow nearshore waters, but were distributed over
wider areas and were more frequently encountered in areas where Chilean dolphins were rare. Conversely, Peale's dolphins were never sighted in several bays
and channels considered to be Chilean dolphin core habitat. Modeling the occurrence of both species in relation to each other allowed a closer examination of
factors distinguishing their preferred habitat. Chilean dolphins generally preferred
areas closer to rivers and with a strong estuarine influence and higher local productivity than Peale's dolphins. Location (latitude, longitude) was also important,
suggesting that there were other, geographically linked, factors that had not been
included in the model. Additional observations showed that Chilean dolphins
were regularly found in areas with strong tidal currents (also Ribeiro et al. 2007),
whereas the preferred coastline of Peale's dolphins often featured small patches of
kelp forests or a mixed rocky–sandy bottom. The patterns in habitat selection may
reflect differences in diet and foraging strategies between the species, although
their actual diet composition remains unknown.

Analysis of the movement patterns of 42 individually identified Chilean dolphins
revealed temporally stable small-scale site fidelity and short alongshore movements
(Table 15.2). There were also individual differences in ranging patterns and site
preference, suggesting some spatial partitioning along environmental and social
parameters within the population. The overall population of Chilean dolphins in the
study area was small (estimated at 55–65 dolphins using mark–recapture analysis)
and resident year-round. In contrast, individually identifiable Peale's dolphins
showed limited site fidelity within years, but were regularly re-sighted in different
years and likely ranged beyond the chosen study areas. Thus, the small population
of Chilean dolphins seemed to depend on abundant and predictable prey sources
within a few productive channels, whereas a slightly larger and wider ranging population of Peale's dolphins (estimated at 65–95 individuals using mark–recapture
analysis) possibly exploited less predictable and/or more patchily distributed prey.

South Pacific

Lagenorhynchus and *Cephalorhynchus* dolphins also occur in sympatry along the Pacific coast of South America (Figure 15.1), but most information stems from incidental observations and strandings (e.g., Goodall et al. 1988a, 1988b, 1997, Aguayo-Lobo et al. 1998). The southern South Pacific coast of the Americas differs dramatically from the shallow and wide continental shelf of the Atlantic coast of South America. The coastline in southern Chile is mountainous and fragments into one of the world's largest systems of convoluted channels, islands, and fjords. The coastal waters are influenced by high loads of freshwater discharge (Dávila et al. 2002), often appear dark (tannin-stained) and deepen quickly to 30 m or more. The continental shelf is narrow, on average only 50 km wide, and deepens abruptly into one of the world's longest deep ocean trenches. The cold nutrient-rich Humboldt (Peruvian) current only reaches the coast at the northern limit of the fjords, off Chiloé Island, 42°S.

Dusky dolphins form large aggregations along the continental shelf of northern Chile, and particularly in the coastal upwelling systems of Peru, but may be less south of 38°S. This range corresponds well with that of the Peruvian anchovy (*E. ringens*), considered a major prey species for dusky dolphins in Peru (McKinnon 1994). Peruvian anchovies (anchovetas) are strongly associated with the plankton-rich waters of the Humboldt Current (range 6°–42°S) and are prone to large-scale fluctuations in distribution and abundance brought about by oceanographic changes in relation to El Niño Southern Oscillation events (Whitehead et al. 1988). Several large groups of 200–1500 dusky dolphins have been sighted close to shore along the outer fjords of Chile (46°S; Pitman and Ballance 1994). Anecdotal sightings (summarized in Aguayo-Lobo et al. 1998) and relatively credible descriptions from local fishermen (F. Viddi, Centro Ballena Azul, Chile, July 2008, personal communication to SH) also suggest that dusky dolphins may occur regularly and in large groups over the narrow continental shelf and along the outer fjords. Intriguingly, dusky dolphins seem to be absent from most of the inner channels and fjord system where Peale's dolphins are the most frequently sighted small cetacean (Heinrich, personal observation, Viddi, personal communication, Pitman and Ballance 1994, Aguayo-Lobo et al. 1998, Gibbons et al. 2002, Dawson and Slooten 2008).

Chilean dolphins also inhabit the southern fjords, with only a slight overlap in range with congeneric Commerson's dolphins in Tierra del Fuego and the Strait of Magellan area. No direct interactions occurred during the few occasions when both congeners were observed in close proximity to each other (Goodall et al. 1988b, Leatherwood et al. 1988). Overall, the two species have largely allopatric ranges that seem to correspond well with the differences in habitat characteristics between the southern South Pacific and South Atlantic coasts (Goodall et al. 1988b). The mechanisms underlying their spatial separation, however, remain unknown. Both species are neritic, occur in a similar depth range (Figure 15.2), in a similar range of water temperature, and have been sighted along various types of coast, near river mouths and in the surf line. Direct competition, restricted diets, or socially mediated factors could be candidates to explain the observed niche separation. There is currently insufficient information available to explore any of these (or other) hypotheses.

The lack of overlap of the two *Cephalorhynchus* species is even more intriguing in light of their large-scale sympatry with Peale's dolphins. Commerson's and Peale's dolphins seem to co-occur widely, have similar diets, and readily form mixed groups with no evidence of competitive interactions (see above). However, direct comparative studies on the two species have yet to be conducted and might reveal more subtle factors that mediate their coexistence. For example, our present information on each species' diet is rather limited and not available for directly sympatric populations.

Sympatric interactions of Chilean dolphins and Peale's dolphins are presently under investigation (Box 15.1, references in Figure 15.2). On a broad scale, the ranges and habitat of both species overlap widely. However, Chilean dolphins are generally patchily distributed and occur in selected bays, channels and fjords (Heinrich, personal observation, F. Viddi, Centro Ballena Azul, Chile, July 2008, personal communication to SH, Goodall et al. 1988b, Aguayo-Lobo et al. 1998, Gibbons et al. 2002, Dawson and Slooten 2008). In areas where detailed studies have been conducted, Chilean and Peale's dolphins show a relatively clear pattern of spatial habitat partitioning (see Box 15.1, Viddi 2003, Heinrich 2006). Unlike for Commerson's dolphins, direct interactions and mixed groups of Chilean and Peale's dolphins are uncommon (Goodall et al. 1988b, Lescrauwaet 1997, Heinrich 2006). For example, during 7 years of dedicated study of both species, only seven incidents of mixed groups (all involving <8 dolphins) and 10 co-occurrences of single species groups within 500 m of each other were observed at Isla Chiloé (total number of groups sighted: Chilean dolphins = 681, Peale's dolphins = 348; Heinrich, unpublished data, Heinrich 2006). Four mixed groups involved mother–calf pairs of both species, and the remaining three mixed groups were foraging/traveling. However, Peale's and Chilean dolphins in the Guaitecas Archipelago (44°S) have been observed to form mixed groups more regularly and engage in cooperative feeding and socializing (Viddi, personal communication). Group sizes for Peale's dolphins are comparable in both areas, but Chilean dolphins seem to form slightly smaller groups in the Guaitecas than at Chiloé (Figure 15.2). Systematic and quantitative comparisons have yet to be made, but initial results promise interesting insights into behavioral plasticity of each species, and intra-specific differences related to habitat type. Sympatric interactions are likely to vary with different environmental conditions and habitats (for examples see Bearzi 2005a); thus, results from local studies should be tested for generality across the wider range of a species. Such comparisons call for systematically conducted and standardized latitudinal studies.

PATTERNS IN SYMPATRIC ECOLOGY

Sympatric mammal species have adopted a variety of strategies to promote coexistence, including differential use of space, temporal separation in space use, and differences in activity patterns and diet (e.g., Arlettaz 1999, Stanford and Nkurunungi 2003, Bearzi and Stanford 2007). This review focused on two sympatric genera whose members occupy a variety of habitat types ranging from the convoluted fjords of southern Chile to the exposed wide shelf waters of southern Africa. While all species exhibit some degree of behavioral, social, and ecological plasticity, a broad pattern of differences among them emerges.

Dusky dolphins show the greatest behavioral plasticity and occur in more disparate habitats than other members of the *Lagenorhynchus* or *Cephalorhynchus* genera. Dusky dolphins range widely, form larger groups, and use varied foraging tactics to exploit relatively unpredictable but profitable resources (see also Chapter 16). When prey resources are sparsely distributed but locally abundant, such as schooling fish in the open ocean, larger groups yield benefits to individual dolphins by increasing individual foraging efficiency through cooperative behaviors and reducing predation risk. Behavioral strategies and grouping patterns of dusky dolphins seem to vary in relation to the openness of the habitat and the behavior of preferred prey species in that area (Gowans et al. 2008).

In contrast, *Cephalorhynchus* dolphins have relatively small movement ranges, small group sizes, and an affinity for coastal shallow waters. In sheltered coastal habitat, such as that occupied by Chilean and Hector's dolphins, resources are more predictable and structured in space and time than in open ocean habitat. Individuals remain within relatively small profitable areas where prey is easily exploited. Predation risk might be relatively low or dolphins might be able to avoid predictable high risk areas (Heithaus and Dill 2002a). The benefits of group living (e.g., finding food more efficiently and avoiding predation) are reduced, but intra-individual competition for finite resources will limit local population size. Thus, dolphins typically form small groups, have small home ranges and form small "resident communities" (sensu Gowans et al. 2008). Small movements, high site fidelity, and a lack of female dispersal are thought to have contributed to the fragmentation and genetic isolation of Hector's dolphin populations (Pichler et al. 1998). More exposed open habitat can result in larger local populations and large temporary aggregations of small groups as has been suggested for Chilean dolphins along the open coast (Goodall et al. 1988b, Perez-Alvarez et al. 2007), Commerson's dolphins off Bahía Engaño, Argentina (Coscarella 2005), and Hector's dolphins off the west coast of New Zealand's South Island (Figure 15.2).

Commerson's and Heaviside's dolphins seem to differ somewhat from their congeners. Heaviside's dolphins exhibit strong diurnal inshore–offshore movements, most likely in relation to the vertical migration and temporally varied availability of demersal prey (Elwen 2008). Commerson's dolphins have been shown to undertake comparatively large alongshore movements (>200 km) over short time periods (i.e., 5 days) several times a year (Coscarella 2005), and diurnal inshore–offshore movements might occur but have yet to be investigated. Such movement patterns are likely driven by a more heterogeneous distribution of prey resources, and require more flexible strategies than exhibited by resident communities in relatively stable sheltered environments (e.g., Chilean dolphins).

Peale's dolphins show a more intermediate pattern, with many characteristics of resident communities (sensu Gowans et al. 2008), such as small group sizes and relatively small range but more diffuse movement patterns. They are known to feed on single prey items usually associated with kelp forests, as well as feeding cooperatively on small surface schooling fish (Lescrauwaet 1997, Schiavini et al. 1997). Interestingly, Peale's and Commerson's dolphins are the only sympatric pair in our compared group known to associate regularly, and in the case of Peale's dolphins to occupy different habitats on both coasts of South America.

The observed inter- and intra-specific variability reflects a complex matrix of ecological influences such as prey distribution, predation risk, and social interactions—as well as evolutionary history (Gowans et al. 2008)—which play an important role in mediating sympatry. Broadly sympatric species with relatively different ecological requirements and behavioral flexibility, such as dusky and Hector's dolphins, rarely interact and are unlikely to compete directly for resources. Species that differ in their behavioral flexibility yet exploit at least some similar resources are likely to exhibit niche partitioning to reduce competition. Dusky and Heaviside's dolphins seem to occupy different niches as they take different prey sizes, or forage at different times of day or at different locations. Species that have regular, non-aggressive interactions and form mixed groups are unlikely to be competing for the same resources. Local resources may be sufficiently abundant to reduce competition, or group formation might be beneficial by increasing foraging efficiency or reducing predation risk. Such patterns have been documented for many mixed species groups (Stensland et al. 2003), including a variety of other dolphin species (Corkeron 1990, Ballance and Pitman 1998). In the group of species reviewed here, only Peale's and Commerson's dolphins were seen to associate regularly. Comparative studies conducted across a range of shared habitats might help shed light on potential factors mediating their direct sympatry.

Comparing sympatric species in different habitats indicates that dolphins seem to avoid resource competition by using dietary, behavioral, and habitat-related specializations (see also Bearzi 2005a). Sympatric ecology can provide interesting insights into the behavioral flexibility, social dynamics, and habitat preferences of different species and how they adapt to different environmental conditions. Species occupying a variety of niches and exhibiting greater behavioral plasticity are likely to be better equipped to adapt to environmental changes.

ACKNOWLEDGMENTS

We are indebted to the editors, Bernd and Melany Würsig, for giving us the opportunity to contribute to this book and for their great patience and encouragement to see this chapter being completed against many odds. Sincere thanks to: Miguel Iñiguez (Fundación Cethus, Argentina) and Francisco Viddi (Centro Ballena Azul, Chile) for generously sharing their unpublished data and knowledge, Adrian Dahood (Texas A&M University) for her great help with sourcing New Zealand dusky dolphin information, Mariano Coscarella (CENPAT, Argentina) for answering last-minute questions, and to Clint Blight (SMRU, UK) for his help with Figure 15.1.

Chapter 16

A Large-brained Social Animal

Heidi C. Pearson[a] and Deborah E. Shelton[b]
[a]*Marine Mammal Research Program, Texas A&M University, Galveston, Texas, USA*
[b]*Ecology and Evolutionary Biology, University of Arizona, Tucson, Arizona, USA*

We are following a single dusky dolphin on a calm day in the outer reaches of our study site in Admiralty Bay, New Zealand. After an hour of casually interspersing traveling with foraging and resting, the dusky begins swimming underwater for longer periods of time, traversing longer distances, until he (she?) joins with a group of five other dolphins. Was the lone dusky whistling underwater to this other group and "planning" to join it, or was it merely coincidence that he found the other dolphins? Throughout the next hour, the group engages in boisterous social activity that includes mating, belly-up rolls, chasing, head-up "spyhops," fluke-ups, and leaps. Group size fluctuates between two and ten as some individuals break off while others join. Three New Zealand fur seals also join in the mix, adding to the excitement. We are not immune to this contagious level of high activity, and express our enthusiasm by gleefully snapping photographs as the duskies leap next to our boat, sometimes appearing as if they will land right in it! Gradually, the activity level dies down, as does the winter sun which slowly starts to slip behind the hills of D'Urville Island. On the boat ride back to camp, we are somewhat awed to know that we were granted an intimate look at a tiny bit of the social life of dusky dolphins. We are also left wondering. What was the cause of this social excitement—was it receptive females, a reunion of "long-lost friends," or simply exuberance? How does this social exuberance relate to an individual's mental life? Could dusky dolphins be so engaged in social activities because they evolved large brains, or vice versa?

Heidi Pearson, Galveston, Texas, September 2008

INTRODUCTION

In 1961, Jane Goodall observed chimpanzees (*Pan troglodytes*) using tools to "fish" for termites in Gombe National Park, Tanzania (van Lawick-Goodall 1968, Goodall 1986). This observation indicated sophisticated thought on the part of the chimpanzee and conflicted with the traditional idea that tool-making is a distinctly human capacity. Documentation of chimpanzee tool use was part of a shift in our understanding of the mental lives of animals that is yet incomplete; scientists are still grappling with how best to study animal cognition, its origins, and its evolutionary implications.

At about the same time as Goodall's report on chimpanzee tool use, the notion of intelligence in cetaceans was also beginning to receive attention. In the early 1960s, neurobiologist John Lilly sought to understand the "extraterrestrial" intelligence of dolphins in captivity. While his methods and conclusions have rightfully generated controversy (summarized in Samuels and Tyack 2000), Lilly is nevertheless credited with first popularizing the idea of dolphin intelligence, particularly with respect to communication (Lilly and Miller 1961, Lilly 1962, 1965). Later that decade, Karen Pryor and colleagues (Pryor et al. 1969) reported on the capacity for deuterolearning ("learning to learn") in captive rough-toothed dolphins (*Steno bredanensis*). The cognitive significance of Pryor's findings was discussed cautiously, however, and it was not until 11 years later, when Louis Herman (1980) reported on more rigorous tests of the complex cognition of captive bottlenose dolphins (*Tursiops truncatus*), that cetacean intelligence started receiving more scientific attention (see Würsig 2009 for summary and further discussion).

Even as these studies shifted the weight of the evidence in favor of intelligence in dolphins, questions remained about the relevance of studies on captive animals to the evolution and function of wild dolphin cognition. However, in 1984, observations of bottlenose dolphins (*Tursiops* sp.) carrying sponges on their rostrums during benthic foraging provided the first evidence of tool use in wild cetaceans (Smolker et al. 1997); importantly, these observations also demonstrated the presence of sophisticated dolphin behavior outside of captivity. Today, a combination of evidence from experimental, field, and theoretical studies is yielding unprecedented opportunities for scientists to formulate and test hypotheses concerning the evolution of animal intelligence in general, and specifically in dolphins.

We use "intelligence" and "cognition" as umbrella terms encompassing the set of neuronal functions that underpin performance in diverse problem-solving tasks. We have in mind tasks that demand causal reasoning, flexibility, imagination, and prospection (Emery and Clayton 2004). In humans, individuals tend to exhibit rather consistent levels of performance across tasks that seem quite different. That is, for "high-g" tasks, individuals who perform well on one task are likely to perform well on another, even though the tasks seem to demand completely different skills. The pattern is known as "general intelligence," and its explanation (in humans) appears to lie in the recruitment of a restricted neuronal system, the lateral frontal cortex, in response to these seemingly diverse tasks (Duncan et al. 2000).

The details of the mechanistic basis of "general intelligence" in humans is an active area of research and may yet reveal much about the significance of this pattern in a comparative context. As a starting point for this chapter, we consider it likely that non-human animals also have "general intelligence" (i.e., positive correlations among distinct problem-solving tasks, arising from recruitment of a restricted neuronal system). However, we do not suggest that the lateral frontal cortex is the location of this restricted neuronal system in all non-human animals.

How can we apply the idea of "general intelligence" to diverse species? As our chapter title suggests, a useful starting point is to scale back from an interest in specific neural structures and instead look at a much coarser metric: brain size. "Large-brained" is a relative concept and depends on the out-group used for the comparison (Box 16.1). Great apes, dolphins, and corvids are all large-brained when compared to their close relatives. These groups have diverse neuroanatomical characteristics, yet their similarly complex behavior indicates that they have convergently evolved a remarkable adaptation: high general intelligence (e.g., Marino 2002, Emery and Clayton 2005, Marino et al. 2007).

BOX 16.1

Convergent Evolution of Intelligence: Brain Size and Neuroanatomy

Heidi C. Pearson and Deborah E. Shelton

Animal intelligence is difficult to measure. Therefore, brain size[1] is often used to indicate a species' capacity to acquire and process information (Jerison 1973, Tartarelli and Bisconti 2006). Jerison (1973) defined the encephalization quotient (EQ = brain weight/($0.12 \times$ body weight$^{0.67}$)) to describe a mammalian species' deviation from the "expected" brain size for its body weight; for example, an EQ of 3 means that a species has a brain that is three times as large as the "average" extant mammal.

Primates and cetaceans are the two most highly encephalized mammalian groups (i.e., those with the highest EQs; Figure 16.1). The dusky dolphin (*Lagenorhynchus obscurus*) EQ is 4.7, which is one of the highest reported for any odontocete (Marino et al. 2004) and higher than any reported for a non-human primate (Jerison 1973). In fact, the dusky dolphin EQ lies between that of a chimpanzee (*Pan troglodytes*, EQ = 2.3) and that of a modern human (*Homo sapiens*, EQ = 7.6; Jerison 1973). Other delphinids (e.g., Pacific white-sided dolphins, *L. obliquidens*, EQ = 5.3; rough-toothed dolphins, *Steno bredanensis*, EQ = 5.0) also have larger EQs than any non-human primate, and the mean EQ in modern odontocetes (mean EQ = 2.6, range = 0.58–4.56) exceeds that of modern non-human anthropoid primates (mean EQ = 2.0, range = 1.02–3.2; Marino 1997, 1998, Marino et al. 2004, Tartarelli and Bisconti 2006).

However, the EQ is a rough way to estimate the variable of interest (intelligence), and may be problematic. First, EQs for mammals >1000 kg are not meaningful because the expected brain–body size relationship is unlikely to be linear (Jerison 1973). Extremely large odontocetes (e.g., sperm whales, *Physeter macrocephalus*) and all mysticetes have undergone dramatic increases in body size without allometric increases in brain size (Marino 2004). Nonetheless, the sperm whale has the largest brain known to have ever evolved, and behavioral evidence indicates that this relatively small but absolutely large brain produces complex behavior (e.g., Whitehead 1996, Rendell and Whitehead 2003).

Second, it is difficult to compare EQs among different-sized animals, as the brain–body weight slope (i.e., the exponent) used for the EQ calculation may change with taxonomic level (Harvey and Krebs 1990, Marino 1998). Third, EQs vary according to the reference group used. Half of all primates have EQs <1 when compared with other primates, but all primates have EQs >1 when compared with

FIGURE 16.1   Representative primate and cetacean species, and their encephalization quotients (EQ). Clockwise from top left: vervet monkey (*Cercopithecus aethiops*, EQ = 2.1), chacma baboon (*Papio ursinus*, EQ = 1.8), "savanna" chimpanzee (*Pan troglodytes verus*, EQ = 2.3 for *Pan troglodytes*), common dolphin (*Delphinus delphis*, EQ = 4.3), bottlenose dolphin (*Tursiops* sp., EQ = 4.0), and dusky dolphin (*Lagenorhynchus obscurus*, EQ = 4.7). Primate EQs are from Jerison (1973); delphinid EQs are from Marino et al. (2004). All photos except chimpanzee are courtesy of Chris Pearson.

other mammals (Harvey and Krebs 1990). Finally, EQs may not be comparable between animals of different relative maturities (Ridgway and Brownson 1984).

Another way to measure relative brain size is to examine the ratio of the neocortex to the rest of the brain. The neocortex plays a role in abstract thought and rule learning, and is thought to be the "seat of cognitive processes" (Dunbar 1992, 1998, Seyfarth and Cheney 2002). If the neocortex is the brain region underpinning the evolution of intelligence, then measuring neocortical enlargement would be a more direct metric of intelligence than the EQ.

Regardless of the method used to assess relative brain size, odontocetes and primates show convergent encephalization and neocortical elaboration, which is coupled with divergent physical and neuroanatomical evolution (Marino 1998, Reiss and Marino 2001, Marino 2004, Hof et al. 2005, Tartarelli and Bisconti 2006). Some of the differences in cetacean (vs. primate) brains include an enlarged cerebellum with well-developed regions for automatic control of complicated movement (e.g., acrobatic swimming); reduction of the limbic system, especially structures related to olfaction; and enhanced development of cerebral regions related to audition (Marino 2004, Tartarelli and Bisconti 2006).

Cetacean brains also contain a complex and enlarged neocortex with a high gyrification index (a measure of neocortex surface area to total brain size) which surpasses that of humans and indicates the presence of neural hardware necessary for complex cognition (Marino 2004). However, the cetacean neocortex is also much thinner than the human neocortex (Marino 2004, Hof et al. 2005, Tartarelli and Bisconti 2006). Nonetheless, a suite of anatomical and behavioral factors should be used to assess cognitive potential, and the combination of large relative brain size, neocortical complexity, and complex behavior and sociality indicate high cognition in cetaceans. Furthermore, these traits indicate that the large cetacean brain is not simply an artifact of the transition of archaeocetes (e.g., *Pakicetus inachus*) from land to water, or the need for echolocation (Marino 2004, 2007, Marino et al. 2004, 2007, 2008, Hof et al. 2005).

Primates and some birds (corvids; African gray parrots, *Psittacus erithacus*) also display convergent cognitive evolution, despite being evolutionarily separated for >150 million years (e.g., Pepperberg 1999, Emery and Clayton 2004, Lefebvre et al. 2004, Emery and Clayton 2005, Prior 2006). Although the absolute size of a bird's brain is restricted by the need for flight, some corvids have the same relative brain size as chimpanzees, and enlargement of the nidopallium and mesopallium in the corvid forebrain may reflect "primate-like" intelligence (Lefebvre et al. 1997, Emery and Clayton 2004). Evidence of complex cognitive behavior lends further support to the convergence between corvids, parrots, and primates. The fact that intelligent behavior may be produced by brains of different structure and organization across diverse taxa (e.g., pongids, delphinids, corvids, parrots) indicates that intelligence may have evolved multiple times via different neurocognitive processes (after Emery and Clayton 2004, Tartarelli and Bisconti 2006).

Several distinct fields are contributing evidence relevant to the task of understanding intelligent behavior in an evolutionary context. Cognitive neurobiologists are using powerful imaging techniques to illuminate the function of brains in real time. Behavioral ecologists are characterizing the behavioral patterns produced by animals in diverse environments, providing insight into the fitness consequences of behavior. Evolutionary biologists are considering the consequences of selection on behavior; through modeling and by mapping brain-related characters onto phylogenetic trees, they are contributing knowledge about evolutionary scenarios that could have led to the observed variation. Paleobiologists are refining ideas about the historical timing and conditions from which large brains emerged, information that can be used to rule out or support particular evolutionary hypotheses.

CHAPTER OVERVIEW

In this chapter we consider what it means to classify dusky dolphins (*Lagenorhynchus obscurus*) as large-brained animals. How are dusky dolphins using their large brains in the marine environment? How and why did this trait evolve in dusky dolphins, and are the reasons essentially similar in other large-brained species?

We begin by describing the evolutionary puzzle of large brains. We then discuss manifestations of dusky dolphin intelligence in terms of flexibility and complexity in feeding and social behavior. We also describe some of these manifestations of intelligence in other species. Next, we summarize several hypotheses for the adaptive value of large brains: ecological intelligence, brain size/environmental change, and the "social brain." In particular, we focus on recent progress in testing the predictions of the social brain hypothesis. Finally, the salient features of the social brain hypothesis lead us to advocate further research in particular areas.

THE EVOLUTIONARY PUZZLE OF LARGE BRAINS

Brain tissue is energetically expensive (Mink et al. 1981, Aiello and Wheeler 1995, Isler and van Schaik 2006), and Darwinian logic dictates that substantial benefits have balanced this energetic cost. Brain functions (i.e., candidate benefits[2]) may be divided into two general categories: controlling body functions and cognition. The portion of the brain that controls body functions scales directly with body size (Box 16.1; Jerison 1973). In contrast, the portion of the brain that functions in cognitively demanding problem-solving tasks does not scale neatly with body size, and only a few lineages have "paid" the *additional* cost of investing in extra brain tissue for high cognition (Jerison 1973). However, we do not intend to give the impression that relative brain size may be viewed as discrete categories of "small" and "large" brains. Rather, relative brain size is best viewed along a continuum, and it is in this context that we refer to "large" and "small" brains throughout this chapter.

The additional cost of large brains is an evolutionary puzzle, and requires an explanation of the benefits that make large brains adaptive under some circumstances. Conventional wisdom dictates that disproportionately large brains are adaptive because they can produce complex and sophisticated behavior. However, relatively simple animals with relatively simple brains can behave in complex ways. Wasps (*Ampulex compressa*) make precise venom injections in their cockroach (*Periplaneta americana*) prey in order to subsequently manipulate the cockroach's behavior (Fouad et al. 1996). Jumping spiders (family Salticidae) execute conditional predatory strategies (Jackson and Pollard 1996). Termites engage in complex colony-level nest-building (Bonabeau et al. 1997).

Examples of such remarkable behavior in insects caution against assuming that complex and sophisticated behavior can *only* arise from large brains. Nevertheless, there seems to be a strong overall relationship between the relative size of a species' brain and the sophistication and flexibility of its behavior. As we discuss examples of behaviors that appear to be manifestations of sophisticated cognition in large-brained animals, we also emphasize examples of behaviors in small-brained animals that seem similar. A cohesive understanding of intelligence in an evolutionary context will require that these kinds of observations be addressed, for they point us towards what characteristics of large brains makes them adaptive. For example, Bshary et al. (2002) point out that fishes as a taxon demonstrate many of the problem-solving skills often discussed with respect to the evolution of primate intelligence (e.g., cooperative hunting, social learning, tool use). They also point out, however, that a diversity of problem-solving skills can be observed within a single primate species, while the same is not true for fishes. Thus, resolving seemingly similar observations of sophisticated behavior in small- and large-brained animals may require taking into account a species' suite of problem-solving skills.

LARGE-BRAINED CETACEANS

Some of the largest and most complex brains to have ever evolved are cetacean brains (Box 16.1). In addition to large brains, cetaceans exhibit several other "indications" of intelligence and complex cognition, including: slow life histories (Box 16.2), reliance on patchy prey resources (Rendell and Whitehead 2001), problem-solving (Herman 1980), innovation (Pryor et al. 1969), tool use (Smolker et al. 1997), vocal and behavioral imitation (Herman 1980, Reiss and McCowan 1993), coordinated or complex behaviors and social interactions

---

BOX 16.2

Large Brains and Slow Life History

Heidi C. Pearson and Deborah E. Shelton

In primates, odontocetes, carnivores, and elephants, large brains are associated with slow life histories (e.g., long lives, high maternal investment, prolonged maternal dependence, long juvenile periods; Deaner et al. 2003, Payne 2003). A slow life history provides time for an individual to develop the cognitive adaptations necessary to respond to socioecological demands (van Schaik and Deaner 2003). Cetaceans and elephants are distinct because they produce precocial young (e.g., able to travel independently straight from birth) yet still have a long developmental period.

A long juvenile period is risky, as juveniles may be more vulnerable to predation and starvation. Thus, benefits obtained during this period must offset costs. An extended juvenile period is not necessary for brain growth, as this is completed before the onset of this period. However, prolonged juvenility provides more time in which to learn the complex foraging and social skills necessary to be a reproductively successful adult (Joffe 1997, Walker et al. 2006).

In corvids, a long developmental period during which juveniles associate with a variety of group members creates increased opportunities for learning (Clayton et al. 2007). In primates, obtaining social knowledge (e.g., regarding dominance relations) during the juvenile period may be crucial for future social success (Pusey 1990, Watts and Pusey 1993). For odontocetes, ecological knowledge of foraging specializations (e.g., Mann and Sargeant 2003) and social knowledge regarding potential alliance partners (Mann 2006) may be obtained during extended juvenile and calf periods. The dusky dolphin also has a slow life history (Cipriano 1992, van Waerebeek and Read 1994) that likely facilitates the attainment of knowledge regarding prey resources, and prospective social and foraging partners.

(e.g., Connor et al. 1992a, Gazda et al. 2005), the ability for mirror self-recognition (Reiss and Marino 2001), culture (Rendell and Whitehead 2001, Whitehead et al. 2004), and the ability to learn artificial codes (Herman 1980).

The comparative approach provides insight into the evolution of large brains and complex cognition, and studies of dolphins make a unique contribution. The most recent common ancestor of dolphins and primates existed about 95 million years ago, and dolphins have followed an evolutionary path distinct from terrestrial mammals for >50 million years (Bromham et al. 1999). Thus, dolphins provide a phylogenetically distinct and highly encephalized group on which brain evolution hypotheses (many of which were developed based on knowledge of great apes) can be tested. A compelling explanation for the evolution of complex cognition must account for which physical, ecological, and social factors are important, and ideas developed based on terrestrial systems should pass muster when tested in the drastically different realm of a marine mammal (Table 16.1). In addition, the about 36-species dolphin family (Delphinidae) allows for more within-family comparisons than is possible in the four-species great ape family (Pongidae; Figure 16.2).

Unlike some better-studied dolphins (e.g. coastal bottlenose dolphins), dusky dolphins are not restricted to inshore habitats, may occur in deep waters up to 2000m, may live in extremely large societies (≤1000 individuals), and may travel long distances (100–1000km; Figure 16.3; Würsig et al. 2007). These differences in ecology and social behavior are significant to many hypotheses about brain evolution. Therefore, differences between dusky dolphins, other dolphins, and terrestrial mammals could serve to falsify or refine brain evolution hypotheses.

Dusky dolphins exhibit several indicators of complex cognition listed by Simmonds (2006): (1) a large brain with a high encephalization quotient (EQ; Marino et al. 2004); (2) a slow life history, with a long lifespan, long period of offspring dependence, and long juvenile period (Box 16.2; Cipriano 1992, van Waerebeek and Read 1994); (3) exploitation of patchy prey resources (Würsig and Würsig 1980); (4) closely coordinated foraging and social behaviors (Würsig and Würsig 1980, Markowitz 2004, Vaughn et al. 2007, Pearson 2008); (5) formation of long-term associations with preferred companions (Markowitz 2004, Pearson 2008); and (6) evidence of culture (Markowitz et al. 2004, Whitehead et al. 2004).

MANIFESTATIONS OF INTELLIGENCE IN DUSKY DOLPHINS

Foraging strategies

Animals are faced with ecological problems such as finding food, avoiding predation, and deciding whether and when to travel. For large-brained animals, intelligence permeates their responses to these problems, and dusky dolphins exhibit indicators of complex cognition in the realm of foraging behavior. In Admiralty Bay, New Zealand dusky dolphins forage diurnally in a coordinated

TABLE 16.1 Distinct features of the oceanic setting and their implications for dolphin socioecology

Feature	Implications	Selected references
Buoyancy	Interactions occur in three dimensions. A low cost of transport may favor group fission–fusion, permit larger home ranges, and increase exposure to socio-ecological variation	Williams et al. 1992, Connor 2000, Rendell and Whitehead 2001
Few physical refuges	Predation pressure may favor sociality among prey species because they cannot physically hide from predators. There is also less opportunity for predators to surprise prey	Norris and Dohl 1980, Norris and Schilt 1988, Connor 2000
Sound propagates well	Complex and long-range auditory communication is possible. Social interactions may occur over wider spatial scales than in the terrestrial environment	Norris and Dohl 1980, Connor et al. 1998, Whitehead 1998, Tyack 2000
Light propagates poorly	Visual communication is important for dolphins, but the use of visual social cues requires close physical proximity	Norris and Dohl 1980, Norris and Schlit 1988, Tyack 2000
Environmental variation is "redder" (i.e., has proportionally greater variance over longer time periods) than on land	Social learning is an efficient means of information transfer and may result in complex cultures	Steele 1985, Rendell and Whitehead 2001,Whitehead 2007
Water has a higher heat capacity than air	Blubber and high-energy (e.g., lipid-rich) diets are necessary for endotherms to cope with the threat of heat loss. The ability to exploit lipid-rich prey may be necessary for large brains to evolve	Castellini 2002, Costa 2002, Connor and Mann 2006
For mammals, breathable air is physically separated from food	This may lead to cooperative fish-herding (to keep fish from escaping by descending)	Würsig and Würsig 1980, Würsig 1986, Connor 2000

manner to herd small schooling fishes (e.g., New Zealand pilchard, *Sardinops neopilchardus*) into tight schools or "prey balls" (Figure 16.4; Markowitz et al. 2004, Vaughn et al. 2007, Pearson 2008). Foraging group size is ≤50 individuals (Pearson 2008), and it appears that individuals take turns herding and feeding

FIGURE 16.2 Species of families Pongidae and Delphinidae exhibit wide variation in morphology, habitat, and social behavior. Family Pongidae is composed of four species including the gorilla (top left, *Gorilla gorilla beringei*) and chimpanzee (top right, *Pan troglodytes verus*). Family Delphinidae is composed of approximately 36 species, including the killer whale (bottom left, *Orcinus orca*) and bottlenose dolphin (bottom right, *Tursiops* sp.). Killer whale and bottlenose dolphin photos are courtesy of Chris Pearson.

FIGURE 16.3 A portion of a large dusky dolphin group off Kaikoura, New Zealand. Many dusky dolphins in New Zealand inhabit the deep waters of the Kaikoura Canyon, forming large groups of ≤1000 individuals (photo courtesy of Chris Pearson).

FIGURE 16.4   Top: Dusky dolphin foraging strategies in Admiralty Bay involve coordination both above and below the surface. Dusky dolphins often feed in conjunction with seabirds; the splash in the background indicates that an Australasian gannet (*Morus serrator*) has just taken a plunge dive towards a school of fish below the surface. Bottom: Coordinated dusky dolphin foraging involves herding small schooling fish such as pilchard (*Sardinops neopilchardus*) into prey balls; dusky dolphins may use the surface of the water as a barrier against which to herd prey balls.

on the prey ball (sensu Würsig 1986). In Golfo San José, Argentina dusky dolphins feed in a similar manner, but feeding group size is much larger (≤300 individuals; Würsig and Würsig 1980).

It is unclear whether or not coordination among feeding dusky dolphins is an indication of advanced intelligence. If by-product benefits (i.e., those that occur incidentally during self-serving individual behavior; Connor 2000) drive the interactions, then the behavioral pattern could occur with or without complex cognition. Similar fish-herding behavior has been reported in yellowtail amberjack (*Seriola lalandei*) and black skipjack (*Euthynnus yaito*; families Carangidae and Scombridae, respectively), and may be explained in terms of by-product benefits (Hiatt and Brock 1948, Schmitt and Strand 1982). Nevertheless, documentation of social bonding in dusky dolphins (Markowitz 2004, Pearson 2008) indicates that a more cognitively demanding form of cooperation—based on a network of individualized, long-term relationships—may occur during feeding.

Another indication of dusky dolphin intelligence is their flexibility in foraging behavior. During the winter, a subset of New Zealand dusky dolphins travels from the cool open waters of Kaikoura to the warmer more sheltered waters of Admiralty Bay (about 275 km north of Kaikoura) where an abundant prey base is seasonally available (Vaughn et al. 2007, Pearson 2008). As a result of these annual movements, some dusky dolphins make a drastic behavioral switch between nocturnal individual foraging on the deep scattering layer (DSL) in Kaikoura (Würsig et al. 2007) and diurnal coordinated foraging on small schooling fishes in Admiralty Bay.

Social strategies

In addition to ecological manifestations of intelligence, dusky dolphins exhibit manifestations of intelligence in the social realm.[3] Dusky dolphin society is composed of groups which frequently change in size and composition according to behavior. In Admiralty Bay, rate of group size increase ("fusion") is highest during foraging, while rate of decrease in group size ("fission") is highest during resting, socializing, and traveling (Pearson in press). In Kaikoura, dusky dolphins typically occur in large groups but also form smaller subgroups composed of mating adults, mothers and calves (nursery groups), or non-sexually active adults (Markowitz 2004, Weir et al. 2008). In Golfo San José, dusky dolphins encounter a variety of associates by switching between small traveling groups and large feeding and socio-sexual groups (Würsig and Würsig 1980, Würsig et al. 1989).

Dusky dolphins maintain relationships within a fluid society, and the nature of these social relationships changes throughout an individual's life (see Markowitz 2004, Weir 2007). The dusky dolphins that travel between Kaikoura and Admiralty Bay may face additional cognitive challenges by maintaining relationships across different physical and social environments. Dusky dolphins are thought to arrive at Admiralty Bay from various points around the South Island (Harlin et al. 2003), so individuals may benefit from remembering past relationships. Some individuals remain bonded as they travel between Kaikoura and Admiralty Bay, associating in foraging groups in Admiralty Bay and in mating groups in Kaikoura (Markowitz 2004, see also Chapter 8).

Watching the dusky dolphin mating system in action leaves one with the impression that it is a chaotic free-for-all, with little role of complex cognition on the part of either sex. Dusky dolphins, like most dolphin species, have a polygynandrous (promiscuous) or multi-mate mating system. The large testes to body mass ratio in most dolphins indicates promiscuity and sperm competition (Connor et al. 2000c). In fact, the testes mass of a dusky dolphin may weigh

8.5% of its total body mass, the highest reported for any odontocete and among the highest of any mammal (Cipriano 1992, van Waerebeek and Read 1994, Pabst et al. 1999; see also Chapter 2).

Closer behavioral examination reveals that complex cognition may play a role in determining mating success in dusky dolphins. Mating often occurs in multi-male single-female groups, so male–male relationships may be important. Maneuverability and speed also appear to be important determinants of male mating success (Markowitz 2004). Could males or females be cooperating with each other to thwart the opposite sex? Further study may reveal a role for complex cognition in the context of mating (see also Chapter 8).

Additional support for complex cognition in dusky dolphins arises from behavioral indicators of social learning. Although social learning is widespread among small- and large-brained creatures (e.g., evident in small-brained organisms such as wood crickets, *Nemobius sylvestris*; Coolen et al. 2005), sustained behavioral traditions are rare, and *multiple* sustained socially transmitted traditions (i.e., culture) are rarer and may be indicative of complex cognition (Whiten and van Schaik 2007). There is evidence for traditional or cultural behavior in New Zealand dusky dolphins, as some of the same individuals annually travel between summer mating grounds in Kaikoura and winter feeding grounds in Admiralty Bay (Markowitz et al. 2004). Molecular markers and field observations indicate that the majority of individuals in Admiralty Bay are males (Harlin et al. 2003, Markowitz et al. 2004, Shelton 2006). Like most cetaceans (Connor et al. 2000c), male dusky dolphins appear to play no role in calf-rearing; thus the seasonal movement between Admiralty Bay and Kaikoura is likely an example of inter-generational or oblique culture (i.e., transmission of knowledge between unrelated members of different generations, with younger males likely learning from older males) rather than vertical culture (i.e., transmission of knowledge from parent(s) to offspring; see Whitehead et al. 2004).

SOME MANIFESTATIONS OF INTELLIGENCE IN OTHER ANIMALS

Foraging strategies

Flexible foraging strategies are present in other large-brained odontocetes, and may occur on the scale of the population, pod, or individual. Pacific white-sided dolphins (*L. obliquidens*) exhibit variability in foraging strategies similar to that of their southern hemisphere relatives, the dusky dolphins. Off California and Japan, Pacific white-sided dolphins forage nocturnally on the deep scattering layer, while off British Columbia, they cooperatively forage on small schooling fishes (Morton 2000, Reeves et al. 2002, Van Waerebeek and Würsig 2002). However, it is unknown if some of the same individuals switch between the two foraging strategies as we see in New Zealand dusky dolphins.

Killer whales (*Orcinus orca*) exhibit pod-specific foraging specializations. Resident killer whales in Prince William Sound, Alaska specialize on salmon (primarily *Oncorhynchus* spp.), while transient killer whales in this same area specialize on marine mammal prey such as harbor seals (*Phoca vitulina*) and Dall's porpoises (*Phocoenoides dalli*; Saulitis et al. 2000).

Female bottlenose dolphins in Shark Bay, Australia display individually distinctive foraging profiles and a female may have up to seven distinct foraging strategies (e.g., bird milling, rooster tailing, bottom grubbing; Mann and Sargeant 2003). Similarly, the foraging repertoire of a chimpanzee may include cooperative hunting of red colobus monkeys (*Colobus badius*), termite-fishing, and careful extraction of fruits from the prickly fronds of oil palm nut trees (Goodall 1986).

Perhaps a more surprising example of sophisticated foraging strategies comes from the Serranidae fish of Ras Mohammad National Park, Egypt. Here, red sea coral groupers (*Plectropomus pessuliferus*) and lunartail groupers (*Variola louti*) solicit giant moray eels (*Gymnothorax javanicus*) to engage in role-specific cooperative hunting (Bshary et al. 2002).

Social strategies

Deducing and remembering individualized social interactions can be cognitively demanding, and large-brained animals often apply their intelligence to these tasks. Corvids "keep track" of social relationships. Pinyon jays (*Gymnorhinus cyanocephalus*) use transitive reasoning to infer their own dominance status relative to strangers by observing interactions between known conspecifics and strangers (Paz-y-Miño et al. 2004). Western scrub jays (*Aphelocoma californica*) engage in anti-theft strategies to protect their caches from non-partners by caching in visually obstructed areas (e.g., in the shade, behind a barrier, far away). If this is not possible, individual jays remember who was watching them during initial caching and later return to the area to hide their food in a different area, unbeknownst to the original observer. Only those individuals who have stolen from others' caches in the past engage in this re-caching behavior (Emery and Clayton 2005, Dally et al. 2006).

Individuals that form coalitions (when one individual supports another during a dispute with a third individual) or alliances (long-term coalitions between the same individuals) must also keep track of changing social dynamics. For example, a chimpanzee seeking to increase his or her dominance rank may strategically switch coalition partners according to daily changes in the social environment (Bearzi and Stanford 2008). Coalitions and alliances have been reported in other cognitively complex species such as spotted hyenas (*Crocuta crocuta*; Engh et al. 2005, Holekamp et al. 2007) and bottlenose dolphins (Connor et al. 1992a), with preliminary evidence for alliance-type relationships in corvids (rooks, *Corvus frugilegus*, and jackdaws, *C. monedula*; Emery et al. 2007). There is currently no strong evidence for coalition or alliance formation in dusky dolphins (but see Chapter 8); however, more data are needed to discern if this is a biological reality or an artifact of insufficient data.

THE ADAPTIVE VALUE OF LARGE BRAINS

Just as observing the behavior of large-brained animals in nature can be challenging, identifying potential causes for the evolution of large brains can also be a daunting task. A variety of hypotheses have been proposed to explain the evolution of large brains (Table 16.2). Below, we focus on three of these hypotheses: ecological intelligence, brain size/evolutionary change, and the social brain.

Ecological intelligence hypothesis

Clutton-Brock and Harvey (1980) were among the first to systematically test an adaptive hypothesis for brain evolution—that large brains evolved for ecological problem-solving in primates. They found that between primate genera, frugivores and those with larger home ranges had larger EQs than folivores and those with smaller home ranges. The "ecological intelligence" explanation for these findings is that frugivores and species with larger home ranges evolved greater information-processing abilities to keep track of spatio temporally variable fruit resources and to "mentally map" a larger area, respectively. However, this and similar "strictly" ecological explanations are insufficient in a broader taxonomic context. Why did primates and not other frugivore-containing lineages require additional brain power to solve essentially similar food acquisition problems? The challenge to explain what is unique about the ecological problems faced by highly encephalized lineages has not been met (Dunbar and Shultz 2007a, 2007b).

Brain size/evolutionary change hypothesis

Strictly ecological explanations for brain evolution have become less popular as researchers have recognized the importance of the interplay between social behavior and ecological problem-solving (e.g., Reader and Laland 2002, Dunbar and Shultz 2007a, 2007b).[4] The "brain size/environmental change" (BSEC) hypothesis posits that large brains evolved because enhanced cognitive abilities increase likelihood of survival in new or altered conditions (Sol et al. 2005). According to this hypothesis, large-brained animals have greater behavioral/foraging flexibility and are better able to innovate and learn from each other, which is selectively favored in changing environments.

Some studies have supported the BSEC hypothesis. In primates, Reader and Laland (2002) found a positive relationship between brain size and innovation frequency. Other studies have reported that the majority of innovations are apparent in the foraging domain (Reader and Laland 2001) and that innovation

TABLE 16.2   Hypotheses regarding the evolution of large brains

Hypothesis	Description	Selected reference(s)
Brain size/ environmental change	Large brains provide increased cognitive benefits (e.g., innovation, flexibility) for animals to survive under altered or new environmental conditions	Sol et al. 2005
Ecological intelligence	Large brains reflect the need for spatiotemporal mapping of patchy food resources	Clutton-Brock and Harvey 1980
Extractive foraging	Large brains are related to the need to manually extract embedded food items; applicable only to taxa which have hands (i.e., primates)	Parker and Gibson 1977, Gibson 1986
Sexual selection (includes the sexual conflict and expensive sexual tissue hypotheses)	Large brains are related to the need to challenge the reproductive investment of and resist being undermined by sexual partners, especially in promiscuously mating societies; sexually selected traits (e.g., costly sexual organs, ornaments, armaments) place energetic constraints on the evolution of large brains	Pitnick et al. 2006, Schillaci 2006
Social learning/ culture	Large brains are related to the need for social learning or culture in solving complex social and ecological problems	van Schaik 2006, Whiten and van Schaik 2007
Social intelligence/ Machiavellian intelligence/social brain	Large brains reflect the need for an individual to keep track of relationships in a complex social environment	Jolly 1966, Humphrey 1976, Byrne and Whiten 1988, Dunbar 1998

rate is highest among individuals that spend more time with others (van Schaik 2006). In birds, species with larger relative brains are better able to respond to novel conditions (e.g., as a result of biological invasion) via increased rates of feeding innovation made possible by increased cognition (Lefebvre et al. 2004, Sol et al. 2005).

Social brain hypothesis

While the BSEC hypothesis has advanced our thinking on the evolution of intelligence, explaining brain evolution in terms of the ecological conse-quences of large brains (e.g., innovative or flexible foraging) remains problem-atic because many small-brained species also evolved in similar environments. In contrast, the social brain/Machiavellian intelligence hypothesis accounts for large brain evolution not by its effects in a particular type of ecological environment, but by the dynamics that occur within social groups.

In *Il Principe* ("The Prince"), Niccolò Machiavelli described shrewd methods by which a prince could acquire and maintain power, thus promoting what he perceived as the greatest moral good—a stable political state (Machiavelli 1532). Byrne and Whiten (1988) used the phrase "Machiavellian intelligence" to describe how intelligence could be used to behave adaptively in social politics. Individuals are faced with the challenge of behaving in their own self-interest within a society of other self-interested individuals. Consequently, the pressure to monitor changing group properties, social hierarchies, and networks of relationships leads to an evolutionary arms race in cleverness. Large brains provide the neural substrate from which Machiavellian attributes arise, and positive feedback leads to the evolution of larger and larger brains. The idea of the importance of intelligence in a social context existed as early as 1966 (Jolly 1966, Humphrey 1976) and was later re-labeled the social intelligence or social brain hypothesis (SBH) to avoid the narrow focus on deceit, manipulation, and punishment that the word "Machiavellian" connotes (e.g., Dunbar 1998, Dunbar and Shultz 2007a).

Social problems may quickly become more complex than many strictly ecological challenges. Consider that a dolphin's blubber is an adaptation to its physical environment. The processes that govern ocean temperature or water conductivity are unaffected by the dolphin's blubber. In contrast, an animal's social environment is evolving. That is, a new adaptation to a particular social environment becomes part of the social environment, changing the original context from which the adaptation arose. The SBH posits that this feedback creates an evolutionary incentive for investment in large brains (Humphrey 1976). A corollary needed for this hypothesis is that there must be a strong incentive for group cohesiveness. (Otherwise, why not go it alone rather than stay and compete in a setting so challenging that an enormous investment in brain tissue is necessary?) In dolphins, strong predation pressure is thought to be the primary factor favoring cohesive groups (Norris and Dohl 1980).

The SBH predicts that brain size increases with group size. In anthropoid primates (Dunbar and Shultz 2007a) and odontocetes (Marino 1997), such a correlation exists and is remarkably robust. However, the relationship does not hold up in other taxa, perhaps indicating a qualitative difference between the societies of anthropoid primates and odontocetes, and those of other groups (Dunbar and Shultz 2007a, 2007b).

Among carnivores, artiodactyls, birds, and bats, brain size is highly correlated with mating system; those with mating pair bonds have larger brains (Dunbar and Shultz 2007a). For example, corvids and parrots form long-term monogamous pair bonds and have larger relative brain sizes than birds that form short-term pair bonds or are promiscuous, polygynous, or polyandrous (Emery et al. 2007). Among ungulates, monogamous species have larger brains than species that are non-social, have harems, or occur in large mixed groups

(Shultz and Dunbar 2006). This indicates the importance of "relationship intelligence," whereby cooperatively coexisting with one's partner enables an individual to better adapt and respond to a complex changing environment (Emery et al. 2007).

Thus, the pair bond appears to be an important influence on social intelligence in birds and many mammals, while in anthropoid primates (and odontocetes) the main influence appears to be the greater social network (summarized in Dunbar and Shultz 2007a, Emery et al. 2007). These data could be explained if social bonds are cognitively demanding, and if only a few lineages (e.g., anthropoid primates, odontocetes) have generalized bond formation to include all group members, not just a mating partner (Shultz and Dunbar 2006, Dunbar and Shultz 2007a).

Understanding brain evolution in the context of social evolution is further complicated because the evolution of cognitive traits can be influenced by several non-genetic inheritance systems (see Jablonka and Lamb 2005). Social learning—a mechanism for behavioral inheritance of a trait—is widespread among animals and can influence evolution (Dugatkin 2000, Whitehead 2007). Social learning or culture is likely to be an efficient way for cetaceans to obtain knowledge, as the high degree of spatiotemporal variability in the marine environment may make individual learning costly (Whitehead 1998). However, the evolutionary effects of social learning and of cultural inheritance remain relatively unexplored (but see Whitehead 1998, Rendell and Whitehead 2001, Avital and Jablonka 2005).

Cetacean culture is readily apparent in the foraging domain. Bottlenose dolphins demonstrate population-specific foraging techniques (Wells 2003), and sympatric resident and transient killer whale pods exhibit dietary differences (Saulitis et al. 2000). However, the greatest potential for culture in cetaceans may be in the acoustic domain (Whitehead 1998). Male humpback whales (*Megaptera novaeangliae*) converging on mating grounds demonstrate horizontal transmission of culture by singing the same song and "keeping up" with the song as it evolves over time (e.g., Clapham 2000, Noad et al. 2000, and references therein). Sperm whales (*Physeter macrocephalus*) and killer whales demonstrate intra-pod specific dialects and inter-pod specific clans (Ford 1991, Whitehead and Weilgart 2000, Yurk et al. 2002, Rendell and Whitehead 2003). Ultimately, the low cost of transport for cetaceans may permit a wide social network as individuals are able to interact with a wide variety of individuals at low energetic cost. The result may be higher cultural complexity in cetaceans than in primates (Rendell and Whitehead 2001).

AN ANSWER TO THE PUZZLE AND FURTHER QUESTIONS

Science is a reticulating endeavor whereby one answer opens more avenues of inquiry. We agree with Dunbar and Shultz (2007a) that an answer to the evolution of large brains likely lies in the nature of social bonds. While ecological unpredictability (e.g., the need to forage on patchy food resources) may have been the initial impetus for the evolution of large brains, the need to solve ecological problems and keep track of relationships within a changing social environment likely fueled the increasing spiral of complex cognition (after Dunbar 2003).

How does the SBH further our understanding of dusky dolphins as large-brained social animals? In the open ocean (e.g., off Kaikoura), large dolphin groups are more than just aggregations of individuals grouped together for predator protection. Further scrutiny indicates the presence of social bond formation and coordinated behavior during foraging (Benoit-Bird et al. 2004; see also Chapter 6) and mating (see Chapter 8). In the more enclosed environments of Admiralty Bay and Golfo San José, social complexity may be heightened as individuals form and reinforce social bonds to facilitate coordinated foraging strategies on prey balls (Vaughn et al. 2007, Pearson 2008). The SBH suggests that for dusky dolphins, a strikingly group-oriented species, the need to keep track of social relationships in a changing environment demands high cognition.

While evidence for intelligence may be apparent in numerous species, one needs to look at the entire "toolbox" of intelligent behaviors that is exhibited by a species. Are multiple examples of intelligent and cognitively complex behaviors displayed within a single species, or are they displayed across the taxon as a whole? An individual chimpanzee, bottlenose dolphin, or dusky dolphin may exhibit an entire suite of intelligent behavior, used to survive in complex ecological and social environments. Collectively, this repertoire of interdependent behaviors indicates complex cognition.

However, for dusky dolphins and other large-brained and cognitively complex species, more studies are needed to generate a consensus about the evolution of "braininess," especially with respect to the SBH. We see three general topics that should be explored further in relation to the SBH. First, the issue of how to observe and document the content and quality of the social interactions that form the basis of relationships (Hinde 1976) should continue to be addressed. That is, when individuals repeatedly interact, *what* are they doing and *how* are they doing it? The "gambit of the group" (i.e., assuming grouped individuals are interacting; Whitehead and Dufault 1999) allows for overall descriptions of a species' social patterns, but understanding the cognitive basis of animal bonds will require field studies that delve deeper into the nature of animal relationships. Complementing efforts to document animal relationships in nature are controlled experiments that illuminate the mechanistic connections between feeling, thinking, and behaving. The mechanisms linking large, complex brains and particular types of social relationships should be explored further in relation to the SBH.

Second, the hierarchical nature of selection should be explored with respect to dolphin societies, particularly for pelagic and "semi-pelagic" species. Although dolphin societies are not as cohesive as "superorganisms" (e.g., eusocial insects, for which group-level selection is clearly important), they nonetheless often show remarkable within-group coordination and cooperation. The importance of sociality in dolphin life is evident throughout the taxon (only some river dolphins are relatively solitary; Reeves et al. 2002) and is often a striking feature of our own species' experience of dolphins. It is difficult to read of pilot whale (*Globicephala* spp.) mass strandings (e.g., Norris and Dohl 1980, Norris and Schilt 1988, Connor 2000) or of the "friendliness" of lone bottlenose dolphins (e.g., Bearzi and Stanford 2008) without gaining an appreciation of how fundamentally social these beings are.

Of course, social behavior does not necessarily indicate that group-level selection has played an important role in dolphin evolution. The idea of group-level selection as a mechanism in nature has had a turbulent history (see Bradley 1999, Krause and Ruxton 2002, Wilson and Wilson 2007) and has often been dismissed without proper consideration. However, a cursory dismissal of group-level processes would compromise further exploration of the SBH as well as exploration of the effects of cultural inheritance on brain evolution. Hierarchically nested levels of selection may ultimately determine how closely the interests of individuals align with the interests of groups of individuals, and social behavioral patterns as well as their neural underpinnings should be understood in this context.

Finally, the issue of how to disentangle cause and effect in systems with feedbacks should be addressed further. The path analysis used by Dunbar and Shultz (2007a, 2007b) and ancestral character state reconstruction of sociality and brain size (Perez-Barberia et al. 2007) both seem to be promising approaches. Additionally, discerning the timescale of relevant ecological and evolutionary processes is critical for establishing the importance of feedback between the two processes.

FUTURE RESEARCH DIRECTIONS

New and continued dusky dolphin field studies will further our understanding of how these animals respond to a complex socioecological environment. Continued photo-identification and comparison between Kaikoura and Admiralty Bay (as an example, although detailed studies in other parts of the world would be very welcome as well) may indicate which individuals are responsible for the spread of the "traditional" seasonal movement between these two areas. Detailed behavioral observations may yield information on interactions (i.e., the "quality" of relationships), and may reveal the presence of individualized foraging repertoire as reported for bottlenose dolphins (Mann and Sargeant 2003); females in nursery groups are likely candidates for this behavior. In Kaikoura, mothers in nursery groups may be under pressure to exploit prey resources other than the DSL in order to remain in protective, shallow, nearshore waters, and/or to avoid separation from calves while diving to feed on the DSL (see Weir 2007; Weir et al. 2008).

Observations of juveniles may indicate which socioecologic skills are developed during this protracted life stage, while observations of mating groups may reveal the presence of male alliances as reported for other delphinids. Acoustic studies will be invaluable, as we have yet to uncover how dusky dolphins make use of their acoustic domain to solve socioecologic problems.

In primates, although much research has been conducted on intelligence in Old World monkeys (catarrhines), more data are needed for New World monkeys (platyrrhines). Behavioral data on wild capuchin (*Cebus* spp.) and squirrel (*Saimiri* spp.) monkeys are especially warranted, as these primate genera have the highest EQs (*Cebus* range = 2.5–4.8, *Saimiri* range = 2.4–2.8; Jerison 1973) among non-human primates, yet relatively little is known about their social cognition (Perry et al. 2004).

Still, much more is known about the evolution of intelligence in primates as a whole compared to any other taxon. Perhaps due to our anthropocentric ways, studies of intelligence have traditionally been conducted on primates. However, our understanding of the evolution of cognition and social learning will be furthered by conducting field-based, cross-taxa comparisons between phylogenetically distant groups such as primates, cetaceans, elephants, and corvids (Logan and Pepper 2007). Finally, as the field of cetology advances and closes the gap of 30 years behind which it lags the field of primatology (Connor et al. 2000b), we may yet see odontocetes being used as one important "litmus test" in comparative studies of cognition and sociality.

ACKNOWLEDGMENTS

H. Pearson was supported by the Tom Slick Graduate Fellowship from Texas A&M University, and a Graduate Research Assistantship from Texas A&M University at Galveston. D. Shelton was supported by a graduate fellowship from Science Foundation Arizona. Both were supported in the field by Earthwatch, the National Geographic Society, the Marlborough District Council, and New Zealand's Department of Conservation. We thank S. Lynn, L. Marino, J. Pepper, D. Reiss and B. Würsig for helpful comments on an earlier draft of this manuscript.

NOTES

1. For a discussion of the many ways in which brain size can or should be measured see Harvey and Krebs (1990), Deaner et al. (2000), Reader and Laland (2002), and van Schaik and Deaner (2003).
2. Hypotheses in which the adaptive value of large brains is *not* related to the mechanisms of thinking (e.g., the heat-generating hypothesis; Manger 2006) should not be ruled out a priori. We do not discuss this type of hypothesis further, however, as we agree with Marino and colleagues (2007, 2008) that current evidence supports the idea that the brain is an adaptation for thinking.
3. There is no neat separation between the "ecological" and "social" problems of life. We use the categories for convenience, recognizing that they are not mutually exclusive.
4. Another hypothesis is that large brains are sexually selected. However, this hypothesis predicts large (sexually selected) brains in polygamous species, which actually have relatively small brains (Pitnick et al. 2006, Schillaci 2006, Dunbar and Shultz 2007a).

Chapter 17

Social Creatures in a Changing Sea: Concluding Remarks

Bernd Würsig

Marine Mammal Research Program, Texas A&M University, Galveston, Texas, USA

All across the world, in every kind of environment and region known to man, increasingly dangerous weather patterns and devastating storms are abruptly putting an end to the long-running debate over whether or not climate change is real. Not only is it real, it's here, and its effects are giving rise to a frighteningly new global phenomenon: the man-made natural disaster.

Barack Obama, April 3, 2006

We have learned that dusky dolphins are adaptable social predators that need to factor in their own vulnerabilities as potential prey in order to survive, unlike "top" predators such as lions (*Panthera leo*), killer whales (*Orcinus orca*), and sperm whales (*Physeter macrocephalus*). We know a bit of how they do this, finding and securing prey while avoiding predators, but we do not know all. We also realize that their sociality is not just for eating and to avoid being eaten. Their matrices of associations surely also deal with who gets to mate with whom, how well young are prepared for their roles as social creatures as adults, and how much of sociality has to do with the longer term potentials that we have come to regard as learning that leads to culture. But in this "long-term social" realm, we are merely guessing, our postulates would not survive the rigors of a peer review journal for lack of data, and we can merely hope that more work will shed light on delphinid—including dusky dolphin—society structures, behaviors, and the intriguing question of culture (for example, Whitehead 2009).

Gowans et al. (2008) presented a framework of delphinid social living that essentially boils down to this: large-brained slowly reproducing and long-lived dolphins can be thought of as being (1) social carnivores that integrate their prey search and prey acquisition in a finely tuned cooperative manner, like (for example) wild dogs *(Lycaon pictus)* (Creel and Creel 1995), (2) potential prey that flock or school or group (we can use the mammalian example of grouping, with some sociality, of migrating and non-migrating caribou, *Rangifer tarandus*) and that need to live in herds largely to keep from being eaten (Ballard et al. 1997), and (3) social animals like the great apes that have the brain capacities and longer term social matrices to form relationships, bonds of friendship and enmity, and the abilities to scheme for themselves and their young to survive and reproduce (de Waal 1998). Perhaps (but here I am guessing again), among the odontocete cetaceans, harbor (*Phocoena phocoena*) and finless porpoises (*Neophocaena phocaenoides*) tend to emphasize points 1 and 2 above, while killer and sperm whales emphasize points 1 and 3. The highly social herds or "parties" (sensu Chapter 16) of small dolphins surely need all three. They "are" social carnivores, ungulates, and primates, all rolled into one.

Dusky dolphins have a fission–fusion society whereby dolphins "easily" split (or fission) into (sub)groups of one dozen or fewer animals, as nursery groups, mating groups, mixed age and sex groups, and foraging groups. They again "easily" join (or fuse), and we realize that this lability of group structure allows them to be in a group size most efficient for particular behavioral needs: the small mixed age and sex group, with many of these covering a large area, may be most efficient for finding schooling prey in daytime; the fused group, or large school may be best to herd, stop lateral movement of, and pin a large prey ball against the surface. Nursery groups may move away from the large group to be in protected waters away from boisterous males, and to "hide" from sharks and killer whales in the shallows. And so on for other situations and group sizes.

As we encounter fission–fusion, and begin to investigate individual behavioral interactions, we realize that much socializing is going on, at almost all times. Only when dolphins are heavily resting do they seem to descend into a state akin to schooling fishes, where individuality is subdued, and all pay slight attention to each other through a type of sensory integration system (Norris and Schilt 1988). But during non-resting times, social dolphins interact in myriad ways, such as (this is an example only) briefly pairing off for affiliative rubbing or sexual activity. A third may join, another leave. We obtain the impression that these animals know each other as individuals, and we guess that they "remember" each other as part of the large school or herd and re-acquaint with each other after a time of fission. If so, this matrix of relationships may help to explain their very large brains, useful for more than a reaction to cold water (Manger 2006), but instead built for the complicated social interactions of which we see tantalizing hints (Marino et al. 2008).

Dusky dolphins are doing quite well off the east coast of South Island, New Zealand, where my colleagues and I have spent so much time. I have called them a "robust" species, not at all in danger of population or species level extinction. My colleagues and I have even argued (and stick to that argument) that it is worthwhile to study ecosystems and species that are presently "robust," so that we obtain baselines of information in relatively pristine environments (Würsig et al. 2002). Not all environmentally conscious research needs to be on problem areas and species, in desperate attempts to learn enough to put out the fire of imminent extirpation or extinction (note the sad case of the riverine baiji, *Lipotes vexillifer*, apparently recently extinct, Turvey et al. 2007).

Due in part to data presented in this book, I am cautiously revising my thinking about dusky "robustness." Duskies off Peru are still beleaguered by directed human takes (Van Waerebeek and Würsig 2009 provide a brief summary), although no one knows how significant this is; off Argentina at least some are suffering from indirect killing in nets set for fishes, and—again—we do not know the actual extent of human predation (for example, Dans et al. 2003a, 2003b). In the Marlborough Sounds, wintering duskies are excluded from efficient use of area by mussel farms and—once more—we do not really know the potential or realized impact on dolphin society (Markowitz et al. 2004; Whitehead et al. 2004).

There is a larger worry, present in at least that area where we have the longest database of information, from 1972 onwards, off the Patagonian coast of central Argentina. The worry stems from the fact that the kind of large-scale aggregations of dolphins and seabirds during bait balling that Melany and I described during the early to mid 1970s (Würsig and Würsig 1980) simply "hardly ever" occur anymore. Dolphin groups are smaller, bait ball feeding events are scarcer, and dolphin presence in the two bays of study appears to be reduced. This may be due to human overfishing of anchovy stocks, or perhaps due to incidental kills of dolphins, or a combination of factors. It could even be due to natural factors, cycles of food availability that do not need to be pinned on actions by humans. "The ecology has changed," and we do not know exactly how, and why.

We know that we are in the unhappy midst of a major and frighteningly rapid global climate change (IPCC 2007), with concomitant increase in water temperatures and acidity, and loss of biodiversity in many parts of the ocean (for example, Behrenfeld 2006; effects on marine mammals summarized by Moore 2009). It is not enough to merely assume that marine inhabitants, including dolphins, can move to areas of better resources, as niches are not so easily exchanged, and potentially cultural habits of certain areas for feeding, hiding from predators, calving/nursing, etc., are not easily given up (see Whitehead et al. 2004). A recent long-term ecological "top-down" study of deep water cetaceans indicates that species diversity tracks closely with surface temperature, and loss of diversity is predicted especially for the tropics (Whitehead et al. 2008). Such studies may give us better insight on large-scale effects of climate change, including on dusky dolphins off Patagonia, but even if we have a better grasp of potential and realized effects, solving the global climate change conundrum will be a hard task indeed.

References

Abrams, P.A. (1995) Implications of dynamically variable traits for identifying, classifying, and measuring direct and indirect effects in ecological communities. *American Naturalist* 146: 112–134.

Abrams, P.A. (2007) Defining and measuring the impact of dynamic traits on interspecific interactions. *Ecology* 88: 255–256.

Acevedo-Gutiérrez, A. (2002) Group behavior. Pages 537–544 *in* W.F. Perrin, B. Würsig and J.G.M. Thewissen, eds. *Encyclopedia of Marine Mammals*. Academic Press, San Diego, CA.

Acevedo-Gutiérrez, A. and N. Parker (2000) Surface behavior of bottlenose dolphins is related to spatial arrangement of prey. *Marine Mammal Science* 16: 287–289.

Acevedo-Gutiérrez, A., D.A. Croll and B.R. Tershy (2002) High feeding costs limit dive time in the largest whales. *Journal of Experimental Biology* 205: 1747–1753.

Aguayo, L.A. (1975) Progress report on small cetacean research in Chile. *Journal of the Fisheries Research Board of Canada* 32: 1123–1144.

Aguayo-Lobo, A., D. Torres Navarro and J. Acevedo Ramírez (1998) Los mamíferos marinos de Chile: I. Cetacea. *Serie Científica INACH* 48: 19–159.

Aguilar, A., A. Borrell and P.J.H. Reijnders (2002) Geographical and temporal variation in levels of organochlorine contaminants in marine mammals. *Marine Environmental Research* 53: 425–452.

Aiello, L.C. and P. Wheeler (1995) The expensive-tissue hypothesis: the brain and the digestive system in human and primate evolution. *Current Anthropology* 36: 199–221.

Alcock, J. (1998) *Animal Behaviour: An Evolutionary Approach*, 6th edn. Sinauer Associates, Sunderland MA.

Alcock, J. and J.K. Kemp (2005) The scramble competition mating system of the sphecid wasp *Palmodes praestans* (Kohl). *Journal of Natural History* 39: 2809–2814.

Alexander, R.D. (1974) The evolution of social behavior. *Annual Review of Ecological Systems* 5: 325–383.

Altmann, J. (1974) Observational study of Behaviour: sampling methods. *Behaviour* 49: 227–267.

Angelescu, V. (1982) Ecología trófica de la anchoíta del Mar Argentino (Engraulidae, *Engraulis anchoita*). Parte II. Alimentación, comportamiento y relaciones tróficas en el ecosistema. *Contribuciones del Instituto Nacional de Investigación y Desarrollo Pesquero (INIDEP)* 409: 83.

Angelescu, V. and A. Anganuzzi (1986) Resultados sobre la alimentación de la anchoíta (*Engraulis anchoita*) en el área explorada por el B/I "Shinkai Maru" durante las campañas VI (21/09/78–12/10/78) y VIII (20/11/78–19/12/78) en el Mar Epicontinental Argentino. *Contribuciones del Instituto Nacional de Investigación y Desarrollo Pesquero (INIDEP)* 383: 281–298.

The Dusky Dolphin: Master Acrobat off Different Shores

Angelescu, V. and L.B. Prenski (1987) Ecología trófica de la merluza común del Mar Argentino (Merlucciidae, *Merluccius hubbsi*). Parte 2. Dinámica de la alimentación analizada sobre la base de las condiciones ambientales, la estructura y las evaluaciones de los efectivos en su área de distribución. *Contribuciones del Instituto Nacional de Investigación y Desarrollo Pesquero (INIDEP)* 561: 205.

Anonymous (1875). John Edward Gray, F.R.S. *Nature* 11: 368.

Araabi, B.N., N. Kehtarnavaz, T. McKinney, G. Hillman and B. Würsig (2000) A string matching computer-assisted system for dolphin photo-identification. *Annals of Biomedical Engineering* 28: 1269–1279.

Arias-Schreiber, M. (1995) Informe nacional sobre el estado de los mamiferos marinos en el Peru. IMPARPE. Unpublished report to the Comision Permanente del Pacifico and the United Nations Environment Programme.

Arlettaz, R. (1999) Habitat selection as a major resource partitioning mechanism between the two sibling bat species *Myotis myotis and Myotis blythii. Journal of Animal Ecology* 68.

Armitage, K.B. and O.A. Schwartz (2000) Social enhancement of fitness in yellow-bellied marmots. *Proceedings of the National Academy of Sciences of the United States of America* 97: 12149–12152.

Arnold, K. and A. Whiten (2003) Grooming interactions among the chimpanzees of the Budongo Forest, Uganda: tests of five explanatory models. *Behaviour* 140: 519–552.

Asseburg, C., J. Harwood, J. Matthiopoulos and S. Smout (2006) The functional response of generalist predators and its implications for the monitoring of marine ecosystems. Pages 262–274 *in* I. Boyd, S. Wanless and C.J. Camphuysen, eds. *Top Predators in Marine Ecosystems. Their role in monitoring and management.* Cambridge University Press, Cambridge, UK.

Au, W.W.L. (1993) *The Sonar of Dolphins.* Springer-Verlag, New York.

Au, W.W.L. (2002) Echolocation. Pages 358–367 *in* W.F. Perrin, B. Würsig and J.G.M. Thewissen, eds. *Encyclopedia of Marine Mammals.* Academic Press, San Diego, CA.

Au, W.W.L. (2004) A comparison of the sonar capabilities of bats and dolphins. Pages xiii–xxvii *in* J.A. Thomas, C.F. Moss and M. Vater, eds. *Echolocation in Bats and Dolphins.* University of Chicago Press, Chicago, IL.

Au, W.W.L. and K.J. Benoit-Bird (2003) Automatic gain control in the echolocation system of dolphins. *Nature* 423: 861–863.

Au, W.W.L. and D.L. Herzing (2003) Echolocation signals of wild Atlantic spotted dolphin (*Stenella frontalis*). *Journal of the Acoustical Society of America* 113: 598–604.

Au, W.W.L. and B. Würsig (2004) Echolocation signals of dusky dolphins (*Lagenorhynchus obscurus*) in Kaikoura, New Zealand. *Journal of the Acoustical Society of America* 115: 2307–2313.

Au, W.W.L., D.A. Carder, R.H. Penner and B.L. Scronce (1985) Demonstration of adaptation in beluga whale echolocation signals. *Journal of the Acoustical Society of America* 77: 230–726.

Au, W.W.L., R.H. Penner and C.W. Turl (1987) Propagation of beluga echolocation signal. *Journal of the Acoustical Society of America* 82: 807–813.

Au, W.W.L., J. Pawloski, P.E. Nachtigall, M. Blonz and R. Gisner (1995) Echolocation signal and transmission beam pattern of a false killer whale (*Pseudorca crassidens*). *Journal of the Acoustical Society of America* 98: 51–59.

Au, W.W.L., A.N. Popper and R.R. Fay (2000) *Hearing by Whales and Dolphins.* Springer-Verlag, New York, NY.

Au, W.W.L., J.K.B. Ford, J.K. Horne and K.A. Allman (2003) Echolocation signals of free-ranging killer whales (*Orcinus orca*) and modeling of foraging for chinook salmon (*Oncorhynchus tshawytscha*). *Journal of the Acoustical Society of America* 115: 901–909.

Aubauer, R. (1995) *Korrelationsverfahren zur Flugbahnverfolgung echoortender Fledemause.* Unpublished Dipl.-Ing. Technishen Hochschule Darmstadt, Darmstadt.

Aubone, A., S. Bezzi, R. Castrucci, C. Dato, P. Ibañez, G. Irusta, M. Pérez, M. Renzi, B. Santos, N. Scarlato, M. Simonazzi, L. Tringali and F. Villarino (1999) Merluza (*Merluccius hubbsi*) Secretaría de Agricultura, Ganadería, Pesca y Alimentación, Mar del Plata (CD-Rom version). Pages 27–35 *in* J.L. Cajal and L.B. Prenski, eds. *Diagnóstico de los recursos pesqueros de la República Argentina: 1999.* Instituto Nacional de Investigación y Desarrollo Pesquero.

Avilés, L., J.A. Fletcher and A.D. Cutter (2004) The kin composition of social groups: trading group size for degree of altruism. *American Naturalist* 164: 132–144.

Avital, E. and E. Jablonka (2005) *Animal Traditions: Behavioural inheritance in evolution.* Cambridge University Press, Cambridge, UK.

Ayers, D., M.P. Francis, L.H. Griggs and S.J. Baird (2004). Fish bycatch in New Zealand tuna longline fisheries, 2000–01 and 2001–02. *New Zealand Fisheries Assessment Report 2004/46.* 47 pp.

Azevedo, A.F., S.C. Viana, A.M. Oliveira and M.V. Sluys (2005) Group characteristics of marine tucuxis (*Sotalia fluviatilis*) (Cetacea: Delphinidae) in Guanabara Bay, south-eastern Brazil. *Journal of the Marine Biological Association of the United Kingdom* 85: 209–212.

Aznar, F.J., J.A. Balbuena, M. Fernández and J.A. Raga (2001) Living together: the parasites of marine mammals. Pages 385–419 *in* P. Evans and J.A. Raga, eds. *Marine Mammals: Biology and conservation.* Kluwer Academic/Plenum Publishers. New York.

Baird, R.W. (2000) The killer whale: foraging specializations and group hunting. Pages 127–153 *in* J. Mann, R.C. Connor, P.L. Tyack and H. Whitehead, eds. *Cetacean Societies: field studies of dolphins and whales.* University of Chicago Press, Chicago, IL.

Baird, R.W. and L.M. Dill (1995) Occurrence and behaviour of transient killer whales: seasonal and pod-specific variability, foraging behaviour, and prey handling. *Canadian Journal of Zoology* 73: 1300–1311.

Baird, R.W., P.A. Abrams and L.M. Dill (1992) Possible indirect interactions between transient and resident killer whales: implications for the evolution of foraging specializations in the genus *Orcinus. Oecologia* 89: 125–132.

Baird, R.W., A.D. Ligon, S.K. Hooker and A.M. Gorgone (2001) Subsurface and night-time behaviour of pan tropical spotted dolphins in Hawai'i. *Canadian Journal of Zoology* 79: 988–996.

Baird, R.W., M.B. Hanson and L.M. Dill (2005) Factors influencing the diving behaviour of fish-eating killer whales: sex differences and diel and interannual variation in diving rates. *Canadian Journal of Zoology* 83: 257–267.

Bakeman, R. and J.M. Gottman (1997) *Observing Interaction: An introduction to sequential analysis*, 2nd edn. Cambridge University Press, Cambridge, UK.

Baker, A.N. (1972a) Reproduction, early life history and age-growth relationships of the New Zealand Pilchard, *Sardinops neopilchardus* (Steindachner). *Fisheries Research Bulletin* No. 5. Fisheries Research Division, New Zealand Marine Department. 64 pp.

Baker, A.N. (1972b) New Zealand whales and dolphins. *Tuatara* 20: 1–49.

Baker, A.N. (1977) Spectacled porpoise, *Phocoena dioptrica*, new to the subantarctic Pacific Ocean. *New Zealand Journal of Marine and Freshwater Research* 11: 401–406.

Baker, A.N. (1983) *Whales and Dolphins of New Zealand and Australia: an identification guide*. Victoria University Press, Wellington.

Baker, A.N., A.N.H. Smith and F.B. Pichler (2002) Geographical variation in Hector's dolphin: recognition of new subspecies of *Cephalorhynchus hectori*. *Journal of the Royal Society of New Zealand* 32: 713–727.

Bakun, A. (1996) *Patterns in the ocean: Ocean processes and marine population dynamics*. California Sea Grant College System, National Oceanic and Atmospheric Administration, in cooperation with Centro de Investigaciones Biológicas del Noroeste, La Paz, Mexico, 323 pp.

Balbuena, J.A. and J.A. Raga (1994) Intestinal helminths as indicators of segregation and social structure of pods of long-finned pilot whales (*Globicephala melaena*) off the Faroe Islands. *Canadian Journal of Zoology* 72: 443–448.

Balbuena, J.A., F.J. Aznar, M. Fernandez and J.A. Raga (1995) The use of parasites as indicators of social structure and stock identity of marine mammals. Pages 133–139 *in* A.S. Blix, L. Wallaoe and O. Ulltanf, eds. *Whales, Seals, Fish and Man*. Elsevier Science, Amsterdam.

Balino, B.M. and D.L. Aksnes (1993) Winter distribution and migration of the sound scattering layers, zooplankton and micronekton, in Masfjorden, western Norway. *Marine Ecology Progress Series* 102: 35–50.

Ballance, L.T. and R.L. Pitman (1998) Cetaceans of the western tropical Indian Ocean: distribution, relative abundance, and comparison with cetacean communities of two other tropical ecosystems. *Marine Mammal Science* 14: 429–459.

Ballance, L.T., R.L. Pitman and P.C. Fiedler (2006) Oceanographic influences on seabirds and cetaceans of the eastern tropical Pacific: a review. *Progress in Oceanography* 69: 360–390.

Ballard, W.B., L.A. Ayres, P.R. Krausman, D.J. Reed and S.G. Fancy (1997) Ecology of wolves in relation to a migratory caribou herd in northwest Alaska. *Wildlife Monographs* 135: 5–47.

Banks, J.D., K.F. Levine, J. Syvanen, J. Theis and A. Gilson (1995) DNA tissue gender typing, in *Standard Operating Procedure of the Wildlife Forensic Laboratory of the California Department of Fish and Game*. Sacramento, CA.

Bannister, J.L., C.M. Kemper and R.M. Warneke (1996) *The Action Plan for Australian Cetaceans*. Wildlife Australia Endangered Species Program Project No. 380. Australian Nature Conservation Agency. 242 pp.

Barnard, K.H. (1954) *A Guide Book to South African Whales and Dolphins*. South African Museum Guide, 4, Cape Town, 33 pp.

Barr, K. (1997) The impacts of marine tourism on the behaviour and movement patterns of dusky dolphins (*Lagenorhynchus obscurus*) at Kaikoura, New Zealand. M.Sc. thesis, Dunedin, New Zealand: University of Otago. 97 pp.

Barr, K. and L. Slooten (1998) Effects of tourism on dusky dolphins at Kaikoura. *Proceedings of the International Whaling Commission SC/50/WW* 10: 1–30.

Barr, K. and E. Slooten (1999) Effects of tourism on dusky dolphins at Kaikoura. *Conservation Advisory Science Notes*, no. 229, Department of Conservation, Wellington, New Zealand, 28 pp.

Barrett, T., M. Blixenkrone-Møller, G. Di Guardo, M. Domingo, P. Duignan, A. Hall, L. Mamaev and A.D.M.E. Osterhaus (1995) Morbilliviruses in aquatic mammals: report on round table discussion. *Veterinary Microbiology* 44: 261–265.

Bastida, R. and D. Rodriguez (2005) *Marine Mammals of Patagonia and Antarctica* (English edition), Vasquez Mazzini Editores, Buenos Aires, Argentina.

Bastida, R., V. Lichtschein and R.N.P. Goodall (1988) Food habits of *Cephalorhynchus commersonii* off Tierra del Fuego. Pages 143–160 *in* R.L. Brownell Jr. and G.P. Donovan, eds. *Biology of the Genus Cephalorhynchus.* International Whaling Commission, Cambridge.

Beale, C.M. and P. Monaghan (2004a) Human disturbance: people as predation-free predators? *Journal of Applied Ecology* 41: 335–343.

Beale, C.M. and P. Monaghan (2004b) Behavioural responses to human disturbance: a matter of choice. *Animal Behaviour* 68: 1065–1069.

Bean, D. and J.M. Cook (2001) Male mating tactics and lethal combat in the nonpollinating fig wasp *Sycoscapter australis. Animal Behaviour* 62: 535–542.

Bearzi, M. (2005a) Dolphin sympatric ecology. *Marine Biology Research* 1: 165–175.

Bearzi, M. (2005b) Habitat partitioning by three species of dolphins in Santa Monica Bay, California. *Bulletin of the Southern California Academy of Science* 104: 113–124.

Bearzi, M. and C.B. Stanford (2007) Dolphins and African apes: comparisons of sympatric socio-ecology. *Contributions to Zoology* 76: 235–254.

Bearzi, M. and C.B. Stanford (2008) *Beautiful Minds: The parallel lives of great apes and dolphins.* Harvard University Press, Cambridge, MA.

Beauchamp, G. (2007) Exploring the role of vision in social foraging: what happens to group size, spacing, aggression and habitat use in birds and mammals that forage at night? *Biological Reviews* 82: 511–525.

Beauchamp, G. and R. McNeil (2003) Vigilance in greater flamingos foraging at night. *Ethology* 109: 511–520.

Bednekoff, P.A. and S.L. Lima (1998) Randomness, chaos and confusion in the study of antipredator vigilance. *Trends in Ecology and Evolution* 13: 284–287.

Beentjes, M.P., B. Bull, R.J. Hurst and N.W. Bagley (2002) Demersal fish assemblages along the continental shelf and upper slope of the east coast of the South Island, New Zealand. *New Zealand Journal of Marine and Freshwater Research* 36: 197–223.

Behrenfeld, M.J., R.T. O'Malley, D.A. Siegel, C.R. McCain, J.L. Sarmiento, G.C. Feldman, A.J. Milligan, P.G. Falkowski, R.M. Letelie and E.S. Boss (2006) Climate-driven trends in contemporary ocean productivity. *Nature* 444: 752–755.

Bejder, L. and A. Samuels (2003) Evaluating impacts of nature-based tourism on cetaceans. Pages 229–256 *in* N. Gales, M. Hindell and R. Kirkwood, eds. *Marine Mammals: Fisheries, tourism and management issues.* CSIRO Publishing, London.

Bejder, L., S.M. Dawson and J.A. Harraway (1999) Responses of Hector's dolphins to boats and swimmers in Porpoise Bay, New Zealand. *Marine Mammal Science* 15: 738–750.

Bejder, L., A. Samuels, H. Whitehead and N. Gales (2006) Interpreting short-term behavioural responses to disturbance within a longitudinal perspective. *Animal Behaviour* 72: 1149–1158.

Benoit-Bird, K.J., B. Würsig and C.J. McFadden (2004) Dusky dolphin (*Lagenorhynchus obscurus*) foraging in two different habitats: active acoustic detection of dolphins and their prey. *Marine Mammal Science* 20: 215–231.

Benoit-Bird, K.J., A.D. Dahood and B. Würsig (2009) Using active acoustics to compare predator-prey behavior in two marine mammal species. *Marine Ecology Progress Series*. doi: 10.3354/meps07793.

Bernard, H.J. and A.A. Hohn (1989) Difference in feeding habits between pregnant and lactating spotted dolphins (*Stenella attenuata*). *Journal of Mammalogy* 70: 211–215.

Berón-Vera, B., S.N. Pedraza, J.A. Raga, A. Gil, E.A. Crespo, M. Koen Alonso and R.N.P. Goodall (2001) Gastrointestinal helminths of Commerson's dolphins *Cephalorhynchus commersonii* from Central Patagonia and Tierra del Fuego. *Diseases of Aquatic Organisms* 47: 201–208.

Berón-Vera, B., E.A. Crespo, J.A. Raga and M. Fernández (2007) Parasite communities of common dolphins (*Delphinus delphis*) from Patagonia: the relation with host distribution and diet and comparison with sympatric hosts. *Journal of Parasitology* 93: 1056–1060.

Berón-Vera, B., E.A. Crespo and J.A. Raga Parasites of some little known cetacean species from the Southwestern Atlantic Ocean, Patagonia, Argentina. *Journal of Parasitology* (in press).

Berta, A. and J.L. Sumich (1999) *Marine Mammals: Evolutionary Biology*. Academic Press, San Diego, CA.

Bertin, A. and F. Cezilly (2003) Sexual selection, antennae length and the mating advantage of large males in *Asellus aquaticus*. *Journal of Evolutionary Biology* 16: 698–707.

Bertram, B.C.R. (1978) *Living in Groups: Predators and prey*. Blackwell, Oxford.

Best, P.B. (1976) Tetracycline marking and the rate of growth layer formation in the teeth of a dolphin (*Lagenorhynchus obscurus*). *South African Journal of Science* 72: 216–218.

Best, P.B. (2007) *Whales and Dolphins of the Southern African Subregion*. Cambridge University Press, Cape Town.

Best, P.B. and R.B. Abernethy (1994) Heaviside's dolphin (*Cephalorhynchus heavisidii*). Pages 289–310, 415–416 *in* S.H. Ridgway and R. Harrison, eds. *Handbook of Marine Mammals*. Academic Press, London.

Best, P.B. and G.J.B. Ross (1977) Exploitation of small cetaceans off southern Africa. *Reports of the International Whaling Commission* 27: 494–497.

Best, P.B. and G.J.B. Ross (1984) Live-capture fishery for small cetaceans in South African waters. *Reports of the International Whaling Commission* 34: 615–618.

Bezzi, S., M. Renzi, M. Pérez, G. Cañete, G. Irusta and H. Lassen (1995) Evaluación y estrategias de manejo del recurso merluza. Page 32 *in VI Congreso Latinoamericano de Ciencias del Mar (COLACMAR)*, resúmen 88. Mar del Plata.

Bicca-Marques, J.C. and P.A. Garber (2004) Use of spatial, visual, and olfactory information during foraging in wild nocturnal and diurnal anthropoids: a field experiment comparing *Aotus, Callicebus,* and *Saguinus*. *American Journal of Primatology* 62: 171–187.

Bierman, W.H. and E.J. Slipjer (1948) Remarks on the species of the genus *Lagenorhynchus*, Pt. II. *Proceedings of the Koninklijke Nederlandse Akademie van Wetenschappen* 51: 127–133.

Blaxter, J.H.S. (1974) The role of light in the vertical migration of fish—a review *in Proceedings of Light as an Ecological Factor II: The 16th Symposium of the British Ecological Society*. British Ecological Society.

Blumstein, D.T. and J.C. Daniel (2007) *Quantifying Behavior the JWatcher Way.* Sinauer Associates, Sunderland, MA.

Boesch, C. and H. Boesch (1989) Hunting behavior of wild chimpanzees in the Taï National Park. *American Journal of Physical Anthropology* 78: 547–573.

Boesch, C. and K. Boesch-Ackerman (2000) *The Chimpanzees of the Taï Forest. Behavioural ecology and evolution.* Oxford University Press, Oxford.

Boinski, S., L. Kauffman, A. Westoll, C.M. Stickler, S. Cropp and E. Ehmke (2003) Are vigilance, risk from avian predators and group size consequences of habitat structure? A comparison of three species of squirrel monkey (*Saimiri oerstedii, S. boliviensis,* and *S. sciureus*). *Behaviour* 140: 1421–1467.

Bon, R., C. Rideau, J.C. Villaret and J. Joachim (2001) Segregation is not only a matter of sex in Alpine ibex, *Capra ibex ibex. Animal Behaviour* 62: 495–504.

Bonabeau, E., G. Theraulaz, J.L. Deneubourg, S. Aron and S. Camazine (1997) Self-organization in insects. *Trends in Ecology and Evolution* 12: 188–193.

Bonner, W.N. (1982) *Seals and Man, a Study of Interactions.* Sea Grant Publication. University of Washington Press, Seattle, WA.

Boren, L.J., N.J. Gemmell and K.J. Barton (2002) Controlled approaches as an indicator of tourist disturbance on New Zealand fur seals (*Arctocephalus forsteri*). *Australian Mammalogy (Marine Mammal Special Issue)* 24: 85–96.

Boren, L.J., M. Morrissey, C.G. Muller and N.J. Gemmell (2006) Entanglement in New Zealand fur seals in man-made debris at Kaikoura, New Zealand. *Marine Pollution Bulletin* 52: 442–446.

Bowen, W.D. (1990) Population biology of sealworm (*Pseudoterranova decipiens*) in relation to its intermediate and seal hosts. *Canadian Bulletin of Fisheries and Aquatic Science* 222 viii + 306 pp.

Bowen, W.D. (1997) Role of marine mammals in aquatic ecosystems. *Marine Ecology Progress Series* 158: 267–274.

Bowen, W. and D.B. Siniff (1999) Distribution, population biology and feeding ecology of marine mammals. Pages 423–484 *in* J.E. Reynolds III and S.A. Rommel, eds. *Biology of Marine Mammals.* Smithsonian Institution Press, Washington DC.

Boyd, I.L., C. Lockyer and H.D. Marsh (1999) Reproduction in marine mammals. Pages 218–286 *in* J.E. Reynolds III and S.A. Rommel, eds. *Biology of Marine Mammals.* Smithsonian Institution Press, Washington DC.

Boyd, I., S. Wanless and J. Camphuysen (2006) Introduction. Pages 1–10 *in* I. Boyd, S. Wanless and C.J. Camphuysen, eds. *Top Predators in Marine Ecosystems: Their role in monitoring and management.* Cambridge University Press, London.

Boyd, P., J. Laroche, M. Gall, R. Frew and R.M.L. McKay (1999) Role of iron, light, and silicate in controlling algal biomass in subantarctic waters SE of New Zealand. *Journal of Geophysical Research-Oceans* 104: 13395–13408.

Bradley, B.J. (1999) Levels of selection, altruism, and primate behavior. *The Quarterly Review of Biology* 74: 171–194.

Bradshaw, C.J.A., C. Lalas and C.M. Thompson (2000) Clustering of colonies in an expanding population of New Zealand fur seals (*Arctocephalus forsteri*). *Journal of Zoology* 250: 105–112.

Bräger, S. (1998) Behavioural ecology and population structure of Hector's dolphin *Cephalorhynchus hectori.* Ph.D. thesis, Dunedin, New Zealand: University of Otago.

Bräger, S. and K. Schneider (1998) Near-shore distribution and abundance of dolphins along the West Coast of the South Island, New Zealand. *New Zealand Journal of Marine and Freshwater Research* 32: 105–112.

Bräger, S., S.M. Dawson, E. Slooten, S. Smith, G.S. Stone and A. Yoshinaga (2002) Site fidelity and along-shore range in Hector's dolphin, an endangered marine dolphin from New Zealand. *Biological Conservation* 108: 281–287.

Bräger, S., J.A. Harraway and B.F.J. Manly (2003) Habitat selection in a coastal dolphin species (*Cephalorhynchus hectori*). *Marine Biology* 143: 233–244.

Brear, K., J.D. Currey, M.C.S. Kingsley and M. Ramsay (1993) The mechanical design of the tusk of the narwhal (*Monodon monoceros:* Cetacea). *Journal of Zoology* 230: 411–423.

Breed, G.A., W.D. Bowen, J.I. McMillan and M.L. Leonard (2006) Sexual segregation of seasonal foraging habitats in a non-migratory marine mammal. *Proceedings of the Royal Society B: Biological Sciences* 273: 2319–2326.

Bro-Jørgensen, J. (2002) Overt female mate competition and preference for central males in a lekking antelope. *Proceedings of the National Academy of Sciences of the USA* 99: 9290–9293.

Bro-Jørgensen, J. (2003) The significance of hotspots to lekking topi antelopes (*Damaliscus lunatus*). *Behavioral Ecology and Sociobiology* 53: 324–331.

Bro-Jørgensen, J. (2008) The impact of lekking on the spatial variation in payoffs to resource-defending top bulls, *Damaliscus lunatus*. *Animal Behaviour* 75: 1229–1234.

Bro-Jørgensen, J. and S.M. Durant (2003) Mating strategies of topi bulls: getting in the centre of attention. *Animal Behaviour* 65: 585–594.

Bromham, L., M.J. Phillips and D. Penny (1999) Growing up with dinosaurs: molecular dates and the mammalian radiation. *Trends in Ecology and Evolution* 13: 113–118.

Brown, C.R. and M.B. Brown (1996) *Coloniality in the Cliff Swallow: The effect of group size on social behavior*. University of Chicago Press, Chicago, IL.

Brown, J.L. (1975) *The Evolution of Behaviour*. W.W. Norton & Co., Inc., NY.

Brown, J.S., J.W. Laundre and M. Gurung (1999) The ecology of fear: optimal foraging, game theory, and trophic interactions. *Journal of Mammalogy* 80: 385–399.

Brown, N.C. (2000) The dusky dolphin, *Lagenorhynchus obscurus*, off Kaikoura, New Zealand: a long-term comparison of behaviour and habitat use. M.Sc. thesis, Auckland, New Zealand: University of Auckland. 153 pp.

Brown, P.C., S.J. Painting and K.L. Cochrane (1991) Estimates of phytoplankton and bacterial biomass and production in the northern and southern Benguela ecosystems. *South African Journal of Marine Science* 11: 537–564.

Brownell, Jr, R.L. (1974) Small odontocetes of the Antarctic. *in* V.C. Bushnell, ed. *Antarctic Mammals. Antarctic Map Folio Series, Folio 18*. American Geographic Society, NY.

Brownell, Jr, R.L. and F. Cipriano (1999) Dusky dolphin *Lagenorhynchus obscurus* (Gray, 1828). Pages 85–104 *in* S.H. Ridgway and R. Harrison, eds. *Handbook of Marine Mammals, volume 6: The second book of dolphins and the porpoises*. Academic Press, San Diego, CA.

Brownell, Jr., R.L. and J.G. Mead (1989) Taxonomic status of the delphinid (Mammalia: Cetacea) *Tursiop panope* Philippi, 1895. *Proceedings of the Biological Society of Washington* 102: 532–534.

Bshary, R., W. Wickler and H. Fricke (2002) Fish cognition: a primate's eye view. *Animal Cognition* 5: 1–13.

Buchanan, F.C., M.K. Friesen, R.P. Littlejohn and J.W. Clayton (1996) Microsatellites from the beluga whale *Delphinapterus leucas*. *Molecular Ecology* 5: 571–575.

Buchholtz, E.A., E.M. Wolkovich and R. Cleary (2005) Vertebral osteology and complexity in *Lagenorhynchus acutus* (Delphinidae) with comparison to other delphinoid genera. *Marine Mammal Science* 21: 411–428.

Buschmann, A.H., D.A. Lopez and A. Medina (1996) A review of the environmental effects and alternative production strategies of marine aquaculture in Chile. *Aquacultural Engineering* 15: 397–421.

Bush, A.O., K.D. Lafferty, J.M. Lotz and A.W. Shostak (1997) Parasitology meets ecology in its own terms: Margolis *et al.*, revisited. *Journal of Parasitology* 83: 575–583.

Bush, A.O., J.C. Fernández, G.W. Esch and J.R. Seed (2001) *Parasitism: The diversity and ecology of animal parasites*. Cambridge University Press, Cambridge, UK.

Butler, D.J. (2003) Possible impacts of marine farming of mussels *(Perna canaliculus)* on king shags *(Leucocarbo carunculatus) DOC Science Internal Series* 111. Department of Conservation, Wellington, New Zealand, 29 pp.

Byrne, R.W. and A. Whiten (1988) *Machiavellian intelligence. Social expertise and the evolution of intellect in monkeys, apes, and humans*. Oxford University Press, Oxford, UK.

Calder, III, W.A. (1982) The pace of growth: an allometric approach to comparative embryonic and post-embryonic growth. *Journal of Zoology,* 198: 215–225.

Caldwell, D.K. and M.C. Caldwell (1971) Sounds produced by two rare cetaceans stranded in Florida. *Cetology* 4: 1–5.

Cárdenas, J.C., M.E. Stutzin, J.A. Oporto, C. Cabello and D. Torres (1986) Manual de identificacion de los cetaceos chilenos. Project WW-445. World Wildlife Fund, U.S., and Comite Nacional Pro Defensa de la Fauna y Flora (Chile). 102 pp.

Carey, F.G. and J.V. Scharold (1990) Movements of blue sharks (*Prionace glauca*) in depth and course. *Marine Biology* 106: 329–342.

Carey, F.G., J.M. Teal and J.W. Kanwisher (1981) The visceral temperatures of mackerel sharks (Lamnidae). *Physiological Zoology* 54: 334–344.

Carr, M.E. and E.J. Kearns (2003) Production regimes in four Eastern Boundary Current systems. *Deep-Sea Research II* 50: 3199–3221.

Cassens, I., K. Van Waerebeek, P.B. Best, E.A. Crespo, J. Reyes and M. Milinkovitch (2003) The phylogeography of dusky dolphins (*Lagenorhynchus obscurus*): a critical examination of network methods and rooting procedures. *Molecular Ecology* 12: 1781–1792.

Cassens, I., K. Van Waerebeek, P.B. Best, A. Tzika, A.L. Van Helden, E.A. Crespo and M. Milinkovitch (2005) Evidence for male dispersal along the coasts but no migration in pelagic waters in dusky dolphins (*Lagenorhynchus obscurus*). *Molecular Ecology* 14: 107–121.

Castellini, M. (2002) Thermoregulation. Pages 1245–1250 *in* W.F. Perrin, B. Würsig and J.G.M. Thewissen, eds. *Encyclopedia of Marine Mammals*. Academic Press, San Diego, CA.

Castley, J.G., V.G. Cockcroft and G.I.H. Kerley (1991) A note on the stomach contents of fur seals *Arctocephalus pusillus* beached on the south-east coast of South Africa. *South African Journal of Marine Science* 11: 573–577.

Cawthorn, M.W. (1988) Recent observations of Hector's dolphin, *Cephalorhynchus hectori*, in New Zealand. *Report to the International Whaling Commission. Special Issue* 9: 303–314.

Chapman, P.D., E.K. Pikitch, E.A. Babcock and M.S. Shivj (2007) Deep diving and diel changes in vertical habitat use by Caribbean reef sharks *Carcharinus perezi*. *Marine Ecology Progress Series* 344: 271–275.

Cipriano, F.W. (1992) Behavior and occurrence patterns, feeding ecology and life history of dusky dolphins (*Lagenorhynchus obscurus*) off Kaikoura, New Zealand. Ph.D. thesis, University of Arizona. 216 pp.

Cipriano, F. (1997) Antitropical distributions and speciation in dolphins of the genus *Lagenorhynchus*: a preliminary analysis. Pages 305–316 *in* A.E. Dizon, S.J. Chivers and W.F. Perrin, eds. *Molecular Genetics of Marine Mammals*. International Whaling Commission, Cambridge, MA.

Cipriano, F., D. Blair and J. McKenzie (1985) Methods for dissection and metazoan parasite examination of small cetaceans. *Mauri Ora* 12: 159–169.

Cipriano, F., B. Würsig and M. Würsig (1989) Diurnal variation in dive-times and movements of dusky dolphins feeding on deep scattering layer-associated prey. *American Zoologist* 29: 68A.

Clapham, P.J. (2000) The humpback whale: seasonal feeding and breeding in a baleen whale. Pages 173–198 *in* J. Mann, R.C. Connor, P.L. Tyack and H. Whitehead, eds. *Cetacean Societies: Field studies of dolphins and whales*. University of Chicago Press, Chicago, IL.

Clarke, G.W. (1970) Light conditions in the sea in relation to the diurnal vertical migrations Pages 41–50 *in Proceedings of an International Symposium on Biological Sound Scattering in the Ocean*. Maury Center for Oceanic Science, Washington DC.

Clarke, R. (1962) Whale observation and whale marking off the coast of Chile in 1958 and from Ecuador towards and beyond the Galapagos Islands in 1959. *Norsk Hvalfangst-Tidende* 7: 265–287.

Clayton, N.S., J.M. Dally and N.J. Emery (2007) Social cognition by food-caching corvids. The western scrub-jay as a natural psychologist. *Philosophical Transactions of the Royal Society of London. Series B, Biological Sciences* 362: 507–522.

Clua, É. and F. Grosvalet (2001) Mixed-species feeding aggregation of dolphins, large tunas and seabirds in the Azores. *Aquatic Living Resources* 14: 11–18.

Clutton-Brock, T.H. and P.H. Harvey (1980) Primates, brains and ecology. *Journal of Zoology* 190: 323–390.

Clutton-Brock, T.H., S.D. Albon, R.J. Gibson and F.E. Guinness (1979) The logical stag: adaptive aspects of fighting in red deer (*Cervus elaphus L*). *Animal Behaviour* 27: 211–225.

Cockcroft, V.G. and G.J.B. Ross (1990) Observations on the early development of a captive bottlenose dolphin calf. Pages 461–478 *in* S. Leatherwood and R.R. Reeves, eds. *The Bottlenose Dolphin*. Academic Press, San Diego, CA.

Collado-May, L.J., I. Agnarsson and D. Wartzok (2007) Phylogenetic review of tonal sound production in whales in relation to sociality. *BMC Evolutionary Biology* 7: 136.

Committee on the Alaska Groundfish Fishery and Steller Sea Lions, National Research Council (2003) *The Decline of the Steller Sea Lion in Alaskan Waters: Untangling food webs and fishing nets*. The National Academies Press, Washington DC.

Compagno, L.J.V. (1984) Sharks of the world. An annotated and illustrated catalogue of shark species known to date, Part 1. *FAO Fisheries Synopsis,* 125 pp.

Connor, R.C. (2000) Group living in whales and dolphins. Pages 199–218 *in* J. Mann, R.C. Connor, P.L. Tyack and H. Whitehead, eds. *Cetacean Societies: Field studies of dolphins and whales*. The University of Chicago Press, Chicago, IL.

Connor, R.C. (2007) Dolphin social intelligence: complex alliance relationships in bottlenose dolphins and a consideration of selective environments for extreme brain size evolution in mammals. *Philosophical Transactions of the Royal Society of London. Series B, Biological Sciences* 362: 587–602.

Connor, R. and J. Mann (2006) Social cognition in the wild: Machiavellian dolphins. Pages 329–367 *in* S. Hurley and M. Nudds, eds. *Rational Animals*. Oxford University Press, Oxford, UK.

Connor, R.C., R.A. Smolker and A.F. Richards (1992a) Dolphin alliances and coalitions. Pages 415–443 *in* A.H. Harcourt and F.B.M. De Waal, eds. *Coalitions and Alliances in Humans and Other Animals*. Oxford University Press, Oxford.

Connor, R.C., R.A. Smolker and A.F. Richards (1992b) Two levels of alliance formation among male bottlenose dolphins (*Tursiops* sp.). *Proceedings of the National Academy of Sciences* 89: 987–990.

Connor, R.C., A.F. Richards, R.A. Smolker and J. Mann (1996) Patterns of female attractiveness in Indian Ocean bottlenose dolphins. *Behaviour* 133: 37–69.

Connor, R.C., J. Mann, P.L. Tyack and H. Whitehead (1998) Social evolution in toothed whales. *Trends in Ecology and Evolution* 13: 228–232.

Connor, R.C., M.R. Heithaus and L.M. Barre (1999) Super-alliances of bottlenose dolphins. *Nature* 397: 571–572.

Connor, R.C., R.S. Wells, J. Mann and A.J. Read (2000a) The bottlenose dolphin: social relationships in a fission-fusion society. Pages 91–126 *in* J. Mann, R.C. Connor, P.L. Tyack and H. Whitehead, eds. *Cetacean Societies: Field studies of dolphins and whales*. University of Chicago Press, Chicago, IL.

Connor, R.C., J. Mann, P.L. Tyack and H. Whitehead (2000b) The social lives of whales and dolphins. Pages 1–6 *in* J. Mann, R.C. Connor, P.L. Tyack and H. Whitehead, eds. *Cetacean Societies: Field studies of dolphins and whales*. University of Chicago Press, Chicago, IL.

Connor, R.C., A.J. Read and R. Wrangham (2000c) Male reproductive strategies and social bonds. Pages 247–269 *in* J. Mann, R.C. Connor, P.L. Tyack and H. Whitehead, eds. *Cetacean Societies: Field studies of dolphins and whales*. University of Chicago Press, Chicago, IL.

Constantine, R. (1996) Distribution of dusky dolphins (*Lagenorhynchus obscurus*) in New Zealand waters. Unpublished report for the Department of Conservation, Auckland, New Zealand. 6 pp.

Constantine, R. (1999) The effects of tourism on marine mammals: A review of literature relevant to managing the industry in New Zealand. *Science for Conservation Series,* 106. Department of Conservation, Wellington, New Zealand, 60 pp.

Constantine, R. (2001) Increased avoidance of swimmers by wild bottlenose dolphins (*Tursiops truncatus*) due to long-term exposure to swim-with-dolphin tourism. *Marine Mammal Science* 17: 689–702.

Constantine, R. and C.S. Baker (1997) Monitoring the commercial swim-with-dolphin operations in the Bay of Islands, New Zealand. *Science and Research Series*, No. 104. Department of Conservation, Wellington, New Zealand, 54 pp.

Constantine, R., I. Visser, D. Buurman, R. Buurman and B. McFadden (1998) Killer whale (*Orcinus orca*) predation on dusky dolphins (*Lagenorhynchus obscurus*) in Kaikoura, New Zealand. *Marine Mammal Science* 14: 324–330.

Constantine, R., D.H. Brunton and T. Dennis (2004) Dolphin-watching tour boats change bottlenose dolphin (*Tursiops truncatus*) behaviour. *Biological Conservation* 117: 299–307.

Coolen, I., O. Dangles and J. Casas (2005) Social learning in noncolonial insects? *Current Biology* 15: 1931–1935.

Cope, E.D. (1866) Third contribution to the history of the Balaenidae and Delphinidae. *Proceedings of the Academy of Natural Sciences of Philadelphia* 18: 293–300.

Cope, E.D. (1876) Fourth contribution to the history of existing Cetacea. *Proceedings of the Academy of Natural Sciences of Philadelphia* 28: 129–139.

Corcuera, J., F. Monzon, E.A. Crespo, A. Aguilar and J.A. Raga (1994) Interactions between marine mammals and coastal fisheries of Necochea and Claromecó (Buenos Aires Province, Argentina). Gillnets and Cetaceans. *International Whaling Commission Special Issue* 15: 269–281.

Corkeron, P.J. (1990) Aspects of the behavioral ecology of inshore dolphins *Tursiops truncatus* and *Sousa chinensis* in Moreton Bay, Australia. Pages 285–293 *in* S. Leatherwood and R.R. Reeves, eds. *The Bottlenose Dolphin*. Academic Press, San Diego, CA.

Coscarella, M. (2005) Ecología, comportamiento y evaluación del impacto de embarcaciones sobre manadas de tonina overa *Cephalorhynchus commersonii* en Bahía Engano, Chubut. Ph.D. thesis, Buenos Aires, Argentina: Universidad de Buenos Aires. 225 pp.

Coscarella, M.A., S.L. Dans, E.A. Crespo and S.N. Pedraza (2003) Potential impact of dolphin watching unregulated activities in Patagonia. *Journal of Cetacean Research and Management* 5: 77–84.

Costa, D.P. (2002) Energetics. Pages 387–394 *in* W.F. Perrin, B. Würsig and J.G.M. Thewissen, eds. *Encyclopedia of Marine Mammals*. Academic Press, San Diego, CA.

Courbis, S. (2007) Effect of spinner dolphin presence on the level of swimmer and vessel activity in Hawaiian Bays. *Tourism in Marine Environment* 4: 1–14.

Crawford, R.J.M., E. Goya, J.-P. Roux and C.B. Zavalaga (2006) Comparison of assemblages and some life-history traits of seabirds in the Humboldt and Benguela systems. *African Journal of Marine Science* 28: 553–560.

Creel, S. and D. Christianson (2008) Relationships between direct predation and risk effects. *Trends in Ecology and Evolution* 23: 194–201.

Creel, S. and N.M. Creel (1995) Communal hunting and pack size in African wild dogs, *Lycaon pictus*. *Animal Behaviour* 50: 1325–1339.

Creel, S. and J.A. Winnie (2005) Responses of elk herd size to fine-scale spatial and temporal variation in the risk of predation by wolves. *Animal Behaviour* 69: 1181–1189.

Crespi-Abril, A., N.A. García, E.A. Crespo and M.A. Coscarella (2003) Marine mammals consumption by broadnose sevengill shark *Notorynchus cepedianus* in the northern and central Patagonian Shelf. *Latin American Journal of Aquatic Mammals* 2: 101–107.

Crespo, E.A. and M.A. Hall (2001) Interactions between *aquatic mammals* and humans in the context of ecosystem management. Pages 463–490 *in* P.G.H. Evans and J.A. Raga, eds. *Marine Mammals: Biology and Conservation*. Kluwer Academic/Plenum Publishers, New York.

Crespo, E.A., J. Corcuera and A. Lopez Cazorla (1994) Interactions between marine mammals and fisheries in some fishing areas of the coast of Argentina. Gillnets and Cetaceans. *International Whaling Commission Special Issue* 15: 283–290.

Crespo, E.A., S.N. Pedraza, M. Coscarella, N.A. García, S.L. Dans, M. Iñiguez, L.M. Reyes, M.K. Alonso, A.C.M. Schiavini and R. González (1997a) Distribution and school size of dusky dolphins, *Lagenorhynchus obscurus* (Gray, 1828), in the southwestern South Atlantic. *Oceanic Reports of the International Whaling Commission,* 47: 693–697.

Crespo, E.A., S.N. Pedraza, S.L. Dans, M. Koen Alonso, L.M. Reyes, N.A. García, M. Coscarella and A.C.M. Schiavinni (1997b) Direct and indirect effects of the high seas fisheries on the marine mammal populations in the northern and central Patagonian Coast. *Journal of Northwest Atlantic Fisheries Science* 22: 189–207.

Crespo, E.A., M. Koen Alonso, S.L. Dans, N.A. García, S.N. Pedraza, M. Coscarella and R. González (2000) Incidental catch of dolphins in mid-water trawls for southern anchovy off Patagonia. *Journal of Cetacean Research and Management* 2: 11–16.

Croft, D.P., L.J. Morrell, A.S. Wade, C. Piyapong, C.C. Ioannou, J.R.G. Dyer, B.B. Chapman, W. Yan and J. Krause (2006) Predation risk as a driving force for sexual segregation: a cross-population comparison. *American Naturalist* 167: 867–878.

Crovetto, A., J. Lamilla and G. Pequeno (1992) *Lissodelphis peronii,* Lacepede 1804 (Delphinidae, Cetacea) within the stomach contents of a sleeping shark, *Somniosus cf pacificus,* Bigelow and Schroeder 1944, in Chilean waters. *Marine Mammal Science* 8: 312–314.

Cruickshank, R.A. and S.G. Brown (1981) Recent observations and some historical records of southern right whale dolphins *Lissodelphis peronii. Fisheries Bulletin of South Africa* 15: 109–122.

Cushing, D.H. (1973) *The Deep Scattering Layer.* Pergamon Press, Oxford.

Dahood, A.D. and B. Würsig, Z. Vernon, I. Bradshaw, D. Buurman and L. Buurman (2008) Tour operator data illustrate long term dusky dolphin (*Lagenorhynchus obscurus*) occurrence patterns near Kaikoura, New Zealand. *Proceedings of the International Whaling Commission* SC/60/WW2: 1–19, June 2008.

Dailey, M.D. and W.K. Vogelbein (1991) Parasite fauna of three species of Antartic whales with reference to their use as potential stock indicators. *Fishery Bulletin, U.S.* 89: 355–365.

Dailey, M.D. and K.A. Otto (1982) Parasites as biological indicators of the distribution and diets of marine mammals common to the eastern pacific. *National Marine Fisheries Service Center Administrative Report* LJ-82–13C. Available from SWFSC, P.O. Box 271, La Jolla, CA 92038, USA. 44 pp.

Dally, J.M., N.J. Emery and N.S. Clayton (2006) Food-caching western scrub-jays keep track of who was watching when. *Science* 312: 1662–1665.

Dammhahn, M. and P.M. Kappeler (2005) Social system of *Microcebus berthae,* the world's smallest primate. *International Journal of Primatology* 26: 407–435.

Dans, S.L., E.A. Crespo, N.A. Garcia, L.M. Reyes, S.N. Pedraza and M. Koen Alonso (1997a) Incidental mortality of Patagonian dusky dolphins in mid-water trawling: retrospective effects from the early 80's. *Reports of the International Whaling Commission* 47: 699–704.

Dans, S.L., E.A. Crespo, S.N. Pedraza and M.K. Alonso (1997b) Notes on the reproductive biology of female dusky dolphins (*Lagenorhynchus obscurus*) off the Patagonia coast. *Marine Mammal Science* 13: 303–307.

Dans, S.L., L.M. Reyes, S.N. Pedraza, J.A. Raga and E.A. Crespo (1999) Gastrointestinal helminths of the dusky dolphin, *Lagenorhynchus obscurus* (Gray, 1828), off Patagonia, in the Southwestern Atlantic. *Marine Mammal Science* 15: 649–660.

Dans, S.L., M. Koen Alonso, S.N. Pedraza and E.A. Crespo (2003a) Incidental catch of dolphins in trawling fisheries off Patagonia, Argentina: can populations persist? *Ecological Applications* 13: 754–762.

Dans, S.L., M. Koen Alonso, E.A. Crespo, S.N. Pedraza and N.A. García (2003b) Interactions between marine mammals and high seas fisheries in Patagonia under an integrated approach. Pages 100–115 *in* N. Gales, M. Hindell and R. Kirkwood, eds. *Marine Mammals: Fisheries; tourism and management issues.* CSIRO Publishing.

Dans, S.L., E.A. Crespo, S.N. Pedraza, M. Degrati and G. Garaffo (2008) Dusky dolphin and tourist interaction: impact on diurnal feeding behavior. *Marine Ecology Progress Series* 369: 287–296.

Darwin, C. (1859) *On the Origin of Species by Means of Natural Selection, or the preservation of favoured races in the struggle for life.* John Murray, London.

Darwin, C.R. et al. (1843a) Report of a Committee appointed "to consider the rules by which the nomenclature of Zoology may be established on a uniform and permanent basis." *Report of the British Association for the Advancement of Science, Manchester 1842* 12: 105–121.

Darwin, C.R. et al. (1843b) Series of propositions for rendering the nomenclature of zoology uniform and permanent, being the Report of a Committee for the consideration of the subject appointed by the British Association for the Advancement of Science. *Annals and Magazine of Natural History, including Zoology, Botany, and Geology* 11: 259–275.

Davidson, R.J., S.P. Courtney, I.R. Millar, D.A. Brown, N.A. Deans, P.R. Clerke, J.C. Dix, P.F. Lawless, S.J. Mavor and S.M. McRae (1995) *Ecologically important marine, freshwater, island and mainland areas from Cape Soucis to Ure River, Marlborough, New Zealand: recommendations for protection.* Occasional Publication No. 16. Department of Conservation, Nelson/Marlborough Conservancy 286 pp.

Davies, J.L. (1963a) The antitropical factor in cetacean speciation. *Evolution* 17: 107–116.

Davies, J.L. (1963b) *Whales and Seals of Tasmania.* Tasmanian Museum and Art Gallery, Hobart 32 pp.

Dávila, P.M., D. Figueroa and E. Muller (2002) Freshwater input into the coastal ocean and its relation with the salinity distribution off austral Chile (35–55°S). *Continental Shelf Research* 22: 521–534.

Dawber, C. (2004) *Lines in the Water. A history of greenshell mussel farming in New Zealand.* River Press, Picton, New Zealand.

Dawkins, M.S. (1980) *Animal Suffering: The science of animal welfare.* Chapman and Hall, London.

Dawson, S.M. (1988) The high frequency sounds of free-ranging Hector's dolphin, *Cephalorhynchus hectori. Reports of the International Whaling Commission* Special Issue 9: 339–341.

Dawson, S.M. (1991) Incidental catch of Hector's dolphin in inshore gillnets. *Marine Mammal Science* 7: 283.

Dawson, S.M. (2002) *Cephalorhynchus* dolphins. Pages 200–203 *in* W.F. Perrin, B. Würsig and J.G.M. Thewissen, eds. *Encyclopedia of Marine Mammals.* Academic Press, San Diego, CA.

Dawson, S. and E. Slooten (2005) Management of gillnet bycatch of cetaceans in New Zealand. *Journal of Cetacean Research and Management* 7: 59–64.

Dawson, S., E. Slooten, S. DuFresne, P. Wade and D. Clement (2004) Small-boat surveys for coastal dolphins: Line-transect surveys for Hector's dolphins (*Cephalorhynchus hectori*). *Fishery Bulletin* 102: 441–451.

Dawson, S., and L. Slooten (2008) A platform-of-opportunity survey for cetaceans, especially Chilean dolphins, *Cephalorhynchus eutropia,* in the Chilean fjords. *Proceedings of the International Whaling Commission* SC/60/SM11: 1–12.

De Bufrénil V., A. Dziedzic and D. Robineau (1989) Répartition et déplacements des Dauphin de Commerson (*Cephalorhynchus commersonii*) dans un golfe des iles Kerguelen; données du marquage individuel. *Canadian Journal of Zoology* 76: 516–521.

De Haro, J.C. and M.A. Iñíguez (1997) Ecology and behaviour of the Peale's dolphin, *Lagenorhynchus australis* (Peale, 1848) at Carbo Virgenes in Patagonia, Argentina. *Reports of the International Whaling Commission* 47: 723–727.

De Kock, A.C., P.B. Best, V. Cockcroft and C. Bosma (1994) Persistent organochlorine residues in small cetaceans from the east and west coasts of southern Africa. *The Science of the Total Environment* 154: 153–162.

De Waal, F.B.M. (1998) *Chimpanzee Politics: Power and sex among apes, revised edition.* Johns Hopkins University Press, Baltimore, MD.

De Waal, F.B.M. and A.H. Harcourt (1992) Coalitions and alliances: a history of ethological research. Pages 1–22 *in* A.H. Harcourt and F.B.M. De Waal, eds. *Coalitions and Alliances in Humans and Other Animals.* Oxford University Press, Oxford.

Deaner, R.O., C.L. Nunn and C.P. Van Schaik (2000) Comparative tests of primate cognition: different scaling methods produce different results. *Brain, Behavior and Evolution* 55: 44–52.

Deaner, R.O., R.A. Barton and C.P. Van Schaik (2003) Primate brains and life histories: renewing the connection. Pages 233–265 *in* P.M. Kappeler and M.E. Pereira, eds. *Primate Life Histories and Socioecology.* University of Chicago Press, Chicago, IL.

Degrati, M. (2002) Patrón de actividad de las manadas de delfín oscuro *Lagenorhynchus obscurus* en el Golfo Nuevo durante la temporada primavera-verano. B.Sc. thesis, Puerto Madryn, Chubut, Argentina: Universidad Nacional de la Patagonia. 71 pp.

Degrati, M., S.L. Dans, S.N. Pedraza, E.A. Crespo and G.V. Garaffo (2003) Feeding behavior of dusky dolphins in Golfo Nuevo, *in* Abstracts of the 5th Sea Science Conference, Mar del Plata, Buenos Aires, Argentina.

Degrati, M., S.L. Dans, S.N. Pedraza, E.A. Crespo and G.V. Garaffo (2008) Diurnal behaviour of dusky dolphins, *Lagenorhynchus obscurus,* in Golfo Nuevo. *Journal of Mammalogy* 89: 1241–1247.

Dehn, M.M. (1990) Vigilance for predators—detection and dilution effects. *Behavioral Ecology and Sociobiology* 26: 337–342.

Demaster, D.P., C.W. Fowler, S.L. Perry and M.F. Richlen (2001) Predation and competition: the impact of fisheries on marine-mammal populations over the next one hundred years. *Journal of Mammalogy* 82: 641–651.

Department Of Conservation. (1995). Guidelines for ecological investigations of proposed marine farm areas, Marlborough Sounds. Report prepared for Marlborough District Council by Department of Conservation, Nelson/Marlborough Conservancy. Occasional Publication No. 25. 21 pp.

Department Of Conservation (1999) *Dolphin watching at Kaikoura, guidelines for boat operators* Fact Sheet no. 79. Nelson Marlborough Conservancy, Department of Conservation.

Department of the Environment and Heritage (DEH) (2006) *Blue, Fin and Sei Whale Recovery Plan 2005–2010* GPO Box 787, Canberra ACT 2601, Australia. Department of the Environment and Heritage. 11 pp.

Deutsch, S. (2008) Development and social learning of young dusky dolphins. M.Sc. thesis, College Station, TX: Texas A&M University. 87 pp.

Dill, L.M. (1987) Animal decision-making and its ecological consequences – The future of aquatic ecology and behaviour. *Canadian Journal of Zoology* 65: 803–811.

Dill, L.M., M.R. Heithaus and C.J. Walters (2003) Behaviorally mediated indirect interactions in marine communities and their conservation implications. *Ecology* 84: 1151–1157.

Donati, G., A. Bollen, S.M. Borgognini-Tarli and J.U. Ganzhorn (2007) Feeding over the 24-h cycle: dietary flexibility of cathemeral collared lemurs (*Eulemur collaris*). *Behavioral Ecology and Sociobiology* 61: 1237–1251.

Duchesne, P., C. Étienne and L. Bernatchez (2006) PERM: a computer program to detect structuring factors in social units. *Molecular Ecology Notes* 6: 965–967.

Duffield, D.A. and R.S. Wells (1991) The combined application of chromosome protein and molecular data for the investigation of social unit structure and dynamics in *Tursiops truncatus*. Pages 155–170 *in* A.R. Hoelzel (ed). *Genetic Ecology of Whales and Dolphins: Incorporating the proceedings of the workshop on the genetic analysis of cetacean populations*. International Whaling Commission.

Duffy, C.A.J. and D.A. Brown (1994) Recent observations of marine mammals and a leatherback turtle (*Dermochelys coriacea*) in the Marlborough Sounds, New Zealand, 1981–1990 Occasional Publication No. 9. Nelson, New Zealand Department of Conservation. 52 pp.

Dugatkin, L.A. (2000) *The Imitation Factor: Evolution and beyond the gene*. The Free Press, New York.

Dugatkin, L.A., M. Mesterton-Gibbons and A.I. Houston (1992) Beyond the prisoner's dilemma: toward models to discriminate among mechanisms of cooperation in nature. *Trends in Evolution and Ecology* 7: 202–205.

Duignan, P.J., N.J. Gibbs and G.W. Jones (2003) Autopsy of cetaceans incidentally caught in fishing operations 1997/98, 1999/2000, and 2000/01. *DOC Science Internal Series 119*: 66 pp.

Duignan, P.J., N.J. Gibbs and G.W. Jones (2004) Autopsy of cetaceans incidentally caught in commercial fisheries, and all beachcast specimens of Hector's dolphins, 2001/02. *DOC Science Internal Series* 176: 28.

Dunbar, R.I.M. (1992) Neocortex size as a constraint on group size in primates. *Journal of Human Evolution* 20: 469–493.

Dunbar, R. (1997) The monkeys' defence alliance. *Nature* 386: 555–557.

Dunbar, R.I.M. (1998) The social brain hypothesis. *Evolutionary Anthropology* 6: 178–190.

Dunbar, R.I. (2003) The social brain: mind, language, and society in evolutionary perspective. *Annual Review of Anthropology* 32: 163–181.

Dunbar, R.I.M. and S. Shultz (2007a) Evolution in the social brain. *Science* 317: 1344–1347.

Dunbar, R.I.M. and S. Shultz (2007b) Understanding primate brain evolution. *Philosophical Transactions of the Royal Society of London. Series B, Biological Sciences* 362: 649–658.

Duncan, J., R.J. Seitz, J. Kolodny, D. Bor, H. Herzog, A. Ahmed, F.N. Newell and H. Emslie (2000) A neural basis for general intelligence. *Science* 289: 457–460.

Duprey, N.M.T. (2007) Dusky dolphin (*Lagenorhynchus obscurus*) behavior and human interactions: implications for tourism and aquaculture. M.Sc. thesis, College Station, TX: Texas A&M University. 64 pp.

Duprey, N.M.T., J.S. Weir and B. Würsig (2008) Effectiveness of a voluntary code of conduct in reducing vessel traffic around dolphins. *Ocean and Coastal Management* 51: 632–637.

Dyck, M.G. and R.K. Baydack (2004) Vigilance behaviours of polar bears (*Ursus maritimus*) in the context of wildlife viewing activities in Churchill, Manitoba, Canada. *Biological Conservation* 116: 343–350.

Eberle, M. and P.M. Kappeler (2004) Sex in the dark: determinants and consequences of mixed male mating tactics in *Microcebus murinus,* a small solitary nocturnal primate. *Behavioral Ecology and Sociobiology* 57: 77–90.

Eberle, M., M. Perret and P.M. Kappeler (2007) Sperm competition and optimal timing of matings in *Microcebus murinus. International Journal of Primatology* 28: 1267–1278.

Ebert, D.A. (1991) Diet of the sevengill shark *Notorynchus cepedianus* in the temperate coastal waters of southern Africa. *South African Journal of Marine Science* 11: 565–572.

Ebert, D.A. (2002) Ontogenetic changes in the diet of the sevengill shark (*Notorynchus cepedianus*). *Marine and Freshwater Research* 53: 517–523.

Ebert, D.A. (2003) *The Sharks, Rays and Chimaeras of California.* University of California Press, CA.

Edmunds, M. (1974) *Defence in Animals.* Longman, New York.

Eibl-Eibesfeldt, I. (1961) The fighting behavior of animals. *Scientific American* 205: 112–128.

Eklov, P. (1992) Group foraging versus solitary foraging efficiency in piscivorous predators – the perch, *Perca fluviatilis,* and pike, *Esox-lucius,* patterns. *Animal Behaviour* 44: 313–326.

Elgar, M.A. (1989) Predator vigilance and group size in mammals and birds—a critical review of the empirical evidence. *Biological Reviews of the Cambridge Philosophical Society* 64: 13–33.

Elwen, S.H. (2008) The distribution, movements and abundance of Heaviside's dolphins in the nearshore waters of the Western Cape, South Africa. Ph.D. thesis, Pretoria: University of Pretoria. 208 pp.

Elwen, S.H., M.A. Meyer, P.B. Best, P.G.H. Kotze, M. Thornton and S. Swanson (2006) Range and movements of female Heaviside's dolphins (*Cephalorhynchus heavisidii*) as determined by satellite linked telemetry. *Journal of Mammalogy* 87: 866–877.

Emery, N.J. and N.S. Clayton (2004) The mentality of crows: convergent evolution of intelligence in corvids and apes. *Science* 306: 1903–1907.

Emery, N.J. and N.S. Clayton (2005) Evolution of the avian brain and intelligence. *Current Biology* 15: R946–R950.

Emery, N.J., A.M. Seed, A.M.P. von Bayern and N.S. Clayton (2007) Cognitive adaptations of social bonding in birds. *Philosophical Transactions of the Royal Society of London. Series B, Biological Sciences* 362: 489–505.

Emlen, S.T. and L.W. Oring (1977) Ecology, sexual selection and the evolution of mating systems. *Science* 197: 215–223.

Encarnacao, J.A., U. Kierdorf, D. Holweg, U. Jasnoch and V. Wolters (2005) Sex-related differences in roost-site selection by Daubenton's bats *Myotis daubentonii* during the nursery period. *Mammal Review* 35: 285–294.

Engh, A.L., E.R. Siebert, D.A. Greenberg and K.E. Holekamp (2005) Patterns of alliance formation and postconflict aggression indicate spotted hyaenas recognize third-party relationships. *Animal Behaviour* 69: 209–217.

Enright, J.T. (1979) The why and when of up and down. *Limnology and Oceanography* 24: 788–791.

Estes, J.A. and D.O. Duggins (1995) Sea otters and kelp forests in Alaska—generality and variation in a community ecological paradigm. *Ecological Monographs* 65: 75–100.

Estes, J.A., M.T. Tinker, T.M. Williams and D.F. Doak (1998) Killer whale predation on sea otters linking oceanic and nearshore ecosystems. *Science* 282: 473–476.

Estes, J.A., D.P. Demaster, D.F. Doak, T.M. Williams and R.L. Brownell Jr. (2006) *Whales, Whaling, and Ocean Ecosystems*. University of California Press, Berkeley, CA.

Evans, P.G.H. (1982) Associations between seabirds and cetaceans: a review. *Mammal Review* 12: 187–206.

Evans, W.E. (1975) Distribution, differentiation of populations, and other aspects of the natural history of *Delphinus delphis* (Linnaeus) in the northeastern Pacific. Ph.D. dissertation, Los Angeles: University of California.

Fertl, D. and B. Würsig (1995) Coordinated feeding by Atlantic spotted dolphins (*Stenella frontalis*) in the Gulf of Mexico. *Aquatic Mammals* 21: 3–5.

Fertl, D., A. Acevedo-Gutierrez and F.L. Darby (1996) A report of killer whales (*Orcinus orca*) feeding on a carcharhinid shark in Costa Rica. *Marine Mammal Science* 12: 606–611.

Findlay, K.P., P.B. Best, G.J.B. Ross and V.G. Cockroft (1992) The distribution of small odontocete cetaceans off the coasts of South Africa and Namibia. *South African Journal of Marine Science* 12: 237–270.

Fischer, F. and K.E. Linsenmair (1999) The territorial system of the kob antelope (*Kobus kob kob*) in the Comoé National Park, Côte d'Ivoire. *African Journal of Ecology* 37: 386–399.

Fisher, R.A. (1930) *The Genetical Theory of Natural Selection*. Dover, New York.

Fitzgibbon, C.D. (1990) Why do hunting cheetahs prefer male gazelles? *Animal Behaviour* 40: 837–845.

Fitzgibbon, C.D. (1994) The costs and benefits of predator inspection behavior in Thomson's gazelles. *Behavioral Ecology and Sociobiology* 34: 139–148.

Fleming, P.A. and S.W. Nicholson (2004) Sex differences in space use, body condition and survivorship during the breeding season in the Namaqua rock mouse, *Aethomys namaquensis*. *African Zoology* 39: 123–132.

Flower, W.H. (1883) On the characters and divisions of the family Delphinidae. *Proceedings of the Zoological Society of London*: 466–513.

Flower, W.H. (1885) *List of the Specimens of Cetacea in the Zoological Department of the British Museum*. Taylor and Francis, London 36 pp.

Foellmer, M.W. and D.J. Fairbairn (2005) Competing dwarf males: sexual selection in an orb-weaving spider. *Journal of Evolutionary Biology* 18: 629–641.

Ford, J.K.B. (1991) Vocal traditions among resident killer whales (*Orcinus orca*) in coastal waters of British Columbia. *Canadian Journal of Zoology* 69: 1454–1483.

Ford, J.K.B., G.M. Ellis, L.G. Barrett-Lennard, A.B. Morton, R.S. Palm and K.C. Balcomb (1998) Dietary specialization in two sympatric populations of killer whales (*Orcinus orca*), in coastal British Columbia and adjacent waters. *Canadian Journal of Zoology* 76: 1456–1471.

Fouad, K., F. Libersat and W. Rathmayer (1996) Neuromodulation of the escape behavior of the cockroach *Periplaneta americana* by the venom of the parasitic wasp *Ampulex compressa*. *Journal of Comparative Physiology A* 178: 91–100.

Francis, M.P., J.D. Stevens and P.R. Last (1988) New records of *Somniosus* (Elasmobranchii, Squalidae) from Australasia, with comments on the taxonomy of the genus. *New Zealand Journal of Marine and Freshwater Research* 22: 401–409.

Fraser, F.C. (1948) Cetaceans. Pages 203–350 *in* J.R. Norman and F.C. Fraser, eds. *Giant Fishes, Whales, and Dolphins*, 2nd edn. Putnam, London.

Fraser, F.C. (1966) Comments on the Delphinoidea. Pages 7–31 *in* K.S. Norris, ed. *Whales, Dolphins, and Porpoises*. University of California Press, Berkeley.

Fraser, F.C. and P.E. Purves (1960) Hearing in cetaceans. *Bulletin of the British Museum of Natural History, Zoology* 7: 1–140.

Fréon, P., J. Alheit, E.D. Barton, S. Kifani and P. Marchesiello (2006) Modelling, forecasting and scenarios in comparable upwelling ecosystems: California, Canary and Humboldt. Pages 185–220 *in* V. Shannon, G. Hempel, P. Malanotte-Rizzoli, C. Moloney and J. Woods, eds. *Benguela: Predicting a large marine ecosystem. Large marine ecosystems*, volume 14. Elsevier, Amsterdam.

Frid, A. and L.M. Dill (2002) Human-caused disturbance stimuli as a form of predation risk. *Conservation Ecology* 6: 11–27.

Gago, F.J. and R.C. Ricord (2005) *Symbolophorus reversus:* a new species of lanternfish from the eastern Pacific (Myctophiformes: Myctophidae). *Copeia* 2005: 138–145.

Gall, M., A. Ross, J. Zeldis and J. Davis (2000) Phytoplankton in Pelorus Sound: food for mussels. *Water and Atmosphere* 8: 1–16.

Gallardo, A. (1912) El delfin *Lagenorhynchus fitzroyi* (Waterhouse) Flower capturado en Mar del Plata. *Anales del Museo Nacional de Historia Natural, Buenos Aires* 23: 391–398.

Gallardo, A. (1913) Notas sobre la anatomia del aparato espiracular, laringe y hioides de dos delfines: *Phocaena dioptrica* Lahille y *Lagenorhynchus fitzroyi* (Waterhouse) Flower. *Anales del Museo Nacional de Historia Natural, Buenos Aires* 24: 235–246.

Gambell, J. (1999) The International Whaling Commission and the contemporary whaling debate. Pages 179–198 *in* J.R. Twiss and R.R. Reeves, eds. *Conservation and Management of Marine Mammals*. Smithsonian Institution, Washington, DC.

Garaffo, G.V., S.L. Dans, S.N. Pedraza, E.A. Crespo and M. Degrati (2007) Habitat use by dusky dolphin in Patagonia: how predictable is their location? *Marine Biology* 152: 165–177.

García De La Rosa, S.B. (1998) Estudio de las interrelaciones tróficas de dos elasmobranquios del Mar Argentino, en relación con las variaciones espacio-temporales y ambientales: *Squalus acanthias* (Squalidae) y *Raja flavirostris* (Rajidae). Ph.D. thesis, Facultad de Ciencias Exactas y Naturales, Universidad Nacional de Mar del Plata. 215 pp.

García De La Rosa S.B. and F. Sánchez (1997) Alimentación de *Squalus acanthias* y predación sobre merluza *Merluccius hubbsi* en el Mar Argentino entre 34°47′–47°S. *Revista de Investigación y Desarrollo Pesquero* 11: 119–133.

Gardiner, J.M. and J. Atema (2007) Sharks need the lateral line to locate odor sources: rheotaxis and eddy chemotaxis. *Journal of Experimental Biology* 210: 1925–1934.

Gaskin, D.E. (1968a) Distribution of Delphinidae (Cetacea) in relation to sea surface temperatures off eastern and southern New Zealand. *New Zealand Journal of Marine and Freshwater Research* 2: 527–534.

Gaskin, D.E. (1968b) The New Zealand Cetacea. *Fisheries Research Bulletin* No. 1 (New Series): 1–92.

Gaskin, D.E. (1972) *Whales, Dolphins and Seals with Special Reference to the New Zealand Region*. Heinemann Educational Books, London.

Gaskin, D.E. (1982) *The Ecology of Whales and Dolphins*. Heineman Educational Books, London.

Gautier-Hion, A., R. Quris and J. Gautier (1983) Monospecific versus polyspecific life: a comparative study of foraging and antipredation tactics in a community of *Cercopithecus* monkeys. *Behavioral Ecology and Sociobiology* 12: 325–335.

Gazda, S.K., R.C. Connor, R.K. Edgar and F. Cox (2005) A division of labour with role specialization in group-hunting bottlenose dolphins (*Tursiops truncatus*) off Cedar Key, Florida. *Proceedings of the Royal Society of London. Series B* 272: 135–140.

Gerson, H.B. and J.P. Hickie (1985) Head scarring on male narwhals (*Monodon monoceros*): evidence for aggressive tusk use. *Canadian Journal of Zoology* 63: 2083–2087.

Gibbons, J.E., C. Venegas, L. Guzmán, G. Pizzaro, D. Boré and P. Gálvez (2002) *Programa de monitoreo de pequenos cetaceos en areas selectas de la XII Region* Final Project report Proyecto FIP-IT/99–28 Universidad de Magallanes, Punta Arenas, Chile 155 pp.

Gibson, D.I. and E.A. Harris (1979) The helminth-parasites of cetaceans in the collection of the British Museum (Natural History). *Investigations on Cetacea* 10: 309–324.

Gibson, K.R. (1986) Cognition, brain size and the extraction of embedded food resources. Pages 93–104 *in* J.G. Else and P.C. Lee, eds. *Primate Ontogeny, Cognition and Social Behaviour*. Cambridge University Press, Cambridge, UK.

Gibson, Q.A. and J. Mann (2008a) Early social development in wild bottlenose dolphins: sex differences, individual variation, and maternal influence. *Animal Behaviour* 76: 375–387.

Gibson, Q.A. and J. Mann (2008b) The size, composition and function of wild bottlenose dolphin (*Tursiops* sp.) mother-calf groups in Shark Bay, Australia. *Animal Behaviour* 76: 389–405.

Gilby, I. and R. Wrangham (2008) Association patterns among wild chimpanzees (*Pan troglodytes schweinfurthii*) reflect sex differences in cooperation. *Behavioral Ecology and Sociobiology* 62: 1831–1842.

Gill, P.C., G.J.B. Ross, W.H. Dawbin and H. Wapstra (2000) Confirmed sightings of dusky dolphins (*Lagenorhynchus obscurus*) in southern Australian waters. *Marine Mammal Science* 16: 452–459.

Gill, T. (1865) On two new species of Delphinidae, from California, in the Smithsonian. *Proceedings of the Academy of Natural Sciences of Philadelphia* 17: 177–178.

Gittleman, J.L. (1985) Carnivore body size: ecological and taxonomic correlates. *Oecologia* 67: 540–554.

Goldstein, H.E. (1988) Estudios comparativos de los hábitos alimentarios y de los nichos tróficos de dos peces costeros: la brótola (*Urophysis brasiliensis*) y el mero (*Acanthistius brasilianus*). Publicaciones de la Comisión Técnica Mixta del Frente Marítimo 4: 151–161.

Gomenido, M., A.H. Harcourt and E.R.S. Roldan (1998) Sperm competition in mammals 826 pp. Pages 467–775 *in* T. Birkhead and A.P. Møller, eds. *Sperm Competition and Sexual Selection*. Academic Press, San Diego, CA.

Gonzalez-Solis, J., J.P. Croxall and A.G. Wood (2000) Sexual dimorphism and sexual segregation in foraging strategies of northern giant petrels, *Macronectes halli*, during incubation. *Oikos* 90: 390–398.

Goodall, J. (1986) *The Chimpanzees of Gombe. Patterns of Behavior*. Harvard University Press, Cambridge, MA.

Goodall, R.N.P. (1978) Report on the small cetaceans stranded on the coasts of Tierra del Fuego. *Scientific Reports of the Whales Research Institute, Tokyo* 30: 197–230.

Goodall, R.N.P. (1989) The last whales of Tierra del Fuego. *Oceanus* 32: 89–95.

Goodall, R.N.P. (1994) Commerson's dolphin *Cephalorhynchus commersonii* (Lacepede, 1804). Pages 241–268 *in* S.H. Ridgway and R. Harrison, eds. *Handbook of Marine Mammals*. Academic Press, London.

Goodall, R.N.P. (1997) Review of sightings of the hourglass dolphin, *Lagenorhynchus cruciger*, in the South American Sector of the Antarctic and Sub-Antarctic. *Reports of the International Whaling Commission* 47: 1001–1013.

Goodall, R.N.P. (2002) Peale's dolphin. Pages 890–894 *in* W.F. Perrin, B. Würsig and J.G.M. Thewissen, eds. *Encyclopedia of Marine Mammals*. Academic Press, San Diego, CA.

Goodall, R.N.P. and I. Cameron (1980) Exploitation of small cetaceans off Southern South America. *Reports of the International Whaling Commission* 30: 445–450.

Goodall, R.N.P. and A.R. Galeazzi (1985) A review of the food habits of the small cetaceans of the Antarctic and the sub-Antarctic. Pages 566–572 *in* W.R. Siegfried, P.R. Condy and R.M. Laws, eds. *Antarctic Nutrient Cycles and Food Webs*. Springer-Verlag, Berlin.

Goodall, R.N.P., A.R. Galeazzi, S. Leatherwood, K.W. Miller, I.S. Cameron, R.K. Kastelein and A.P. Sobral (1988a) Studies of commerson's dolphins, *Cephalorhynchus commersonii*, off Tierra del Fuego, 1976–1984, with a review of information on the species in the South Atlantic. Pages 3–70 *in* R.L. Brownell Jr. and G.P. Donovan, eds. *Biology of the Genus Cephalorhynchus*. International Whaling Commission, Cambridge.

Goodall, R.N.P., K.S. Norris, A.R. Galeazzi, J.A. Oporto and I.S. Cameron (1988b) On the Chilean dolphin, *Cephalorhynchus eutropia* (Gray, 1846). Pages 197–257 *in* R.L. Brownell Jr. and G.P. Donovan, eds. *Biology of the Genus Cephalorhynchus. Reports of the International Whaling Commission Special Issue 9*. International Whaling Commission, Cambridge.

Goodall, R.N.P., A.C.M. Schiavini and C. Fermani (1994) Net fisheries and net mortality of small cetaceans off Tierra del Fuego, Argentina. Pages 295–306 *in* W.F. Perrin, G.P. Donovan and J. Barlow, eds. *Gillnets and Cetaceans. Reports of the International Whaling Commission Special Issue 15*. International Whaling Commission, Cambridge.

Goodall, R.N.P., J.C. De Haro, F. Fraga, M.A. Iñíguez and K.S. Norris (1997) Sightings and behaviour of Peale's dolphins, *Lagenorhynchus australis*, with notes on dusky dolphins, *L. obscurus*, off southernmost South America. *Reports of the International Whaling Commission* 47: 757–775.

Gowans, S. and H. Whitehead (1995) Distribution and habitat partitioning by small odontocetes in the Gully, a submarine canyon on the Scotian Shelf. *Canadian Journal of Zoology* 73: 1599–1608.

Gowans, S., B. Würsig and L. Karczmarski (2008) The social structure and strategies of delphinids: Predictions based on an ecological framework. *Advances in Marine Biology* 53: 195–294.

Graham, J. (2004) Movement patterns and bioenergetics of the shortfin mako shark. California Sea Grant College Program. Research Profiles. Paper PPFisheries04_03. 4 pp. http://repositories.cdlib.org/csgc/rp/PPFisheries04_03. Accessed March 2008.

Grange, K. and R. Cole (1997) Mussel farming impacts. *Aquaculture Update* 17: 1–3.

Grant, J., A. Hatcher, D.B. Scott, P. Pocklington, C.T. Schafer and G.V. Winters (1995) A multidisciplinary approach to evaluating impacts of shellfish aquaculture on benthic communties. *Estuaries* 18: 124–144.

Graw, B. and M. Manser (2007) The function of mobbing in cooperative meerkats. *Animal Behaviour* 74: 507–517.

Gray, J.E. (1828) *Spicilegia Zoologica; original figures and short systematic descriptions of new and unfigured animals*. Treüttel, Würtz and Co. and W. Wood, London.

Gray, J.E. (1846a) On the British Cetacea. *The Annals and Magazine of Natural History (London)* 17: 82–85.

Gray, J.E. (1846b) Mammalia, birds. Pages 1–53 *in* J. Richardson and J.E. Gray, eds. *The Zoology of the Voyage of the HMS Erebus and Terror under the command of James Clark Ross, during the years 1839–1843*, Volume 1. E.W. Janson, London.

Gray, J.E. (1850) *Catalogue of the specimens of Mammalia in the collection of the British Museum, Part 1, Cetacea*. Trustees of the British Museum, London.

Gray, J.E. (1864) On the Cetacea which have been observed in the seas surrounding the British Islands. *Proceedings of the Zoological Society of London* 1864: 195–248.

Gray, J.E. (1866a) *Catalogue of Seals and Whales in the British Museum*. Trustees of the British Museum, London.

Gray, J.E. (1866b) Notes on the skulls of dolphins, or bottlenose whales, in the British Museum. *Proceedings of the Zoological Society of London* 1866: 211–216.

Gray, J.E. (1868a) *Synopsis of the Species of Whales and Dolphins in the Collection of the British Museum*. B. Quartch, London.

Gray, J.E. (1868b) Notice of *Clymene similis*, a new dolphin sent from the Cape by Mr. Layard. *Proceedings of the Zoological Society of London* 1868: 146–149.

Gray, J.E. (1871) *Supplement to the Catalogue of Seals and Whales in the British Museum*. Trustees of the British Museum, London.

Griffin, D.R., F.A. Webster and C.R. Michael (1960) The echolocation of flying insects by bats. *Animal Behaviour* 8: 141–154.

Grimwood, I.R. (1969) Notes on the distribution and status of some Peruvian mammals (1968). *New York Zoological Society Special Publication* 21.

Grinnell, J., C. Packer and A.E. Pusey (1995) Cooperation in male lions: kinship, reciprocity, or mutualism? *Animal Behaviour* 49: 95–105.

Gubbins, C., B. McCowan, S.K. Lynn, S. Hooper and D. Reiss (1999) Mother-infant spatial relations in captive bottlenose dolphins, *Tursiops truncatus*. *Marine Mammal Science* 15: 751–765.

Guerra, C., K. Van Waerebeek, G. Portflitt and G. Luna (1987) Presencia de cetáceos frente a la segunda region de Chile. *Estudios Oceanológicos* 6: 87–96.

Guinet, C. and J. Bouvier (1995) Development of intentional stranding hunting techniques in killer whale (*Orcinus orca*) calves at Crozet Archipelago. *Canadian Journal of Zoology* 73: 27–33.

Guo, S.W. and E.A. Thompson (1992) Performing the exact test of Hardy-Weinberg proportion for multiple alleles. *Biometrics* 48: 361–372.

Gurnell, J., L.A. Wauters, D. Preatoni and G. Tosi (2001) Spacing behaviour, kinship, and population dynamics of grey squirrels in a newly colonized broadleaf woodland in Italy. *Canadian Journal of Zoology* 79: 1533–1543.

Gygax, L. (2002) Evolution of group size in the dolphins and porpoises: interspecific consistency of intraspecific patterns. *Behavioral Ecology* 13: 583–590.

Haldane, J.B.S. (1954) An exact test for randomness of mating. *Journal of Genetics* 52: 631–635.

Hamilton, W.D. (1964a) The genetical evolution of social behaviour I. *Journal of Theoretical Biology* 7: 1–16.

Hamilton, W.D. (1964b) The genetical evolution of social behaviour II. *Journal of Theoretical Biology* 7: 17–52.

Hamilton, W.D. (1970) Selfish and spiteful behaviour in an evolutionary model. *Nature* 228: 1218–1220.

Hamilton, W.D. (1971) Geometry of the selfish herd. *Journal of Theoretical Biology* 31: 295–311.

Hamilton, W.D. and M. Zuk (1982) Heritable true fitness and bright birds: a role for parasites? *Science* 218: 384–387.

Haney, J.F., A. Craggy, K. Kimball and F. Weeks (1990) Light control of evening vertical migrations by *Chaoborus punctipennis* larvae. *Limnology and Oceanography* 35: 1068–1078.

Harcourt, A.H., P.H. Harvey, S.G. Larson and R.V. Short (1981) Testis weight, body weight and breeding system in primates. *Nature* 293: 55–57.

Harlin, A.D. (2004) Molecular systematics and phylogeography of *Lagenorhynchus obscurus* derived from nuclear and mitochondrial loci. Ph.D. dissertation, Galveston, TX: Texas A&M University.

Harlin, A.D., B. Würsig, C.S. Baker and T.M. Markowitz (1999) Skin swabbing for genetic analysis: Application to dusky dolphins (*Lagenorhynchus obscurus*). *Marine Mammal Science* 15: 409–425.

Harlin, A.D., T.M. Markowitz, C.S. Baker, B. Würsig and R.L. Honeycutt (2003) Genetic structure, diversity and historical demography of New Zealand's dusky dolphin (*Lagenorhynchus obscurus*). *Journal of Mammalogy* 84: 702–717.

Harlin-Cognato, A.D. and R.L. Honeycutt (2006) Multi-locus phylogeny of dolphins in the subfamily Lissodelphininae: character synergy improves phylogenetic resolution. *BMC Evolutionary Biology* 6: 1–16.

Harlin-Cognato, A.D., T. Markowitz, B. Würsig and R.L. Honeycutt (2007) Multi-locus phylogeography of the dusky dolphin (*Lagenorhynchus obscurus*): passive dispersal via the west-wind drift or response to prey species and climate change? *BMC Evolutionary Biology* 7: 131.

Harrison, N.M., M.J. Whitehouse, D. Heinemann, P.A. Prince, G.L. Hunt Jr. and R.R. Veit (1991) Observations of multispecies seabird flocks around South Georgia. *Auk* 108: 801–810.

Harrison, R.J., R.L. Brownell Jr. and R.C. Boice (1972) Reproduction and gonadal appearances in some odontocetes. Pages 361–429 *in* R.J. Harrison (ed). *Functional Anatomy of Marine Mammals*, Volume 1. Academic Press, London and New York.

Harvell, C.D., K. Kim, J.M. Burkholder, R.R. Colwell, P.R. Epstein, D.J. Grimes, E.E. Hofmann, E.K. Lip, A.D.M.E. Osterhaus, R.M. Overstreet, J.W. Porter, G.W. Smith and G.R. Vasta (1999) Emerging marine diseases—climate links and anthropogenic factors. *Science* 285: 1505–1510.

Harvey, P.H. and J.R. Krebs (1990) Comparing brains. *Science* 249: 140–146.

Hatakeyama, Y. and H. Soeda (1990) Observation of porpoises taken in salmon gillnet fisheries. Pages 269–281 *in* J.A. Thomas and R. Kastelein, eds. *Sensory Abilities of Cetaceans*. Plenum Press, New York.

Hector, J. (1873) On the whales and dolphins of the New Zealand seas. *Transactions and Proceedings of the New Zealand Institute* 5: 154–171.

Hector, J. (1878) Notes on the whales of the New Zealand seas. *Transactions and Proceedings of the New Zealand Institute* 10: 331–343 (Note that although this paper was presented in 1877, it was not published until the following year).

Heinrich, S. (2006) Ecology of Chilean dolphins and Peale's dolphins at Isla Chiloé, southern Chile. Ph.D. thesis, St Andrews, UK: University of St Andrews. 239 pp.

Heithaus, M.R. (2001a) Predator–prey and competitive interactions between sharks (order Selachii) and dolphins (suborder Odontoceti): a review. *Journal of Zoology* 253: 53–68.

Heithaus, M.R. (2001b) The biology of tiger sharks, *Galeocerdo cuvier*, in Shark Bay, Western Australia: sex ratio, size distribution, diet, and seasonal changes in catch rates. *Environmental Biology of Fishes* 61: 25–36.

Heithaus, M.R. (2004) Predator–prey interactions. Pages 487–521 *in* J.C. Carrier, J. Musick and M.R. Heithaus, eds. *The Biology of Sharks and their Relatives*. CRC Press, Oxford.

Heithaus, M.R. and L.M. Dill (2002a) Food availability and tiger shark predation risk influence bottlenose dolphin habitat use. *Ecology* 83: 480–491.

Heithaus, M.R. and L.M. Dill (2002b) Feeding strategies and tactics. Pages 412–422 *in* W.F. Perrin, B. Würsig and J.G.M. Thewissen, eds. *Encyclopedia of Marine Mammals*. Academic Press, San Diego, CA.

Heithaus, M.R. and L.M. Dill (2006) Does tiger shark predation risk influence foraging habitat use by bottlenose dolphins at multiple spatial scales? *Oikos* 114: 257–264.

Heithaus, M.R., A. Frid, A.J. Wirsing and B. Worm (2008) Predicting ecological consequences of marine top predator declines. *Trends in Ecology and Evolution* 23: 202–210.

Hennig, W. (1996) *Phylogenetic Systematics*. University of Illinois Press, Urbana, IL.

Herman, L.M. (1980) Cognitive characteristics of dolphins. Pages 363–429 *in* L.M. Herman (ed). *Cetacean Behavior: Mechanisms and functions*. Robert E. Krieger Publishing Company, Malabar, FL.

Hershkovitz, P. (1966) Catalog of living whales. *United States National Museum Bulletin* No. 246. Smithsonian Institution, Washington, DC.

Herzing, D.L. (1988) A quantitative description and behavioral association of a burst-pulsed sound, the squawk, in captive bottlenose dolphins, *Tursiops truncatus*. M.Sc. thesis, San Francisco, CA: San Francisco State University. 87 pp.

Herzing, D.L. (1996) Vocalizations and associated underwater behavior of free-ranging Atlantic spotted dolphins, *Stenella frontalis*, and bottlenose dolphins, *Tursiops truncatus*. *Aquatic Mammals* 22: 61–79.

Herzing, D.L. (2004) Social and nonsocial uses of echolocation in free-ranging *Stenella frontalis* and *Tursiops truncatus*. Pages 404–410 *in* J.A. Thomas, C.F. Moss and M. Vater, eds. *Echolocation in Bats and Dolphins*. University of Chicago Press, Chicago, IL.

Herzing, D.L., K. Moewe and B.J. Brunnick (2003) Interspecific interactions between Atlantic spotted dolphins, *Stenella frontalis*, and bottlenose dolphins, *Tursiops truncatus*, on Great Bahama Bank, Bahamas. *Aquatic Mammals* 29: 335–341.

Hiatt, R.W. and V.E. Brock (1948) On the herding of prey and the schooling of black skipjack, *Euthynnus yaito* Kashinouye. *Pacific Science* 2: 297–298.

Hillix, W.A. and D.M. Rumbaugh (2004) *Animal Bodies, Human Minds: Ape, dolphin, and parrot language skills*. Springer-Verlag, New York, NY.

Hillman, G.R., B. Würsig, G.A. Gailey, N. Kehtarnavaz, A. Drobyshevsky, B.N. Araabi, H.D. Tagare and D.W.Weller (2003) Computer-assisted photo-identification of individual marine vertebrates: a multi-species system. *Aquatic Mammals* 29: 117–123.

Hinde, R. (1976) Interactions, relationships and social structure. *Man* 11: 1–17.

Hof, P.R., R. Chanis and L. Marino (2005) Cortical complexity in cetacean brains. *The Anatomical Record Part A* 287A: 1142–1152.

Hohmann, G. and B. Fruth (2003) Intra- and inter-sexual aggression by bonobos in the context of mating. *Behaviour* 140: 1389–1413.

Holekamp, K.E., S.T. Sakai and B.L. Lundrigan (2007) The spotted hyena (*Crocuta crocuta*) as a model system for study of the evolution of intelligence. *Journal of Mammalogy* 88: 545–554.

Holts, D.B. and D.W. Bedford (1993) Horizontal and vertical movements of the shortfin mako shark, *Isurus oxyrinchus*, in the southern California bight. *Australian Journal of Marine and Freshwater Research* 44: 901–909.

Hoyt, E. (2001) *Whale Watching 2001: Worldwide tourism numbers, expenditures, and expanding socioeconomic benefits*. International Fund for Animal Welfare, Yarmouth Port, MA, 158 pp.

Hoyt, E. (2005) *Marine Protected Areas for Whales, Dolphins and Porpoises*. Earthscan, Sterling, VA.

Hrdy, S.B. (1977) *The Langurs of Abu*. Harvard University Press, Cambridge, MA.

Hulbert, L.B., M.F. Sigler and C.R. Lunsford (2006) Depth and movement behavior of the Pacific sleeper shark in the north-east Pacific Ocean. *Journal of Fish Biology* 69: 406–425.

Humphrey, N.K. (1976) The social function of intellect. Pages 303–317 *in* P.P.G. Bateson and R.A. Hinde, eds. *Growing Points in Ethology: Based on a conference sponsored by St. John's College and King's College, Cambridge*. Cambridge University Press, Cambridge, UK.

Huntingford, F. and A. Turner (1987) *Animal Conflict*. Chapman and Hall, London.

Iilende, T., T. Strømme and E. Johnsen (2001) Dynamics of the pelagic component of Namibian hake stocks. A.I.L. Payne, S.C. Pillar, R.J.M. Crawford, eds. *A Decade of Namibian Fisheries Science. South African Journal of Marine Science* 23: 337–346.

Imber, M.J. (1975) Lycoteuthid squids as prey of petrels in New Zealand seas. *New Zealand Journal of Marine and Freshwater Research* 9: 483–492.

Iñíguez, M.A. and J.C. de Haro (1994) Preliminary reports of feeding habits of the Peale's dolphins (*Lagenorhynchus australis*) in southern Patagonia. *Aquatic Mammals* 2: 35–37.

Inman, A.J. and J. Krebs (1987) Predation and group living. *Trends in Ecology and Evolution* 2: 31–32.

International Union for the Conservation of Nature and Natural Resouces (IUCN). (2008) *2008 IUCN Red List of Threatened Species.* www.iucnredlist.org. Accessed November 1, 2008.

International Whaling Commission (IWC) (1994) Report of the workshop on mortality of cetaceans in passive fishing nets and traps. *Reports of the International Whaling Commission* Special Issue 15.

Ioannou, C.C. and J. Krause (2008) Searching for prey: the effects of group size and number. *Animal Behaviour* 75: 1383–1388.

IPCC (Intergovernmental Panel on Climate Change) (2007) *Fourth Assessment Report.* http://www.ipcc.ch/.

Isler, K. and C.P. van Schaik (2006) Metabolic costs of brain size evolution. *Biology Letters* 2: 557–560.

Isvaran, K. (2005) Female grouping best predicts lekking in blackbuck (*Antilope cervicapra*). *Behavioral Ecology and Sociobiology* 57: 283–294.

Isvaran, K. and Y. Jhala (2000) Variation in lekking costs in blackbuck (*Antilope cervicapra*): relationship to lek-territory location and female mating patterns. *Behaviour* 137: 547–563.

Ivanovic, M.L. and N.E. Brunetti (1994) Food and feeding of *Illex argentinus*. *Antarctic Science* 6: 185–193.

Jablonka, E. and M.J. Lamb (2005) *Evolution in Four Dimensions: Genetic, epigenetic, behavioral, and symbolic variation in the history of life.* MIT Press, Cambridge, MA.

Jackson, R.R. and S.D. Pollard (1996) Predatory behavior of jumping spiders. *Annual Review of Entomology* 41: 287–308.

Jarman, P.J. (1974) The social organization of antelope in relation to their ecology. *Behaviour* 48: 215–267.

Jefferson, T.A., P.J. Stacey and R.W. Baird (1991) A review of killer whale interactions with other marine mammals—Predation to coexistence. *Mammal Review* 21: 151–180.

Jefferson, T.A., S. Leatherwood and M.A. Webber (1993) *Marine Mammals of the World: FAO species identification guide.* United Nations Food and Agriculture Organization, Rome.

Jefferson, T.A., M.A. Webber and R.L. Pitman (2008) *Marine Mammals of the World: a comprehensive guide to their identification.* Academic Press, New York.

Jerison, H.J. (1973) *Evolution of the Brain and Intelligence.* Academic Press, New York.

Joffe, T.H. (1997) Social pressures have selected for an extended juvenile period in primates. *Journal of Human Evolution* 32: 593–605.

Johnson, A. and A. Acevedo-Gutiérrez (2007) Regulation compliance and harbor seal (*Phoca vitulina*) disturbance. *Canadian Journal of Zoology* 85: 290–294.

Johnson, M.V. (1948) Sound as a tool in marine ecology, from data on biological noises and the deep scattering layer. *Journal of Marine Research* 7: 443–458.

Jolly, A. (1966) Lemur social behavior and primate intelligence. *Science* 153: 501–506.

Jones, P.D., D.J. Hannah, S.J. Buckland, T. van Maanen, S.V. Leathem, S. Dawson, E. Slooten, A.van Helden and M. Donoghue (1999) Polychlorinated dibenzo-p-dioxins, dibenzofurans and polychlorinated biphenyls in New Zealand cetaceans. *Journal of Cetacean Research and Management* Supplement 1: 157–167.

Kalinowski, S.T., A.P. Wagner and M.L. Taper (2006) ML-RELATE: a computer program for maximum likelihood estimation of relatedness and relationship. *Molecular Ecology Notes* 6: 576–579.

Kaminski, G., S. Brandt, E. Baubet and C. Baudoin (2005) Life-history patterns in female wild boars (*Sus scrofa*): Mother-daughter postweaning associations. *Canadian Journal of Zoology* 83: 474–480.

Kamminga, C. (1988) Echolocation signal types of odontocetes. Pages 9–22 *in* P.E. Nachtigall and P.W.B. Moore, eds. *Animal Sonar: Processes and performance.* Plenum Press, New York.

Kamminga, C. and H. Wiersma (1981) Investigations of cetacean sonar II. Acoustical similarities and differences in odontocete sonar signals. *Aquatic Mammals* 8: 41–62.

Karczmarski, L. (1999) Group dynamics of humpback dolphins (*Sousa chinensis*) in the Algoa Bay region, South Africa. *Journal of Zoology (London)* 249: 283–293.

Karczmarski, L., V.G. Cockcroft and A. McLachlan (2000a) Habitat use and preferences of Indo-Pacific humpback dolphins *Sousa chinensis* in Algoa Bay, South Africa. *Marine Mammal Science* 16: 65–79.

Kasamatsu, F., G. Joyce, P. Ensor and J. Mermoz (1990) Current occurrence of cetacea in the southern hemisphere minke whale assessment cruises, 1978/79–1987/88. Paper SC/42/015, presented to the International Whaling Commission Scientific Committee.

Kastelein, R.A., C.A. Van Der Elst, H.K. Tennant and P.R. Wiepkema (2000) Food consumption and growth of a female dusky dolphin (*Lagenorhynchus obscurus*). *Zoo Biology* 19: 131–142.

Kasuya, T. (1977) Age determination and growth of the Baird's beaked whale with a comment on the fetal growth rate. *Scientific Reports of the Whales Research Institute, Tokyo* 29: 1–20.

Kellogg, R. (1928) The history of whales—their adaptation to life in the water. *The Quarterly Review of Biology* 3: 29–76.

Kellogg, R. (1941) On the identity of the porpoise *Sagmatias amblodon*. *Field Museum of Natural History and Zoology* 27: 293–311.

Kelly, C.D., L.F. Bussiere and D.T. Gwynne (2008) Sexual selection for male mobility in a giant insect with female-biased size dimorphism. *American Naturalist* 172: 417–423.

Kenagy, G.J. and S.C. Trombulak (1986) Size and function of mammalian testes relative to body size. *Journal of Mammalogy* 67: 1–22.

Ketterson, E.D. and V. Nolan Jr. (1994) Hormones and life histories: an integrative approach. Pages 327–353 *in* L.A. Real, ed. *Behavioral Mechanisms in Evolutionary Ecology.* University of Chicago Press, Chicago, IL.

Klatsky, L.J., R.S. Wells and J.C. Sweeney (2007) Offshore bottlenose dolphins (*Tursiops truncatus*): Movement and dive behavior near the Bermuda Pedestal. *Journal of Mammalogy* 88: 59–66.

Klimley, A.P., S.C. Beavers, T.H. Curtis and S.J. Jorgensen (2002) Movements and swimming behavior of three species of sharks in La Jolla Canyon, California. *Environmental Biology of Fishes* 63: 117–135.

Koeman, J.H. (1973) PCB in mammals and birds in the Netherlands. Pages 35–43 *in* E. Ahnland and A. Akerblom, eds. *PCB Conference II.* National Swedish Environmental Protection Board, Stockholm.

Koeman, J.H., W.S.M. van de Ven, J.J.M. Goeij, P.S. Tijoe and J.L. van Haaften (1975) Mercury and selenium in marine mammals and birds. *The Science for the Total Environment* 3: 279–287.

Koen-Alonso, M. (1999) Estudio comparado de la alimentación entre algunos preda-dores de alto nivel trófico de la comunidad marina del norte y centro de Patagonia. Ph.D. thesis, Argentina: University of Buenos Aires. 182 pp.

Koen-Alonso, M. and P. Yodzis (2005) Multispecies modelling of some components of the marine community of northern and central Patagonia, Argentina. *Canadian Journal of Fisheries and Aquatic Sciences* 62: 1490–1512.

Koen-Alonso, M., E.A. Crespo, S.N. Pedraza, N.A. Garcia and M. Coscarella (2000a) Feeding habits of the southern sea lion *Otaria flavescens* of Patagonia. *Fishery Bulletin* 97: 250–263.

Koen-Alonso, M., S.N. Pedraza, E.A. Crespo, N.A. Garcia and S.L. Dans (2000b) A multispecies approach to the dynamics of some high trophic predators of the north-ern and central Patagonia marine community. Page 47 *in 14th Annual Conference European Cetacean Society.* Cork, Ireland.

Koen-Alonso, M., E. Crespo, N.A. Garcia, S.N. Pedraza and M.A. Coscarella (1998) Diet of dusky dolphins, *Lagenorhynchus obscurus,* in waters off Patagonia, Argentina. *Fishery Bulletin* 96: 366–374.

Koen-Alonso, M., E.A. Crespo, N.A. García, S.N. Pedraza, P.A. Mariotti, B. Berón Vera and N.J. Mora (2001) Food habits of *Dipturus chilensis* (Pisces: Rajidae) off Patagonia, Argentina. *ICES Journal of Marine Science* 58: 288–297.

Koen-Alonso, M., E.A. Crespo, N.A. García, S.N. Pedraza, P.A. Mariotti and N.J. Mora (2002) Fishery and ontogenetic driven changes in the diet of the spiny dog-fish *Squalus acanthias* in Patagonian waters, Argentina. *Environmental Biology of Fishes* 63: 193–202.

Konishi, M. and E.I. Knudsen (1979) Oilbird—hearing and echolocation. *Science* 204: 425–427.

Koslow, J.A. (1979) Vertical migrators see the light? *Limnology and Oceanography* 24: 783–784.

Kotler, B.P. and R.D. Holt (1989) Predation and competition—the interaction of 2 types of species interactions. *Oikos* 54: 256–260.

Kramer, M.O. (1961) Dolphins have the laugh on us . . . as far as speed goes. *South African Yachting* Jan/Feb: 28–30.

Krause, J. and J.G.J. Godin (1995) Predator preferences for attacking particular prey group sizes—consequences for predator hunting success and prey predation risk. *Animal Behaviour* 50: 465–473.

Krause, J. and G. Ruxton (2002) *Living in Groups.* Oxford University Press, London.

Krebs, J.R. and N.B. Davies (1993) *An Introduction to Behavioural Ecology.* Blackwell, London.

Krebs, J.R., J.C. Ryan and E.L. Charnov (1974) Hunting by expectation or optimal foraging—study of patch use by chickadees. *Animal Behaviour* 22: 953–964.

Kruijt, J.P. and J.A. Hogan (1967) Social behavior on the lek in black grouse, *Lyrurus tetrix tetrix. Ardea* 55: 203–240.

Krützen, M., L.M. Barre, L.M. Möller, M.R. Heithaus, C. Simms and W.B. Sherwin (2002) A biopsy system for small cetaceans: Darting success and wound healing in *Tursiops* spp. *Marine Mammal Science* 18: 863–878.

Krützen, M., W.B. Sherwin, R.C. Connor, L.M. Barre, T. Van De Casteele, J. Mann and R. Brooks (2003) Contrasting relatedness patterns in bottlenose dolphins (*Tursiops* sp.) with different alliance strategies. *Proceedings of the Royal Society of London Series B-Biological Sciences* 270: 497–502.

Krützen, M., J. Mann, M.R. Heithaus, R.C. Connor, L. Bejder and W.B. Sherwin (2005) Cultural transmission of tool use in bottlenose dolphins. *Proceedings of the National Academy of Sciences of the USA* 102: 8939–8943.

Kuker, K.J., J.A. Thomson and U. Tscherter (2005) Novel surface feeding tactics of minke whales, *Balaenoptera acutorostrata,* in the Saguenay-St. Lawrence National Marine Park. *Canadian Field Naturalist* 119: 214–218.

La Rosa, T., S. Mirto, A. Marino, V. Alonzo, T.L. Maugeri and A. Mazzola (2001) Heterotrophic bacteria community and pollution indicators of mussel farm impact in the Gulf of Gaeta (Tyrrhenian Sea). *Marine Environmental Research* 52: 301–321.

La Rosa, T., S. Mirto, E. Favaloro, B. Savona, G. Sara, R. Danovaro and A. Mazzola (2002) Impact on the water column biogeochemistry of a Mediterranean mussel and fish farm. *Water Research* 36: 713–721.

Labropoulou, M. and C. Papaconstantinou (2000) Comparison of otolith growth and somatic growth in two macrourid fishes. *Fisheries Research* 46: 177–188.

Laevastu, T. and F. Favorite (1988) *Fishing and Stock Fluctuations.* Fishing News Books, Farnham, UK.

Lafferty, K., S.M. Allesina, C.J. Arim, G. Briggs, De Leo, A.P. Dobson, J.A. Dunne, P.T.J. Johnson, A.M. Kuris, D.J. Marcogliese, N.D. Martinez, J. Memmott, P.A. Marquet, J.P. McLaughlin, E.A. Mordecai, M. Pascual, R. Poulin and D.W. Thieltges (2008) Parasites in food webs: the ultimate missing links. *Ecology Letters* 11: 533–546.

Lahvis, G.P., R.S. Wells, D.W. Kuehl, J.L. Stewart, H.L. Rhinehart and C.S. Via (1995) Decreased lymphocyte responses in free-ranging bottlenose dolphins (*Tursiops truncatus*) are associated with increased concentrations of PCBs in peripheral blood. *Environmental Health Perspectives* 103: 67–72.

Lalli, C.M. and T.R. Parsons (1995) *Biological Oceanography: An introduction.* Butterworth-Heinemann, Boston.

Lammers, M.O., W.W.L. Au and D.L. Herzing (2003a) The broadband social acoustic signaling behavior of spinner and spotted dolphins. *Journal of the Acoustical Society of America* 114: 1629–1639.

Lammers, M.O., W.W.L. Au, R. Aubauer and P.E. Nachtigall (2003b) A comparative analysis of echolocation and burst-pulse click trains in *Stenella longirostris.* Pages 414–419 *in* J. Thomas, C. Moss and M. Vater, eds. *Echolocation in Bats and Dolphins.* University of Chicago Press, Chicago, IL.

Lammers, M.O., M. Schotten and W.W.L. Au (2006) The spatial context of free-ranging Hawaiian spinner dolphins (*Stenella longirostris*) producing acoustic signals. *Journal of the Acoustical Society of America* 119: 1244–1250.

Laundre, J.W., L. Hernandez and K.B. Altendorf (2001) Wolves, elk, and bison: reestablishing the "landscape of fear" in Yellowstone National Park, USA. *Canadian Journal of Zoology* 79: 1401–1409.

Leaper, R., P. Best, T. Branch, G. Donovan, H. Murase and K. Van Waerebeek (2008) Report of review group of data sources on odontocetes in the Southern Ocean: in preparation for IWC/CCAMLR workshop in August 2008. *Proceedings of the International Whaling Commission* SC/60/EM 1, Santiago, Chile.

Learmonth, J.A., C.D. MacLeod, M.B. Santos, G.J. Pierce, H.Q.P. Crick and R.A. Robinson (2006) Potential effects of climate change on marine mammals. *Oceanography and Marine Biology: An Annual Review* 44: 431–464.

Leatherwood, S. and R.R. Reeves (1983) *The Sierra Club Handbook of Whales and Dolphins*. Sierra Club Books, San Francisco, CA.

Leatherwood, S., R.R. Reeves, W.F. Perrin and W.E. Evans (1982) Whales, dolphins, and porpoises of the eastern North Pacific and adjacent Arctic waters: a guide to their identification. NOAA Technical Report, NMFS Circular 444.

Leatherwood, S., R.K. Kastelein and K.W. Miller (1988) Observations of Commerson's dolphin and other cetaceans in southern Chile, January–February (1984). Pages 71–83 *in* R.L. Brownell Jr. and G.P. Donovan, eds. *Biology of the Genus Cephalorhynchus*. International Whaling Commission, Cambridge.

LeBoeuf, B.J. and S. Kaza (1981) *The Natural History of Ano Nuevo*. Boxwood Press, Pacific Grove.

Leduc, R.G., W.F. Perrin and A.E. Dizon (1999) Phylogenetic relationships among the delphinid cetaceans based on full cytochrome b sequences. *Marine Mammal Science* 15: 619–648.

Lee, P.C. (1987) Allomothering among African elephants. *Animal Behaviour* 35: 278–291.

Lefebvre, L., P. Whittle, E. Lascaris and A. Finkelstein (1997) Feeding innovations and forebrain size in birds. *Animal Behaviour* 53: 549–560.

Lefebvre, L., S.M. Reader and D. Sol (2004) Brains, innovations and evolution in birds and primates. *Brain, Behavior and Evolution* 63: 233–246.

Lehner, P.N. (1996) *Handbook of Ethological Methods*, 2nd edn. Cambridge University Press, Cambridge.

Lescrauwaet, A.K. (1997) Notes on the behaviour and ecology of the Peale's dolphin, *Lagenorhynchus australis,* in the Strait of Magellan, Chile. *Reports of the International Whaling Commission* 47: 747–755.

Leslie, Jr., D.M. and K.J. Jenkins (1985) Rutting mortality among male Roosevelt elk. *Journal of Mammalogy* 66: 163–164.

Lesson, R.P. and P. Garnot (1826) *Zoologie in Voyage autour du Monde, Exécuté par Ordre du Roi, sur la Corvette de Sa Majesté, La Coquille, pendant les Années 1822, 1823, 1824 et 1825* Volume 1, Part 1. Arthus Bertrand, Paris.

Lett, C., J. Veitch, C.D. Van Der Lingen and L. Hutchings (2007) Assessment of an environmental barrier to transport of ichthyoplankton from the southern to the northern Benguela ecosystems. *Marine Ecology Progress Series* 347: 247–259.

Lewis, K.B. and P.L. Barnes (1999) Kaikoura Canyon, New Zealand: active conduit from near-shore sediment zones to trench-axis channel. *Marine Geology* 162: 39–69.

Lichter, A. and A. Hooper (1983) *Guia para el reconocimiento de cetáceos del Mar Argentino*. Fundación vida Silvestre Argentina, Buenos Aires.

Liley, S. and S. Creel (2007) What best explains vigilance in elk: characteristics of prey, predators, or the environment? *Behavioral Ecology* 19: 245–254.

Lillie, D.G. (1915) Cetacea. *British Antarctic (Terra Nova) Expedition Natural History Report, Zoology* 1: 85–124.

Lilly, J.C. (1962) Vocal behavior of the bottlenose dolphins. *Proceedings of the American Philosophical Society* 106: 520–529.

Lilly, J.C. (1965) Vocal mimicry in *Tursiops:* ability to match numbers and durations of human vocal bursts. *Science* 147: 300–301.

Lilly, J.C. and A.M. Miller (1961) Vocal exchanges between dolphins. *Science* 134: 1873–1876.

Lima, S.L. (1998) Stress and decision making under the risk of predation: Recent developments from behavioral, reproductive, ecological perspectives. *Advances in the Study of Behavior* 27: 215–290.

Lima, S.L. (2002) Putting predators back into behavioral predator–prey interactions. *Trends in Ecology and Evolution* 17: 70–75.

Lima, S.L. and L.M. Dill (1990) Behavioral decisions made under the risk of predation—a review and prospectus. *Canadian Journal of Zoology* 68: 619–640.

Lincoln, G.A., F. Guinness and R.V. Short (1972) The way in which testosterone controls the social and sexual behavior of the red deer stag (*Cervus elephus*). *Hormones and Behavior* 3: 375–396.

Lloyd, B.D. (2003) Potential effects of mussel farming on New Zealand's marine mammals and seabirds: a discussion paper. Department of Conservation, Wellington, New Zealand. 34 pp.

Logan, C.J. and J.W. Pepper (2007) Social learning is central to innovation, in primates and beyond. *Behavioral and Brain Sciences* 30: 416–417.

Loizaga de Castro, R., M. Degrati, G.V. Garaffo, S.L. Dans and E.A. Crespo (2006) Potencialidad de la técnica de fotoidentificación para la estimación de abundancia del delfín oscuro (*Lagenorhynchus obscurus*) en los Golfos Norpatagónicos, *in Abstracts of the 6th Sea Science Conference*, Puerto Madryn, Chubut, Argentina.

Long, D.J. and R.E. Jones (1996) White shark predation and scavenging on cetaceans in the eastern North Pacific Ocean. Pages 293–307 *in* A.P. Klimley and D.G. Ainley, eds. *Great White Sharks: The biology of* Carcharodon carcharias. Academic Press, New York.

Lorenz, K. (1966) *On Aggression*. Methuen and Co., London.

Lucifora, L.O., R.C. Menni and A.H. Escalante (2005) Reproduction, abundance and feeding habits of the broadnose sevengill shark *Notorynchus cepedianus* in north Patagonia, Argentina. *Marine Ecology Progress Series* 289: 237–244.

Ludwig, J.A. and J.F. Reynolds (1988) *Statistical Ecology*. John Wiley and Sons.

Lukas, D., V. Reynolds, C. Boesch and L. Vigilant (2005) To what extent does living in a group mean living with kin? *Molecular Ecology* 14: 2181–2196.

Lundquist, D. (2007) Behavior and movement of southern right whales—effects of boats and swimmers. M.Sc. thesis, College Station, TX: Texas A&M University. 132 pp.

Lusseau, D. (2003a) Male and female bottlenose dolphins *Tursiops* spp. have different strategies to avoid interactions with tour boats in Doubtful Sound, New Zealand. *Marine Ecology Progress Series* 257: 267–274.

Lusseau, D. (2003b) The effects of tour boats on the behavior of bottlenose dolphins: using Markov chains to model anthropogenic effects. *Conservation Biology* 17: 1785–1793.

Lusseau, D. (2004) The hidden cost of tourism: detecting long-term effects of tourism using behavioral information. *Ecology and Society* 9: 2.

Lusseau, D. (2005) Residency pattern of bottlenose dolphins, *Tursiops* spp. *in* Milford Sound, New Zealand, is related to boat traffic. *Marine Ecology Progress Series* 295: 265–272.

Lusseau, D. (2006) The short-term behavioural reactions of bottlenose dolphins to interactions with boats in Doubtful Sound, New Zealand. *Marine Mammal Science* 22: 802–818.

Lusseau, D., R. Williams, B. Wilson, K. Grellier, T.R. Barton, P.S. Hammond and P.M. Thompson (2004) Parallel influence of climate on the behaviour of Pacific killer whales and Atlantic bottlenose dolphins. *Ecology Letters* 7: 1068–1076.

Luttbeg, B. and J.L. Kerby (2005) Are scared prey as good as dead? *Trends in Ecology and Evolution* 20: 416–418.

Lydekker, R. (1906) *Sir William Flower*. J.M. Dent and Co, London.

Lynn, S.K. and D.L. Reiss (1992) Pulse sequence and whistle production by two captive beaked whales *Mesoplodon* species. *Marine Mammal Science* 8: 299–305.

Machiavelli, N. (1532) *Il Principe*. Available on-line from Project Gutenberg, Chapel Hill, NC. http://www.gutenberg.org/files/1232/1232-h/1232-h.htm. Accessed May 27, 2008.

Madsen, C.J. and L.M. Herman (1980) Social and ecological correlates of cetacean vision and visual appearance. Pages 101–147 *in* L.M. Herman, ed. *Cetacean Behavior*. John Wiley and Sons, New York.

Main, M.B., F.W. Weckerly and V.C. Bleich (1996) Sexual segregation in ungulates: New directions for research. *Journal of Mammalogy* 77: 449–461.

Majluf, P., E.A. Babcock, J.C. Riveros, M.A. Schreiber and W. Alderete (2002) Catch and by-catch of sea birds and marine mammals in the small-scale fishery of Punta San Juan, Peru. *Conservation Biology* 16: 1333–1343.

Mamaev, E.G. and V. N. Burkanov (2006) Killer whales (*Orcinus orca*) and northern fur seals of the Commander Islands: Is it feeding specialization development? Pages 347–351 *in Marine mammals of the Holarctic*. Proceedings of Fourth International Conference, St. Petersburg, Russia.

Manger, P.R. (2006) An examination of cetacean brain structure with a novel hypothesis correlating thermogenesis to the evolution of a big brain. *Biological Reviews* 81: 293–338.

Maniatis, T., E. Fritsch and J. Sambrook (1982) *Molecular Cloning: A laboratory manual*. Cold Spring Harbor Laboratory, Cold Spring Harbor, NY.

Manly, F.J. (2001) *Randomization, Bootstrap, and Monte Carlo Methods in Biology*. Chapman and Hall, London.

Mann, J. (1999a) Behavioral sampling methods for cetaceans: a review and critique. *Marine Mammal Science* 15: 102–122.

Mann, J. (1999b) Recent changes in female dolphin ranging in Red Cliff Bay, off Monkey Mia, Shark Bay. Report to West Australian Department of Fisheries and West Australian Department of Conservation and Land Management, Perth, Australia.

Mann, J. (2006) Establishing trust: socio-sexual behaviour among Indian Ocean bottlenose dolphins and the development of male-male bonds. Pages 107–130 *in* S. Volker and P.L. Vasey, eds. *Homosexual Behaviour in Animals: An evolutionary perspective*. Cambridge University Press, Cambridge, UK.

Mann, J. and H. Barnett (1999) Lethal tiger shark (*Galeocerdo cuvier*) attack on bottlenose dolphin (*Tursiops* sp.) calf: defense and reactions by the mother. *Marine Mammal Science* 15: 568–575.

Mann, J. and C. Kemps (2003) The effects of provisioning on maternal care in wild bottlenose dolphins, Shark Bay, Australia. Pages 305–317 *in* N. Gales, M. Hindell and R. Kirkwood, eds. *Marine Mammals: Fisheries, Tourism and Management Issues*. CSIRO Publishing, Collingwood, Victoria, Australia.

Mann, J. and B. Sargeant (2003) Like mother, like calf: the ontogeny of foraging traditions in wild Indian Ocean bottlenose dolphins (*Tursiops* sp.). Pages 236–266 *in* D.M. Fragaszy and S. Perry, eds. *The Biology of Traditions: Models and evidence*. Cambridge University Press, Cambridge, UK.

Mann, J. and B.B. Smuts (1998) Natal attraction: allomaternal care and mother-infant separations in wild bottlenose dolphins. *Animal Behaviour* 55: 1097–1113.

Mann, J. and B. Smuts (1999) Behavioral development in wild bottlenose dolphin newborns (*Tursiops* sp.). *Behaviour* 136: 529–566.

Mann, J. and J. Watson-Capps (2005) Surviving at sea: ecological and behavioural predictors of calf mortality in Indian Ocean bottlenose dolphins (*Tursiops* sp.). *Animal Behaviour* 69: 899–909.

Mann, J., R.C. Connor, L.M. Barre and M.R. Heithaus (2000) Female reproductive success in bottlenose dolphins (*Tursiops* sp.): life history, habitat, provisioning, and group-size effects. *Behavioral Ecology* 11: 210–219.

Manson, J.H. (1996) Male dominance and mount series duration in Cayo Santiago rhesus macaques. *Animal Behaviour* 51: 1219–1231.

Manzanilla, S.R. (1989) The 1982–1983 El Niño event recorded in dentinal growth layers in teeth of Peruvian dusky dolphins (*Lagenorhynchus obscurus*). *Canadian Journal of Zoology* 67: 2120–2125.

Margalef, R. (1974) *Ecología*. Ediciones Omega, Barcelona.

Marino, L. (1997) What can dolphins tell us about primate evolution? *Evolutionary Anthropology* 5: 81–86.

Marino, L. (1998) A comparison of encephalization between odontocete cetaceans and anthropoid primates. *Brain, Behavior and Evolution* 51: 230–238.

Marino, L. (2002) Convergence of complex cognitive abilities in cetaceans and primates. *Brain, Behavior and Evolution* 59: 21–32.

Marino, L. (2004) Cetacean brain evolution: multiplication generates complexity. *International Journal of Comparative Psychology* 17: 1–16.

Marino, L. (2007) Cetacean brains: how aquatic are they? *The Anatomical Record* 290: 694–700.

Marino, L., D.W. McShea and M.D. Uhen (2004) Origin and evolution of large brains in toothed whales. *The Anatomical Record Part A* 281A: 1247–1255.

Marino, L., R.C. Connor, R.E. Fordyce, L.M. Herman, P.R. Hof, L. Lefebvre, D. Lusseau, B. McCowan, E.A. Nimchinsky, A.A. Pack, L. Rendell, J.S. Reidenberg, D. Reiss, M.D. Uhen, E. Van der Gucht and H. Whitehead (2007) Cetaceans have complex brains for complex cognition. *PLoS Biology* 5: e139.

Marino, L., C. Butti, R.C. Connor, R.E. Fordyce, L.M. Herman, P.R. Hof, L. Lefebvre, D.Lusseau, B.McCowan, E.A.Nimchinsky, A.A. Pack, J.S. Reidenberg, D. Reiss, L. Rendell, M.D. Uhen, E. Van Der Gucht and H. Whitehead (2008) A claim in search of evidence: reply to Manger's thermogenesis hypothesis of cetacean brain structure. *Biological Reviews* 83: 417–440.

Markowitz, T.M. (2004) Social organization of the New Zealand dusky dolphin. Dissertation, Galveston, TX: Texas A&M University. 278 pp.

Markowitz, T.M. and W.J. Markowitz (2007) Tourism effects on dusky dolphins at Kaikoura. KDDTRP Progress Report II. New Zealand Department of Conservation. Nelson Marlborough Conservancy Office, Munro State Building, 186 Bridge Street, Nelson, New Zealand. 21 pp.

Markowitz, T.M. and B. Würsig (2004) Distribution, abundance, and group structure of dusky dolphins inhabiting Admiralty Bay, New Zealand during Winter 2004, with comparisons to previous years. Marlborough District Council and New Zealand Department of Conservation. Nelson Marlborough Conservancy Office, Munro State Building, 186 Bridge Street, Nelson, New Zealand. 31 pp.

Markowitz, T.M., A.D. Harlin and B. Würsig (2003a) Digital photography improves efficiency of individual dolphin identification. *Marine Mammal Science* 19: 217–223.

Markowitz, T.M., A.D. Harlin and B. Würsig (2003b) Digital photo-identification: a reply to Mizroch. *Marine Mammal Science* 19: 608–612.

Markowitz, T.M., A.D. Harlin, B. Würsig and C.J. McFadden (2004) Dusky dolphin foraging habitat: overlap with aquaculture in New Zealand. *Aquatic Conservation: Marine and Freshwater Ecosystems* 14: 133–149.

Martin, A.R. (1986) Feeding associations between dolphins and shearwaters around the Azores Islands. *Canadian Journal of Zoology* 64: 1372–1374.

Martin, P. and P. Bateson (1993) *Measuring Behaviour*, 2nd edn. Cambridge University Press, Cambridge, UK.

Mass, A.M. and A.Y. Supin (2002) Vision. Pages 1280–1293 *in* W.F. Perrin, B. Würsig and J.G.M. Thewissen, eds. *Encyclopedia of Marine Mammals*. Academic Press, San Diego, CA.

Matkin, C.O., L.G. Barrett-Lennard, H. Yurk, D. Ellifrit and A. Trites (2007) Ecotypic variation and predatory behavior among killer whales (*Orcinus orca*) off the eastern Aleutian Islands, Alaska. *Fishery Bulletin* 105: 74–87.

Matsubara, M. (2003) Costs of mate guarding and opportunistic mating among wild male Japanese macaques. *International Journal of Primatology* 24: 1057–1075.

May-Collado, L. and I. Agnarsson (2006) Cytochrome b and Bayesian inference of whale phylogeny. *Molecular Phylogenetics and Evolution* 38: 344–354.

McBride, A.F. and H. Kritzler (1951) Observations on pregnancy, parturition, and post-natal behavior in the bottlenose dolphin. *Journal of Mammalogy* 32: 251–266.

McCann, T.S. (1981) Aggression and sexual activity of male southern elephant seals, *Mirounga leonina*. *Journal of Zoology* 185: 295–310.

McDonald, M.A., J.A. Hildebrand and S.M. Wiggins (2006) Increases in deep ocean ambient noise in the Northeast Pacific west of San Nicolas Island, California. *Journal of the Acoustical Society of America* 120: 711–728.

McDonald, R.A. (2002) Resource partitioning among British and Irish mustelids. *Journal of Animal Ecology* 71: 185–200.

McFadden, C.J. (2003) Behavioral flexibility of feeding dusky dolphins (*Lagenorhynchus obscurus*) in Admiralty Bay, New Zealand. M.Sc. thesis, College Station, Texas: Texas A&M University. 92 pp.

McKinnon, J. (1994) Feeding habits of the dusky dolphin, *Lagenorhynchus obscurus,* in the coastal waters of central Peru. *Fishery Bulletin US* 92: 569–578.

McOmber, L.C. (1999) Mating activity among dusky dolphins (*Lagenorhynchus obscurus*) off Kaikoura, New Zealand. B.Sc. thesis, Hanover, NH: Dartmouth College.

Mead, J.G. and R.L. Brownell Jr. (2005) Order Cetacea. Pages 723–743 *in* D.E. Wilson and D.M. Reeder, eds. *Mammal Species of the World*. Johns Hopkins University Press, Baltimore, MD.

Meeuwis, J.M. and J.R.E. Lutjeharms (1990) Surface thermal characteristics of the Angola-Benguela front. *South African Journal of Marine Science* 9: 261–279.

Mehta, A.V., J.M. Allen, R. Constantine, C. Garrigue, B. Jann, C. Jenner, M.K. Marx, C.O. Matkin, D.K. Mattila, G. Minton, S.A. Mizroch, C. Olavarria, J. Robbins, K.G. Russell, R.E. Seton, G.H. Steiger, G.A. Vikingsson, P.R. Wade, B.H. Witteveen and P.J. Clapham (2007) Baleen whales are not important as prey for killer whales *Orcinus orca* in high-latitude regions. *Marine Ecology-Progress Series* 348: 297–307.

Menni, R. (1986) Shark biology in Argentina: a review. Pages 425–436 *in* T. Uyenyo, R. Arai, T. Taniuchi and K. Matsuura, eds. *Indo Pacific Fish Biology: Proceedings of the Second International Conference on Indo-Pacific Fishes*. Ichtyological Society of Japan.

Menni, R., M.B. Cousseau and A.E. Gosztonyi (1986) Sobre la biología de los tiburones costeros de la provincia de Buenos Aires. *Anales de la Sociedad Científica Argentina* CCXIII: 3–26.

Michaud, R. (2005) Sociality and ecology of the odontocetes. Pages 303–326 *in* K.E. Ruckstuhl and P. Neuhaus, eds. *Sexual Segregation in Vertebrates: Ecology of the two sexes.* Cambridge University Press, Cambridge, UK.

Ministry Of Economic Development (2007) Our blue horizon, He pae kikorangi, the government's commitment to aquaculture. Ministry of Economic Development, 33 Bowen St, Wellington, New Zealand. Available at www.aquaculture.govt.nz.

Ministry of Fisheries (2007a) Blue Sharks (BWS) Fisheries Summary. Ministry of Fisheries Plenary Report. http://services.fish.govt.nz/fishresourcespublic/Plenary2006/BWS_06.pdf. Accessed March 2008.

Ministry of Fisheries (2007b) Mako Shark (MAK) Fisheries Summary. Ministry of Fisheries Plenary Report. http://services.fish.govt.nz/fishresourcespublic/Plenary2007/MAK_07.pdf. Accessed March 2008.

Ministry Of Fisheries (2008) Aquaculture in New Zealand. Ministry of Fisheries, The Terrace, Wellington, New Zealand. Available at www.aquaculture.govt.nz.

Mink, J.W., R.J. Blumenschine and D.B. Adams (1981) Ratio of central nervous system to body metabolism in vertebrates: its constancy and functional basis. *The American Journal of Physiology—Regulatory, Integrative and Comparative Physiology* 241: 203–212.

Mirto, S., T. La Rosa, R. Danovaro and A. Mazzola (2000) Microbial and meiofaunal response to intensive mussel-farm biodeposition in coastal sediments of the western Mediterranean. *Marine Pollution Bulletin* 40: 244–252.

Misund, O.A. (1993) Dynamics of moving masses: variability in packing density, shape, and size among herring, sprat, and saithe schools. *ICES Journal of Marine Science* 50: 146–160.

Mitchell, E. (1970) Pigmentation pattern evolution in delphinid cetaceans: an essay in adaptive coloration. *Canadian Journal of Zoology* 48: 717–740.

Mitchell, E. (1975) Porpoise, dolphin and small whale fisheries of the world: status and problems. *International Union for Conservation of Nature and Natural Resources Monograph* 3: 1–129.

Modig, A.O. (1996) Effects of body size and harem size on male reproductive behaviour in the southern elephant seal. *Animal Behaviour* 51: 1295–1306.

Mohn, R. and W.D. Bowen (1996) Grey seal predation on the eastern Scotian Shelf: modelling the impact on Atlantic cod. *Canadian Journal of Fisheries and Aquatic Science* 53: 2722–2738.

Moore, M.R., W. Vetter, C. Gaus, G.R. Shaw and J.F. Müller (2002) Trace organic compounds in the marine environment. *Marine Pollution Bulletin* 45: 62–68.

Moore, S.E. (2009) Climate change. Pages 238–241 *in* W.F. Perrin, B. Würsig and J.G.M. Thewissen, eds. *Encyclopedia of Marine Mammals*, 2nd edn. Academic Press, San Diego, CA.

Morisaka, T. and R.C. Connor (2007) Predation by killer whales (*Orcinus orca*) and the evolution of whistle loss and narrow-band high frequency clicks in odontocetes. *Journal of Evolutionary Biology* 20: 1439–1458.

Morse, D.H. (1977) Feeding behavior and predator avoidance in heterospecific groups. *Bioscience* 27: 332–339.

Morton, A. (2000) Occurrence, photo-identification and prey of Pacific white-sided dolphins (*Lagenorhynchus obliquidens*) in the Broughton Archipelago, Canada 1984–(1998). *Marine Mammal Science* 16: 80–93.

Moss, C.J. (1988) *Elephant Memories*. William Morrow, New York.

Mougeot, F. and V. Bretagnolle (2000) Predation risk and moonlight avoidance in nocturnal seabirds. *Journal of Avian Biology* 31: 376–386.

Mouzo, F.H., M.L. Garza, J.F. Izquierdo and R.O. Zibecchi (1978) Rasgos de la geología submarina del Golfo Nuevo. *Acta Oceanográfica Argentina* 2: 69–91.

Myers, R.A., J.K. Baum, T.D. Shepherd, S.P. Powers and C.H. Peterson (2007) Cascading effects of the loss of apex predatory sharks from a coastal ocean. *Science* 315: 1846–1850.

Nefdt, R.J.C. and S.J. Thirgood (1997) Lekking, resource defense, and harassment in two subspecies of lechwe antelope. *Behavioral Ecology* 8: 1–9.

Nei, M. (1978) Estimation of average heterozygosity and genetic distance from a small number of individuals. *Genetics* 89: 583–590.

New Zealand Aquaculture Council (2006) *The New Zealand aquaculture strategy*. 24 pp. Creative Design Advertising Limited.

New Zealand Aquaculture Council (2007) New Zealand Aquaculture Council annual report 2006–2007.

Newton-Fisher, N.E. (2006) Female coalitions against male aggression in wild chimpanzees of the Budongo Forest. *International Journal of Primatology* 27: 1589–1599.

Ñiquen, M. and M. Bouchon (2004) Impact of El Niño events on pelagic fisheries in Peruvian waters. *Deep-Sea Research* 51: 563–574.

Noad, M.J., D.H. Cato, M.M. Bryden, M.N. Jenner and K.C.S. Jenner (2000) Cultural revolution in whale songs. *Nature* 408: 537.

Noë, R. (1994) A model of coalition formation among male baboons with fighting ability as the crucial parameter. *Animal Behaviour* 47: 211–213.

Noren, S.R. (2008) Infant carrying behavior in dolphins: costly parental care in an aquatic environment. *Functional Ecology* 22: 284–288.

Noren, S.R., G. Biedenbach and E.F. Edwards (2006) Ontogeny of swim performance and mechanics in bottlenose dolphins (*Tursiops truncatus*). *The Journal of Experimental Biology* 209: 4724–4731.

Norman, J.R. and F.C. Fraser, eds (1948) *Giant Fishes, Whales, and Dolphins*, 2nd edn. Putnam, London.

Norris, K.S. (1994) Predators, parasites, and multispecies aggregations. Pages 287–300 *in* K.S. Norris, B. Würsig, R.S. Wells and M. Würsig, eds. *The Hawaiian Spinner Dolphin*. University of California Press, Berkeley, CA.

Norris, K.S. and T.P. Dohl (1980) Behavior of the Hawaiian spinner dolphin, *Stenella longirostris*. *Fisheries Bulletin US* 77: 821–849.

Norris, K.S. and C.R. Schilt (1988) Cooperative societies in 3-dimensional space—on the origins of aggregations, flocks, and schools, with special reference to dolphins and fish. *Ethology and Sociobiology* 9: 149–179.

Norris, K.S., B. Würsig, R.S. Wells and M. Würsig (1994a) *The Hawaiian Spinner Dolphin*. University of California Press, Berkeley.

Norris, K.S., R.S. Wells and C.M. Johnson (1994b) The visual domain. Pages 141–160 *in* K.S. Norris, B. Würsig, R.S. Wells and M. Würsig, eds. *The Hawaiian Spinner Dolphin*. University of California Press, Berkeley.

Nowacek, S.M., R.S. Wells, E.C.G. Owen, T.R. Speakman, R.O. Flamm and D.P. Nowacek (2004) Florida manatees, *Trichechus manatus latirostris,* response to approaching vessels. *Conservation Biology* 119: 517–523.

O'Shea, T.J. and S. Tanabe (2003) Persistent ocean contaminants and marine mammals: a retrospective overview. Pages 99–134 *in* J.G. Vos, G.D. Bossart, M. Fournier and T.J. O'Shea, eds. *Toxicology of Marine Mammals.* Taylor and Francis, London.

Oftedal, O.T. (1984) Milk composition, milk yield, and energy output at peak lactation: a comparative review. *Symposium of the Zoological Society of London* 51: 33–85.

Ogilvie, S., A.H. Ross and D.R. Schiel (2000) Phytoplankton biomass associated with mussel farms in Beatrix Bay, New Zealand. *Aquaculture* 181: 71–80.

Oliver, W.R.B. (1922) A review of the cetacea of the New Zealand seas. *Proceedings of the Zoological Society, London* 1922: 557–585.

Ollervides, F.J. (2001) Gray whales and boat traffic: movement, vocal, and behavioral responses in Bahia Mgdalena, Mexico. Ph.D. dissertation, College Station, TX: Texas A&M University.

Oosthuizen, W.H. (2001) Progress report on cetacean research, January 2001 to December 2001, with statistical data for the calendar year 2001. Submitted to the Scientific Committee of the International Whaling Commission, 26 April–10 May 2002, Shimonoseki, Japan.

Orams, M. (2004) Why dolphins may get ulcers: considering the impacts of cetacean-based tourism in New Zealand. *Tourism in Marine Environments* 1: 5–16.

Östman, J.S.O. (1994) Social organization and social behavior of Hawai'ian spinner dolphins (*Stenella longirostris*). Ph.D. dissertation, University of California at Santa Cruz. 114 pp.

Ott, R.L. and M. Longnecker (2001) *An Introduction to Statistical Methods and Data Analysis.* Duxbury, Pacific Grove, CA.

Otto, S.B., E.L. Berlow, N.E. Rank, J. Smiley and U. Brose (2008) Predator diversity and identity drive interaction strength and trophic cascades in a food web. *Ecology* 89: 134–144.

Overstrom, N.A. (1983) Association between burst-pulse sounds and aggressive behavior in captive Atlantic bottlenose dolphins (*Tursiops truncatus*). *Zoo Biology* 2: 93–103.

Pabst, D.A., S.A. Rommel and W.A. McLellan (1999) The functional morphology of marine mammals. Pages 15–72 *in* J.E. Reynolds III and S.A. Rommel, eds. *Biology of Marine Mammals.* Smithsonian Institution Press, Washington, DC.

Pace, M.L., J.J. Cole, S.R. Carpenter and J.F. Kitchell (1999) Trophic cascades revealed in diverse ecosystems. *Trends in Ecology and Evolution* 14: 483–488.

Packer, C. (1977) Reciprocal altruism in *Papio anubis. Nature* 265: 441–443.

Packer, C. (1983) Sexual dimorphism: the horns of African antelopes. *Science* 221: 1191–1193.

Parker, S.T. and K.R. Gibson (1977) Object manipulation, tool use and sensorimotor intelligence as feeding adaptations in cebus monkeys and great apes. *Journal of Human Evolution* 6: 623–641.

Parmelee, J.R. and C. Guyer (1995) Sexual differences in foraging behavior of an anoline lizard, *Norops humilis. Journal of Herpetology* 29: 619–621.

Parra, G.J. (2006) Resource partitioning in sympatric delphinids: space use and habitat preferences of Australian snubfin and Indo-Pacific humpback dolphins. *Journal of Animal Ecology* 75: 862–874.

Paul, L.J., P.R. Taylor and D.M. Parkinson (2001) Pilchard (*Sardinops neopilchardus*) biology and fisheries in New Zealand, and review of pilchard (*Sardinops,* Sardina) biology, fisheries, and research in the main world fisheries. *New Zealand Fisheries Assessment Report* 37. Wellington, Ministry of Fisheries. 44 pp.

Paulian, P. (1953) Pinnepèdes, cétacés, oiseaux des îles Kerguelen et Amsterdam. *Mémoires de l'Institut Scientifique de Madagascar, Série A* 8: 111–234.

Pauly, D., A.W. Trites, E. Capulli and V. Christensen (1998) Diet composition and trophic levels of marine mammals. *ICES Journal of Marine Science* 55: 467–481.

Payne, K. (2003) Sources of social complexity in the three elephant species. Pages 57–85 *in* F.B.M. de Waal and P.L. Tyack, eds. *Animal Social Complexity: Intelligence, culture, and individualized societies.* Harvard University Press, Cambridge.

Paz-y-Miño, G., A.B. Bond, A.C. Kamil and R.P. Balda (2004) Pinyon jays use transitive inference to predict social dominance. *Nature* 430: 778–781.

Peale, T.R. (1848) *Mammalia and Ornithology* Volume VIII of the United States Exploring Expeditions during the years 1838–1842. C. Sherman, Philadelphia.

Pearcy, W.G., E.E. Krygier, R. Mesecar and F. Ramsey (1977) Vertical distribution and migration of oceanic micronekton off Oregon. *Deep-Sea Research* 24: 223–245.

Pearson, H.C. (2008) Fission-fusion sociality in dusky dolphins (*Lagenorhynchus obscurus*), with comparisons to other dolphins and great apes. Ph.D. dissertation, Texas A&M University, College Station, TX. 136 pp.

Pearson, H.C. Influences on dusky dolphin (*Lagenorhynchus obscurus*) fission-fusion dynamics in Admiralty Bay, New Zealand. *Behavioral Ecology and Sociobiology* (in press).

Peckarsky, B.L., P.A. Abrams, D.I. Bolnick, L.M. Dill, J.H. Grabowski, B. Luttbeg, J.L. Orrock, S.D. Peacor, E.L. Preisser, O.J. Schmitz and G.C. Trussell (2008) Revisiting the classics: considering nonconsumptive effects in textbook examples of predator-prey interactions. *Ecology* 89: 2416–2425.

Pengelley, E.T. and S.J. Asmundson (1971) Annual biological clocks. *Scientific American* 224: 72–79.

Pepperberg, I.M. (1999) The Alex studies: cognitive and communicative abilities of grey parrots. Harvard University Press, Cambridge, MA.

Perez-Alvarez, M.J., E. Alvarez, A. Aguayo-Lobo and C. Olavarria (2007) Occurrence and distribution of Chilean dolphin (*Cephalorhynchus eutropia*) in coastal waters of central Chile. *New Zealand Journal of Marine and Freshwater Research* 41: 405–409.

Perez-Barberia, F.J., S. Shultz and R.I.M. Dunbar (2007) Evidence for coevolution of sociality and relative brain size in three orders of mammals. *Evolution* 68: 2811–2821.

Perrin, W.F. (1972) Color patterns of spinner porpoises (*Stenella* cf. *S. longirostris*) of the eastern Pacific and Hawaii, with comments on delphinid pigmentation. *Fishery Bulletin* 70: 983–1003.

Perrin, W.F. (1975) Variation of spotted and spinner porpoise (genus *Stenella*) in the eastern tropical pacific and Hawaii. *Bulletin of the Scripps Institute of Oceanography* 21: 206.

Perrin, W.F. (1984) Patterns of geographic variation in small cetaceans. *Acta Zoologica Fennica* 172: 137–140.

Perrin, W.F. (1989) *Dolphins, Porpoises, and Whales. An action plan for the conservation of biological diversity: 1988–1992.* IUCN, Gland, Switzerland.

Perrin, W.F. (1999) Selected examples of small cetaceans at risk. Page 296–310 *in* J.R. Twiss and R.R. Reeves, eds. *Conservation and Management of Marine Mammals.* Smithsonian Institution, Washington, D.C.

Perrin, W. and J.E. Powers (1980) Role of a nematode in natural mortality of spotted dolphins. *Journal of Wildlife Management* 44: 960–963.

Perrin, W.F. and S.B. Reilly (1984) Reproductive parameters of dolphins and small whales of the family Delphinidae. *Reports of the International Whaling Commission* 6: 97–133.

Perrin, W.F., D.B. Holts and R.B. Miller (1977) Growth and reproduction of the eastern spinner dolphin, a geographical form of *Stenella longirostris* in the eastern tropical Pacific. *Fishery Bulletin, US* 75: 725–750.

Perry, S., H.C. Barrett and J.H. Manson (2004) White-faced capuchin monkeys show triadic awareness in their choice of allies. *Animal Behaviour* 67: 165–170.

Petersen, S., and Z. Mcdonell (2006) A by-catch assessment of the Cape horse mackerel *Trachurus trachurus capensis* mid-water trawl fishery off South Africa. *Report available from Birdlife South Africa, P.O. Box 515, Randburg 2125 South Africa.* 30 pp.

Petraitis, P.S. (1979) Likelihood measures of niche breadth and overlap. *Ecology* 60: 703–710.

Petricig, R.O. (1995) Bottlenose dolphins (*Tursiops truncatus*) in Bull Creek, South Carolina. Ph.D. dissertation, University of Rhode Island, Kingston, RI. 298 pp.

Philippi, R.A. (1893) Los delfines de la punta austral de la América del Sur. *Anales del Museo Nacional de Chile. (Zool)* 6: 1–17.

Philippi, R.A. (1896) Los cráneos de los delfines Chilenos. *Anales del Museo Nacional de Chile* 12: 1–18.

Piana, E., L. Orquera, R.N.P. Goodall, A.R. Galeazzi and A.P. Sobral (1985) Cetacean remains in Beagle Channel shell middens. Abstract, Sixth Biennial Conference on the Biology of Marine Mammals, Vancouver, British Columbia, Canada.

Pichler, F.B. and C. Olavarria (2002) Resolving Chilean dolphin (*Cephalorhynchus eutropia,* Gray 1846) synonymy by sequencing DNA extracted from teeth of museum specimens. *Revista de Biología Marina y Oceanografía* 36: 117–121.

Pichler, F.B., S.M. Dawson, E. Slooten and C.S. Baker (1998) Geographic isolation of Hector's dolphin populations described by mitochondrial DNA sequences. *Conservation Biology* 12: 676–682.

Pichler, F.B., D. Robineau, R.N.P. Goodall, M.A. Meyer, C. Olavarria and C.S. Baker (2001) Origin and radiation of southern hemisphere coastal dolphins (genus *Cephalorhynchus*). *Molecular Ecology* 10: 2215–2223.

Pieper, R.E. and B.G. Bargo (1980) Acoustic measurements of a migrating layer of the Mexican lampfish, *Triphoturus mexicanus*, at 102 kilohertz. *Fishery Bulletin* 77: 935–942.

Pinkas, L., M.S. Oliphant and I.L.K. Iverson (1971) Food habits of albacore, bluefin tuna and bonito in California waters. *Fishery Bulletin* 152: 1–105.

Pitman, R.L. and L.T. Ballance (1994) Incidental sightings of cetaceans in the Chilean Fjords during March (1994) Unpublished paper SC/46/O 194 presented to the Scientific Committee for the 44th Annual Meeting of the International Whaling Commission. Paper available from the IWC Secretariat: secretariat@iwcoffice.org. 4 pp.

Pitnick, S., K.E. Jones and G.S. Wilkinson (2006) Mating system and brain size in bats. *Proceedings of the Royal Society B: Biological Sciences* 273: 719–724.

Pompanon, F., A. Bonin, E. Bellemain and P. Taberlet (2005) Genotyping errors: causes, consequences and solutions. *Nature Reviews Genetics* 6: 847–859.

Poole, J.H. (1989) Mate guarding, reproductive success and female choice in African elephants. *Animal Behaviour* 37: 842–849.

Popper, A.N. (1980) Sound emission and detection by delphinids. Pages 1–52 *in* L.M. Herman, ed. *Cetacean Behavior: Mechanisms and function*. Wiley-Interscience, New York.

Poulin, R. (1999) The functional importance of parasites in animal communities: many roles at many levels? *International Journal for Parasitology* 29: 903–914.

Preisser, E.L., D.I. Bolnick and M.F. Benard (2005) Scared to death? The effects of intimidation and consumption in predator–prey interactions. *Ecology* 86: 501–509.

Prenski, L.B. and V. Angelescu (1993) Ecología trófica de la merluza común *(Merluccius hubbsi)* del Mar Argentino. Parte 3. Consumo anual de alimento a nivel poblacional y su relación con la explotación de las pesquerías multiespecíficas. *INIDEP Documento Científico* 1: 118.

Prior, H. (2006) Social intelligence in common magpies: on the brink of "theory of mind". *Journal of Ornithology* 147(Suppl): 233–234.

Pryor, K. and I.K. Shallenberger (1991) Social structure in spotted dolphins (*Stenella attenuata*) in the tuna purse seine fishery in the Eastern Tropical Pacific. Pages 161–198 *in* K. Pryor and K.S. Norris, eds. *Dolphin Societies: Discoveries and puzzles*. University of California Press, Berkeley.

Pryor, K.W., R. Haag and J. O'Reilly (1969) The creative porpoise: training for novel behavior. *Journal of the Experimental Analysis of Behavior* 12: 653–661.

Pulliam, H.R. (1973) Advantages of flocking. *Journal of Theoretical Biology* 38: 419–422.

Pusey, A.E. (1990) Behavioural changes at adolescence in chimpanzees. *Behaviour* 115: 203–246.

Pusineri, C., L. Magnin, J. Meynier, S. Spitz, S. Hassani and V. Ridoux (2007) Food and feeding ecology of the common dolphin (*Delphinus delphis*) in the oceanic Northeast Atlantic and comparison with its diet in neritic areas. *Marine Mammal Science* 23: 30–47.

Pyke, G.H., H.R. Pulliam and E.L. Charnov (1977) Optimal foraging—selective review of theory and tests. *Quarterly Review of Biology* 52: 137–154.

Queller, D.C., J.E. Strassmann and C.R. Hughes (1993) Microsatellites and kinship. *Trends in Ecology and Evolution* 8: 285–288.

Raga, J.A., F.J. Aznar, J.A. Balbuena and M. Fernández (2002) Parasites. Pages 867–875 *in* W.F. Perrin, B. Würsig and H.G.M. Thewissen, eds. *Encyclopedia of Marine Mammals*. Academic Press, San Diego, CA.

Rankin, S. and J. Barlow (2007) Sounds recorded in the presence of Blainville's beaked whales, *Mesoplodon densirostris*, near Hawai'i (L). *Journal of the Acoustical Society of America* 122: 42–45.

Rasmussen, M.H., L.A. Miller and W.W.L. Au (2002) Source levels of clicks from free-ranging white beaked dolphins (*Lagenorhynchus albirostris* Gray 1846) recorded in Icelandic waters. *Journal of the Acoustical Society of America* 111: 1122–1125.

Rasmussen, M.H., M. Lammers, K. Beedholm and L.A. Miller (2006) Source levels and harmonic content of whistles in white-beaked dolphins (*Lagenorhynchus albirostris*). *Journal of the Acoustical Society of America* 120: 510–517.

Raymond, M. and F. Rousset (1995) Genepop (version-1.2)—Population-genetics software for exact tests and ecumenicism. *Journal of Heredity* 86: 248–249.

Read, A.J., K. Van Waerebeek, J.C. Reyes, J.S. McKinnon and L.C. Lehman (1988) The exploitation of small cetaceans in coastal Peru. *Biological Conservation* 46: 53–70.

Reader, S.M. and K.N. Laland (2001) Primate innovation: sex, age and social rank differences. *International Journal of Primatology* 22: 787–805.

Reader, S.M. and K.N. Laland (2002) Social intelligence, innovation, and enhanced brain size in primates. *Proceedings of the National Academy of Sciences of the USA* 99: 4436–4441.

Redfern, J.V., M.C. Ferguson, E.A. Becker, K.D. Hyrenbach, C. Good, J. Barlow, K. Kashner, M.F. Baumgartner, K.A. Forney, L.T. Ballance, P. Fauchald, P. Halpin, T. Hamazaki, A.J. Pershing, S.S. Qian, A. Read, S.B. Reilly, L. Torres and F. Werner (2006) Techniques for cetacean-habitat modeling. *Marine Ecology Progress Series* 310: 271–295.

Reeves, R.R., C. Smeenk, C.C. Kinze, R.L. Brownell, Jr. and J. Lien (1999) White-beaked dolphin, *Lagenorhynchus albirostris* Gray, 1846. Pages 1–30 *in* S.H. Ridgway and R. Harrison, eds. *Handbook of Marine Mammals, Volume 6: The second book of dolphins and the porpoises.* Academic Press, London.

Reeves, R.R., B.S. Stewart, P.J. Clapham and J.A. Powell (2002) *National Audubon Society Guide to Marine Mammals of the World.* Chanticleer Press, New York.

Reeves, R.R., B.D. Smith, E.A. Crespo and G. Notarbartolo di sciara, Giuseppe (compilers) (2003) *Dolphins, Whales and Porpoises: 2002–2010 Conservation Action Plan for the World's Cetaceans.* IUCN/SSC Cetacean Specialist Group, IUCN, Gland, Switzerland and Cambridge, UK. 139 pp.

Reid, K. and J.P. Croxall (2001) Environmental response of upper trophic-level predators reveals a system change in an Antarctic marine ecosystem. *Proceedings of the Royal Society B: Biological Sciences* 268: 377–384.

Reid, K., J.P. Croxall, D.R. Briggs and E.J. Murphy (2005) Antarctic ecosystem monitoring: quantifying the response of ecosystem indicators to variability in Antarctic krill. *ICES Journal of Marine Science* 62: 366–373.

Reiss, D. and B. McCowan (1993) Spontaneous vocal mimicry and production by bottlenose dolphins (*Tursiops truncatus*): evidence for vocal learning. *Journal of Comparative Psychology* 107: 301–312.

Reiss, D. and L. Marino (2001) Mirror self-recognition in the bottlenose dolphin: a case of cognitive convergence. *Proceedings of the National Academy of Sciences of the USA* 98: 5937–5942.

Rendell, L. and H. Whitehead (2001) Culture in whales and dolphins. *Behavioral and Brain Sciences* 24: 309–382.

Rendell, L. and H. Whitehead (2003) Vocal clans in sperm whales (*Physeter macrocephalus*). *Proceedings of the Royal Society B: Biological Sciences* 270: 225–231.

Reyes, J.C. (1996) A possible case of hybridism in wild dolphins. *Marine Mammal Science* 12: 301–307.

Reyes, L.M. (2006) Cetaceans of Central Patagonia, Argentina. *Aquatic Mammals* 32: 20–30.

Reyes, L.M. (2008) Marine conservation programme for central Patagonia, Argentina: setting basis for the creation of marine protected areas. Unpublished final report to the Rufford Small Grants Foundation. Available online from www.ruffordsmallgrants. org. 45 pp.

Ribeiro, S., F.A. Viddi and T.R.O. Freitas (2005) Behavioural responses of Chilean dolphins to boats in Yaldad Bay, Southern Chile. *Aquatic Mammals* 31: 234–242.

Ribeiro, S., F.A. Viddi, J.L. Cordeiro and T.R.O. Freitas (2007) Fine-scale habitat selection of chilean dolphins (*Cephalorhynchus eutropia*): interactions with aquaculture activities in southern Chiloé Island, Chile. *Journal of the Marine Biological Association of the United Kingdom* 87: 119–128.

Ribic, C.A. (1982) Autumn activity of sea otters in California. *Journal of Mammalogy* 63: 702–706.

Rice, D.W. (1998) *Marine Mammals of the World: Systematics and distribution.* The Society for Marine Mammalogy, Special publication 4. 231 pp. Allen Press, Lawrence, KS.

Rice, W.R. (1989) Analyzing tables of statistical tests. *Evolution* 43: 223–225.

Richter, C., S. Dawson and E. Slooten (2003) Sperm whale watching off Kaikoura, New Zealand: effects of current activities on surfacing and vocalisation patterns. *Science for Conservation* 219: 5–78.

Richter, C., S. Dawson and E. Slooten (2006) Impacts of commercial whale watching on male sperm whales at Kaikoura, New Zealand. *Marine Mammal Science* 22: 46–63.

Ridgway, S.H. (2000) Auditory central nervous system of dolphins. Pages 273–293 *in* W.W.L. Au, A.N. Popper and R.R. Fay, eds. *Hearing by Whales and Dolphins.* Springer-Verlag, New York.

Ridgway, S.H. and R.H. Brownson (1984) Relative brain sizes and cortical surface areas in odontocetes. *Acta Zoologica Fennica* 172: 149–152.

Ringelstein, J., C. Pusineri, S. Hassani, L. Meynier, R. Nicolas and V. Ridoux (2006) Food and feeding ecology of the striped dolphin, *Stenella coeruleoalba,* in the oceanic waters of the north-east Atlantic. *Journal of the Marine Biological Association of the United Kingdom* 86: 909–918.

Ripple, W.J. and R.L. Beschta (2004) Wolves, elk, willows, and trophic cascades in the upper Gallatin Range of Southwestern Montana, USA. *Forest Ecology and Management* 200: 161–181.

Ripple, W.J. and R.L. Beschta (2006) Linking a cougar decline, trophic cascade, and catastrophic regime shift in Zion National Park. *Biological Conservation* 133: 397–408.

Rivarola, M., C. Campagna and A. Tagliorette (2001) Demand-driven commercial whalewatching in Península Valdés (Patagonia): conservation implications for right whales. *Journal of Cetacean Research and Management (Special Issue)* 2: 145–151.

Robertson, D.A. (1978) Blue mackerel, pilchard, anchovy, sprat, saury, and lanternfish. *Proceedings of the Pelagic Fisheries Conference, Occasional Publication* 15: 85–89. Fisheries Research Division, New Zealand Department of Conservation.

Robineau, D. (1985) Données Préliminaires sur la répartition du dauphin de commerson *Cephalorhynchus commersonii* aux Îles Kerguelen, en particulier dans le Golfe du Morbihan. (Preliminary information on the distribution of Commerson's dolphin, *Cephalorhynchus commersonii). Biological Conservation* 31: 85–93.

Robineau, D. (1989) Les cétacés des Îles Kerguelen. *Mammalia* 53: 265–278.

Robineau, D. (1990) Les types de cétacés actuels du Museum National d'Histoire Naturelle II. Delphinidae, Phocoenidae. *Bulletin of the Museum National d'Histoire Naturelle, Paris* Series 4, Section A 12: 197–238.

Robineau, D., R.N.P. Goodall, F. Pichler and C.S. Baker (2007) Description of a new subspecies of Commerson's dolphin *Cephalorhynchus commersonii* (Lacépède 1804) inhabiting the coastal waters of the Kerguelen Islands. *Mammalia* 71: 172–180.

Robinson, K.P. and M.J. Tetley (2007) Behavioural observations of foraging minke whales (*Balaenoptera acutorostrata*) in the outer Morey Firth, north-east Scotland. *Journal of the Marine Biological Association of the United Kingdom* 87: 85–86.

Rooney, N., K. McCann, G. Gellner and J.C. Moore (2006) Structural asymmetry and the stability of diverse food webs. *Nature* 442: 265–269.

Rose, B. and A.I.L. Payne (1991) Occurrence and behavior of the southern right whale dolphin *Lissodelphis peronii* off Namibia. *Marine Mammal Science* 7: 25–34.

Rosel, P.E. (2003) PCR-based sex determination in Odontocete cetaceans. *Conservation Genetics* 4: 647–649.

Ross, G.J.B. (1984) The smaller cetaceans of the south east coast of southern Africa. *Annals of the Cape Provincial Museum (Natural History)* 15: 173–410.

Ross, P. (2002) The roll of immunotoxic environmental contaminants in facilitating the emergence of infectious diseases in marine mammals. *Human and Ecological Risk Assessment* 8: 277–292.

Ross, P., R. De Swart, R. Addison, H. Van Loveren, J. Vos and A. Osterhaus (1996a) Contaminant-induced immunotoxicity in harbour seals: wildlife at risk? *Toxicology* 112: 157–169.

Ross, P., R. De Swart, R. Addison, H. Van Loveren, A. Osterhaus and J. Vos (1996b) The immunotoxicity of environmental contaminants to marine wildlife: a review. *Annual Review of Fish Diseases* 6: 151–165.

Roughgarden, J. (1976) Resource partitioning among competing species: A co-evolutionary approach. *Theoretical Population Biology* 9: 388–424.

Rozas, J. and R. Rozas (1999) DnaSP version 3: an integrated program for molecular population genetics and molecular evolution analysis. *Bioinformatics* 15: 174–175.

Ruckstuhl, K.E. and P. Neuhaus (2000) Sexual segregation in ungulates: a new approach. *Behaviour* 137: 361–377.

Ruckstuhl, K.E., M. Festa-Bianchet and J.T. Jorgenson (2003) Bite rates in Rocky Mountain bighorn sheep (*Ovis canadensis*): effects of season, age, sex and reproductive status. *Behavioral Ecology and Sociobiology* 54: 167–173.

Saayman, G.S. and C.K. Tayler (1979) The socioecology of humpback dolphins (*Sousa sp.*). Pages 165–226 *in* H.E. Winn and B.L. Olla, eds. *The Behavior of Marine Animals, Volume 3: Cetaceans.* Plenum Press, New York.

Salinas, F.V. (2006) Breeding behavior and colonization success of the Cuban treefrog *Osteopilus septentrionalis. Herpetologica* 62: 398–408.

Samuels, A. and P.L. Tyack (2000) Flukeprints: a history of studying cetacean societies. Pages 9–44 *in* J. Mann, R.C. Connor, P.L. Tyack and H. Whitehead, eds. *Cetacean Societies: Field studies of dolphins and whales.* University of Chicago Press, Chicago, IL.

Samuels, A., L. Bejder, R. Constantine and S. Heinrich (2003) A review of swimming with wild cetaceans, with a specific focus on the southern hemisphere. Pages 277–303 *in* N. Gales, M. Hindell and R. Kirkwood, eds. *Marine Mammals Fisheries, Tourism and Management Issues.* CSIRO Publishing, Collingwood, Australia.

Samuels, S. and L. Bejder (2004) Chronic interactions between humans and free-ranging bottlenose dolphins near Panama City Beach, Florida, USA. *Journal of Cetacean Research and Management* 6: 69–77.

Sardella, N.H. and J.T. Timi (2004) Parasites of Argentine hake in the Argentine Sea: population and infracommunity structure as evidence for host stock discrimination. *Journal of Fish Biology* 65: 1472–1488.

Sargeant, B.L., J. Mann, P. Berggren and M. Krützen (2005) Specialization and development of beach hunting, a rare foraging behavior, by wild bottlenose dolphins (*Tursiops* sp.). *Canadian Journal of Zoology* 83: 1400–1410.

Saulitis, E., C. Matkin, L. Barrett-Lennard, K. Heise and G. Ellis (2000) Foraging strategies of sympatric killer whale (*Orcinus orca*) populations in Prince William Sound, Alaska. *Marine Mammal Science* 16: 94–109.

Scammon, C.M. and E.D. Cope (1869) On the cetaceans of the western coast of North America. *Proceedings of the Academy of Natural Sciences of Philadelphia* 21: 13–63.

Scarpaci, C., N. Dayanthi and P.J. Corkeron (2003) Compliance with regulations by "swim-with-dolphins" operations in Port Phillip Bay, Victoria. *Environmental Management* 31: 342–347.

Scarpaci, C., S.W. Bigger, P.J. Corkeron and D. Nugegoda (2000) Bottlenose dolphins (*Tursiops truncatus*) increase whistling in the presence of 'swim-with-dolphin' operations. *Journal of Cetacean Research and Management* 2: 183–185.

Schaller, G.B. (1972) *The Serengeti Lion: A study of predator–prey relations,* Wildlife Behavior and Ecology Series. University of Chicago Press, Chicago, IL.

Schenkkan, E.J. (1973) On the comparative anatomy and function of the nasal tract in odontocetes (Mammalia, Cetacea). *Bijdragen tot de Dierkunde* 43: 127–159.

Schiavini, A.C.M., R.N.P. Goodall, A.K. Lescrauwaet and M.K. Alonso (1997) Food habits of the Peale's dolphin, *Lagenorhynchus australis*; review and new information. *Reports of the International Whaling Commission* 47: 827–833.

Schiavini, A., S.N. Pedraza, E.A. Crespo, R. González and S.L. Dans (1999) Abundance of dusky dolphins (*Lagenorhynchus obscurus*) off north and central Patagonia Argentina in spring and a comparison of incidental catch in fisheries. *Marine Mammal Science* 15: 828–840.

Schillaci, M.A. (2006) Sexual selection and the evolution of brain size in primates. *PLoS One* 1: e62.

Schlegel, H. (1841) Beiträge zur Charakteristik der Cetaceen. *Aghandlungen aus dem Gebiete der Zoologie und Vergleichenden Anatomie* 1: 1–44.

Schmitt, R.J. and S.W. Strand (1982) Cooperative foraging by yellowtail, *Seriola lalandei* (Carangidae) on two species of fish prey. *Copeia* 1982: 714–717.

Schmitz, O.J. (2006) Predators have large effects on ecosystem properties by changing plant diversity, not plant biomass. *Ecology* 87: 1432–1437.

Schmitz, O.J., J.H. Grabowski, B.L. Peckarsky, E.L. Preisser, G.C. Trussell and J.R. Vonesh (2008) From individuals to ecosystem function: toward an integration of evolutionary and ecosystem ecology. *Ecology* 89: 2436–2445.

Schroeder, J.P. (1990) Breeding bottlenose dolphins in captivity. Pages 435–446 *in* S. Leatherwood and R.R. Reeves, eds. *The Bottlenose Dolphin*. Academic Press Inc, San Diego, CA.

Schwartzlose, R.A., J. Alheit, A. Bakun, T.R. Baumgartner, R. Cloete, R.J.M. Crawford, W.J. Fletcher, Y. Green-Ruiz, E. Hagen, T. Kawasaki, D. Lluch-Belda, S.E. Lluch-Cota, A.D. MacCall, Y. Matsuura, M.O. Nevárez-Martínez, R.H. Parrish, C. Roy, R. Serra, K.V. Shust, M.N. Ward and J.Z. Zuzunaga (1999) Worldwide large-scale fluctuations of sardine and anchovy populations. *South African Journal of Marine Science* 21: 289–347.

Sclater, W.L. (1901) *The Mammals of South Africa*. Volume 2. R.H. Porter, London.

Sekiguchi, K. (1994) Studies on feeding habits and dietary analytical methods for smaller odontocete species along the southern African coast. Ph.D. thesis, University of Pretoria, South Africa. 259 pp.

Sekiguchi, K., N.T.W. Klages and P.B. Best (1992) Comparative analysis of the diets of smaller odontocete cetaceans along the coast of southern Africa. *South African Journal of Marine Science* 12: 843–861.

Sergio, F., I. Newton, L. Marchesi and P. Pedrini (2006) Ecological justified charisma: preservation of top predators delivers biodiversity conservation. *Journal of Applied Ecology* 43: 1049–1055.

Seyfarth, R. and D.L. Cheney (2002) What are big brains for? *Proceedings of the National Academy of Sciences of the USA* 99: 4141–4142.

Shane, S.H. (1990) Behavior and ecology of the bottlenose dolphin at Sanibel Island, Florida. Pages 245–265 *in* S. Leatherwood and R.R. Reeves, eds. *The Bottlenose Dolphin*. Academic Press, San Diego, CA.

Shane, S.H., R.S. Wells and B. Würsig (1986) Ecology, behavior and social-organization of the bottle-nosed-dolphin—a review. *Marine Mammal Science* 2: 34–63.

Shelton, D.E. (2006) Dusky dolphins in New Zealand: group structure by sex and relatedness, M.Sc. thesis, Texas A&M University, College Station, TX.

Shimek, S.J. and A. Monk (1977) Daily activity of sea otter off the Monterey Peninsula California. *Journal of Wildlife Management* 41: 117–123.

Shinohara, M., X. Domingo-Roura and O. Takenaka (1997) Microsatellites in the bottlenose dolphin *Tursiops truncatus*. *Molecular Ecology* 6: 695–696.

Shultz, S. and R.I.M. Dunbar (2006) Both social and ecological factors predict ungulate brain size. *Proceedings of the Royal Society of London B: Biological Sciences* 273: 207–215.

Siegel, S. and N.J. Castellan (1995) Estadística no paramétrica aplicada a las ciencias de la conducta. 4th edición española, Editorial Trillas S.A., Mexico D.F., Mexico.

Sigler, M.F., L.B. Hulbert, C.R. Lunsford, N.H. Thompson, K. Burek, G. O'Corry-crowe and A.C. Hirons (2006) Diet of Pacific sleeper shark, a potential Steller sea lion predator, in the north-east Pacific Ocean. *Journal of Fish Biology* 69: 392–405.

Sih, A. (1980) Optimal behavior—Can foragers balance 2 conflicting demands? *Science* 210: 1041–1043.

Sih, A. (1987) Predators and prey lifestyles: an evolutionary and ecological overview. Pages 203–224 *in* W.C. Kerfoot and P.A. Sih, eds. *Predation: Direct and indirect impacts on aquatic communities*. University Press of New England, Hanover, NH.

Similä, T. and F. Ugarte (1993) Surface and underwater observations of cooperatively feeding killer whales in northern Norway. *Canadian Journal of Zoology* 71: 1494–1499.

Simmonds, M.P. (2006) Into the brains of whales. *Applied Animal Behaviour Science* 100: 103–116.

Simmonds, M.P. and S.J. Isaac (2007) The impacts of climate change on marine mammals: early signs of significant problems. *Oryx* 41: 19–26.

Simmons, D.G. and J.R. Fairweather, (1998) Towards a tourism plan for Kaikoura. Tourism Research and Education Center Report No. 10, Lincoln University, Christchurch, New Zealand. 47 pp.

Simon, M., F. Ugarte, M. Wahlberg and L.A. Miller (2006) Icelandic killer whales *Orcinus orca* use a pulsed call suitable for manipulating the schooling behaviour of herring *Clupea harengus*. *Bioacoustics* 16: 57–74.

Simpson, G.G. (1945) The principles of classification and a classification of mammals. *Bulletin of the American Museum of Natural History* 85: 1–350.

Simpson, G.G., A. Roe and R.C. Lewontin (1960) *Quantitative Zoology.* Harcourt Brace, New York.

Sims, D.W., J.P. Nash and D. Morritt (2001) Movements and activity of male and female dogfish in a tidal sea lough: alternative behavioral strategies and apparent sexual segregation. *Marine Biology* 139: 1165–1175.

Slooten, E. (2007) Conservation management in the face of uncertainty effectiveness of four options for managing Hector's dolphin bycatch. *Endangered Species Research* 3: 169–179.

Slooten, E. and S.M. Dawson (1988) Studies on Hector's dolphin, *Cephalorhynchus hectori:* a progress report. Pages 325–338 *in* R.L. Brownell and G.P. Donovan, eds. *Biology of the* Genus Cephalorhynchus. International Whaling Commission, Cambridge.

Slooten, E., D. Fletcher and B.L. Taylor (2000) Accounting for uncertainty in risk assessment: Case study of Hector's dolphin mortality due to gillnet entanglement. *Conservation Biology* 14: 1264–1270.

Slooten, E., S.M. Dawson and W.J. Rayment (2004) Aerial surveys for coastal dolphins: abundance of Hector's dolphins off the South Island west coast, New Zealand. *Marine Mammal Science* 20: 477–490.

Smaal, A.C. (1991) The ecology and cultivation of mussels: new advances. *Aquaculture* 94: 245–261.

Smith, B.D. (2002) Susu and Bhulan: *Platanista gangetica* and *P. g. minor.* Pages 1208–1213 *in* W.F. Perrin, B. Würsig and J.G.M. Thewissen, eds. *Encyclopedia of Marine Mammals.* Academic Press, San Diego, CA.

Smolker, R.A., A.F. Richards, R.C. Connor and J.W. Pepper (1992) Sex differences in patterns of association among Indian Ocean bottlenose dolphins. *Behaviour* 123: 38–69.

Smolker, R., A. Richards, R. Connor, J. Mann and P. Berggren (1997) Sponge carrying by dolphins (Delphinidae, *Tursiops* sp.): a foraging specialization involving tool use? *Ethology* 103: 454–465.

Smuts, B.B. and J.M. Watanabe (1990) Social relationships and ritualized greetings in adult male baboons (*Papio cynocephalus anubis*). *International Journal of Primatology* 11: 147–172.

Sol, D., R.P. Duncan, T.M. Blackburn, P. Cassey and L. Lefebvre (2005) Big brains, enhanced cognition, and response of birds to novel environments. *Proceedings of the National Academy of Sciences of the USA* 102: 5460–5465.

Sorice, M.G., C.S. Shaffer and R.B. Ditton (2006) Managing endangered species within the use- preservation paradox: the Florida manatee (*Trichechus manatus latirostris*) as a tourism attraction. *Environmental Management* 37: 69–83.

Spritzer, M.D., D.B. Meikle and N.G. Solomon (2005a) Female choice based on male spatial ability and aggressiveness among meadow voles. *Animal Behaviour* 69: 1121–1130.

Spritzer, M.D., N.G. Solomon and D.B. Meikle (2005b) Influence of scramble competition for mates upon the spatial ability of male meadow voles. *Animal Behaviour* 69: 375–386.

Stander, P.E. (1992) Foraging dynamics of lions in a semiarid environment. *Canadian Journal of Zoology* 70: 8–21.

Stanford, C.B. (2000) The brutal ape vs. the sexy ape? *American Scientist* 88: 110–112.

Stanford, C.B. and J.B. Nkurunungi (2003) Sympatric ecology of chimpanzees and gorillas in Bwindi Impenetrable National Park, Uganda. *International Journal of Primatology* 24: 901–918.

Steele, J.H. (1985) A comparison of terrestrial and marine ecological systems. *Nature* 313: 355–358.

Stensland, E., A. Angerbjörn and P. Berggren (2003) Mixed species groups in mammals. *Mammal Review* 33: 205–223.

Stimpert, A.K., D.N. Wiley, W.W.L. Au, M.P. Johnson and R. Arsenault (2007) 'Megapclicks': acoustic click trains and buzzes produced during night-time foraging of humpback whales (*Megaptera novaeangliae*). *Biology Letters* 3: 467–470.

Stockin, K.A., R.J. Law, P.J. Duignan, G.W. Jones, L. Porter, L. Mirimin, L. Meynier and M.B. Orams (2007) Trace elements, PCBs and organochlorine pesticides in New Zealand common dolphins (*Delphinus* sp.). *The Science of the Total Environment* 387: 333–345.

Stockin, K., D. Lusseau, V. Binedell, N. Wiseman and M. Orams (2008) Tourism affects the behavioural budget of common dolphins (*Delphinus* spp.) in the Hauraki Gulf, New Zealand. *Marine Ecology Progress Series* 355: 287–295.

Stone, G.S., J. Brown and A. Yoshinaga (1995) Diurnal movement patterns of Hector's dolphin as observed from cliff tops. *Marine Mammal Science* 11: 395–402.

Stone, N.G. (1995) Female foraging responses to sexual harassment in the solitary bee *Anthophora plumipes*. *Animal Behaviour* 50: 405–412.

Stonehouse, B. (1965) Marine birds and mammals at Kaikoura. *Proceedings of the New Zealand Ecological Society* 12: 13–20.

Szabo, A. and D. Duffus (2008) Mother-offspring association in the humpback whale, *Megaptera novaeangliae:* following behaviour in an aquatic mammal. *Animal Behaviour* 75: 1085–1092.

Szymanski, M.D., D.E. Bain, K. Kiehl, S. Pennington, S. Wong and K.R. Henry (1999) Killer whale *Orcinus orca* hearing: auditory brainstem response and behavioral audiograms. *Journal of the Acoustical Society of America* 106: 1134–1141.

Taberlet, P., L.P. Waits and G. Luikart (1999) Noninvasive genetic sampling: Look before you leap. *Trends in Ecology and Evolution* 14: 323–327.

Taggart, S.J., A.G. Andrews, J. Mondragon and E.A. Mathews (2005) Co-occurrence of Pacific sleeper sharks *Somniosus pacificus* and harbor seals *Phoca vitulina* in Glacier Bay. *Alaska Fishery Research Bulletin* 11: 113–117.

Tanabe, S. (2002) Contamination and toxic effects of persistent endocrine disrupters in marine mammals and birds. *Marine Pollution Bulletin* 45: 69–77.

Tanabe, S., T. Mori, R. Tatsukawa and N. Miyazaki (1983) Global pollution of marine mammals by PCBs, DDTs, and HCHs (BHCs). *Chemosphere* 12: 1269–1275.

Tanabe, S., H. Iwata and R. Tatsukawa (1994) Global contamination by persistent organochlorines and their ecotoxicological impact on marine mammals. *The Science of the Total Environment* 154: 163–177.

Tartarelli, G. and M. Bisconti (2006) Trajectories and constraints in brain evolution in primates and cetaceans. *Human Evolution* 21: 275–287.

Tavolga, M. and F.S. Essapian (1957) The behavior of the bottle-nosed dolphin *Tursiops truncatus*: mating, pregnancy, parturition, and mother-infant behavior. *Zoologica* 42: 11–31.

Taylor, M.H. and M. Wolff (2007) Trophic modeling of eastern boundary current systems: a review and prospectus for solving the "Peruvian puzzle". *Revista Peruana de Biología* 14: 87–100.

Te Korowai o Te Tai o Marokura (Kaikoura Coastal Marine Guardians). (2008) *Kaikoura coastal marine values and uses: a characterization report.* 104 pp.

Thomas, A.C., P.T. Strub, M.E. Carr and R. Weatherbee (2004) Comparisons of chlorophyll variability between the four major global eastern boundary currents. *International Journal of Remote Sensing* 25: 1443–1447.

Timi, J.T. (2003) Parasites of Argentine anchovy in the south-west Atlantic: latitudinal pattern and their use for discrimination of host populations. *Journal of Fish Biology* 63: 90–107.

Timi, J.T. and R. Poulin (2003) Parasite community structure within and across host populations of a marine pelagic fish: how repeatable is it? *International Journal for Parasitology* 33: 1353–1362.

Timi, J.T., J.L. Luque and N.H. Sardella (2005) Parasites of *Cynoscion guatucupa* along South American Atlantic coasts: evidence for stock discrimination. *Journal of Fish Biology* 67: 1603–1618.

Timmel, G.B. (2005) Effects of human traffic on the movement patterns of Hawaiian spinner dolphins, *Stenella longirostris*, in Kealakekua Bay, Hawai'i. M.Sc. thesis, San Francisco State University, San Francisco, CA.

Tinbergen, N. (1953) *Social Behaviour in Animals with Special Reference to Vertebrates*. Methuen and Co, London.

Tricas, T.C. (1979) Relationships of the blue shark, *Prionace glauca,* and its prey species near Santa Catalina Island, California. *Fishery Bulletin* 77: 175–182.

Trites, A.W. (2002) Predator-prey relationships. Pages 994–997 *in* W.F. Perrin, B. Würsig and H.G.M. Thewissen, eds. *Encyclopedia of Marine Mammals*. Academic Press, San Diego, CA.

Trites, A.W., V. Christensen and D. Pauly (1997) Competition between fisheries and marine mammals for prey and primary production in the Pacific Ocean. *Journal of the Northwest Atlantic Fishery Science* 22: 173–187.

Trivers, R.L. (1971) The evolution of reciprocal altruism. *Quarterly Review of Biology* 46: 35–57.

Trivers, R.L. (1985) *Social Evolution*. Benjamin Cummings, Menlo Park.

True, F.W. (1889) Contribution to the natural history of the cetaceans: a review of the family Delphinidae. *Bulletin of the U.S. National Museum* 36: 1–191.

True, F.W. (1903) On the species of the South American Delphinidae described by Dr. RA Philippi in 1893 and 1896. *Proceedings of the Biological Society of Washington* 16: 133–144.

Turvey, S.T., R.L. Pitman, B.L. Taylor, J. Barlow, T. Akamatsu, L.A. Barrett, X. Zhao, R.R. Reeves, B.S. Stewart, K. Wang, Z. Wei, X. Zhang, L.T. Pusser, M. Richlen, J.R. Brandon and D. Wang (2007) First human-caused extinction of a cetacean species? *Biology Letters* 3: 537–540.

Tyack, P.L. (2000) Functional aspects of cetacean communication. Pages 270–301 *in* J. Mann, R.C. Connor, P.L. Tyack and H. Whitehead, eds. *Cetacean Societies: Field studies of dolphins and whales*. University of Chicago Press, Chicago, IL.

Urick, R.J. (1983) *Principles of Underwater Sound*. McGraw-Hill, New York.

Van Bresse, M.F., K. Van Waerebeek, M. Fleming and T. Barrett (1998) Serological evidence of morbillivirus infection in small cetaceans from the Southeast Pacific. *Veterinary Microbiology* 59: 89–98.

Van Bressem, M.F. and K. Van Waerebeek (1996) Epidemiology of poxvirus in small cetaceans from the eastern South Pacific. *Marine Mammal Science* 12: 371–382.

Van Bressem, M.F., K. Van Waerebeek, A. Garcia-Godos, D. Dekegel and P.P. Pastoret (1994) Herpes-like virus in dusky dolphins, *Lagenorhynchus obscurus*, from coastal Peru. *Marine Mammal Science* 10: 354–359.

Van Bressem, M.F., K. Van Waerebeek and J.A. Raga (1999) A review of virus infections of cetaceans and the potential impact of morbilliviruses, poxviruses and papillomaviruses on host population dynamics. *Diseases of Aquatic Organisms* 38: 53–65.

Van Bressem, M.F., K. Van Waerebeek, U. Siebert, A. Wünschmann, L. Chávez-Lisambart and J.C. Reyes (2000) Genital diseases in the Peruvian dusky dolphin (*Lagenorhynchus obscurus*). *Journal of Comparative Pathology* 122: 266–277.

Van Bressem, M.F., K. Van Waerebeek, P.D. Jepson, J.A. Raga, P.J. Duignan, O. Nielsen, A.P. Di Beneditto, S. Siciliano, R. Ramos, W. Kant, V. Peddemors, R. Kinoshita, P.S. Ross, A. López-Fernandez, K. Evans, E. Crespo and T. Barrett (2001) An insight into the epidemiology of dolphin morbillivirus worldwide. *Veterinary Microbiology* 81: 287–304.

Van Bressem, M.F., J.A. Raga, T. Barrett, S. Siciliano, E. Di Beneditto, Crespo and K. Van Waerebeek (2007a) Microparasites and their potential impact on the population dynamics of small cetaceans from South America: a brief review. *Proceedings of the International Whaling Commission* SC/59/DW8.

Van Bressem, M.F., K. Van Waerebeek, J.C. Reyes, F. Félix, M. Echegaray, S. Siciliano, A.P. Di Beneditto, L. Flach, F. Viddi, I.C. Avila, J. Bolaños, E. Castineira, D. Montes, E. Crespo, P.A.C. Flores, B. Haase, S. de Souza, M. Laeta and A.B. Fragoso (2007b) A preliminary overview of skin and skeletal diseases and traumata in small cetaceans from South American waters. *Proceedings of the International Whaling Commission* SC/59/DW4.

Van Lawick-Goodall, J. (1968) The behaviour of free-living chimpanzees in the Gombe Stream Reserve. *Animal Behaviour Monographs* 1: 161–311.

Van Oosterhout, C., W.F. Hutchinson, D.P.M. Wills and P. Shipley (2004) Micro-checker: software for identifying and correcting genotyping errors in microsatellite data. *Molecular Ecology Notes* 4: 535–538.

Van Schaik, C.P. (2006) Why are some animals so smart? *Scientific American* April: 64–71.

Van Schaik, C.P. and R.O. Deaner (2003) Life history and cognitive evolution in primates. Pages 5–25 *in* F.B.M. de Waal and P.L. Tyack, eds. *Animal Social Complexity: Intelligence, culture, and individualized societies.* Harvard University Press, Cambridge, MA.

Van Schaik, C.P. and J. Van Hooff (1983) On the ultimate causes of primate social systems. *Behaviour* 85: 91–117.

Van Waerebeek, K. (1992a) Records of dusky dolphins, *Lagenorhynchus obscurus* (Gray, 1828) in the eastern South Pacific. *Beaufortia* 43: 45–61.

Van Waerebeek, K. (1992b) Population identity and general biology of the dusky dolphin *Lagenorhynchus obscurus* (Gray 1828) in the southeastern Pacific, Ph.D. thesis, University of Amsterdam.

Van Waerebeek, K. (1993a) Geographic variation and sexual dimorphism in the skull of the dusky dolphin, *Lagenorhynchus obscurus* (Gray, 1828). *Fishery Bulletin US* 91: 754–774.

Van Waerebeek, K. (1993b) External features of the dusky dolphin *Lagenorhynchus obscurus* (Gray 1828) from Peruvian waters. *Estudios Oceanológicos* 12: 37–53.

Van Waerebeek, K. (1993c) Presumed *Lagenorhynchus* skull at Tasmanian Museum reidentified as *Lissodelphis peronii*. *Australian Mammals* 16: 41–43.

Van Waerebeek, K. (1994) A note on the status of the dusky dolphins (*Lagenorhynchus obscurus*) in Peru. *Reports of the International Whaling Commission Special Issue* 15: 525–528.

Van Waerebeek, K. and A.J. Read (1994) Reproduction of dusky dolphins, *Lagenorhynchus obscurus*, from coastal Peru. *Journal of Mammology* 75: 1054–1062.

Van Waerebeek, K. and J.C. Reyes (1990) Catch of small cetaceans at Puscana Port, Central Peru, during 1987. *Biological Conservation* 51: 15–22.

Van Waerebeek, K. and J.C. Reyes (1994a) Interactions between small cetaceans and Peruvian fisheries in 1988/89 and analysis of trends. *Reports of the International Whaling Commission* Special Issue 15: 495–502.

Van Waerebeek, K. and J.C. Reyes (1994b) Post-ban small cetacean takes off Peru: a review. *Reports of the International Whaling Commission* Special Issue 15: 503–520.

Van Waerebeek, K. and B. Würsig (2002) Pacific white-sided dolphin and dusky dolphin *Lagenorhynchus obliquidens* and *L. obscurus*. Pages 859–861 *in* W.F. Perrin, B. Würsig and J.G.M. Thewissen, eds. *Encyclopedia of Marine Mammals*. Academic Press, San Diego, CA.

Van Waerebeek, K. and B. Würsig (2009) Dusky dolphin *Lagenorhynchus obscurus*. Pages 335–338 *in* W.F. Perrin, B. Würsig and J.G.M. Thewissen, eds. *Encyclopedia of Marine Mammals*, 2nd edn. Academic Press, San Diego, CA.

Van Waerebeek, K., J.C. Reyes and B.A. Luscombe (1988) Revisión de la distribución de pequeños cetaceos frente al Perú. Pages 345–351 *in* H. Salzwedel and A. Landa, eds. *Recursos y Dinamica del Ecosistema de Afloramiento Peruano*. Special Volume I. Instituto del Mar del Perú, Callao.

Van Waerebeek, K., J.C. Reyes and J. Alfaro (1993) Helminth parasites and phoronts of dusky dolphins *Lagenorhynchus obscurus* (Gray, 1828) from Peru. *Aquatic Mammals* 19: 159–169.

Van Waerebeek, K., P.J.H. Van Bree and P.B. Best (1995) On the identity of *Prodelphinus petersii* Lütken, 1889 and records of dusky dolphins *Lagenorhynchus obscurus* (Gray, 1828) from the southern Indian and Atlantic Oceans. *South African Journal of Marine Science* 16: 25–35.

Van Waerebeek, K., M.F. Van Bressem, F. Félix, J. Alfaro-Shigueto, A. García-Godos, L. Chávez-Lisambart, K. Ontón, D. Montes and R. Bello (1997) Mortality of dolphins and porpoises in coastal fisheries off Peru and southern Ecuador in 1994. *Biological Conservation* 81: 43–49.

Van Waerebeek, K., J. Alfaro-Shigueto, D. Montes, K. Onton, L. Santillan and M.F. Van Bressem (2002) Fisheries related mortality of small cetaceans in neritic waters of Peru in 1999–2001. *Proceedings of the International Whaling Commission* SC/54/SM10.

Van Waerebeek, K., R. Leaper, A.N. Baker, V. Papastavrou and D. Thiele (2004) Odontocetes of the Southern Ocean Sanctuary. *Proceedings of the International Whaling Commission* SC/56/SOS1.

Vaughn, R.L., D.E. Shelton, L.L. Timm, L.A. Watson and B. Würsig (2007) Dusky dolphin (*Lagenorhynchus obscurus*) feeding tactics and multi-species associations. *New Zealand Journal of Marine and Freshwater Research* 41: 391–400.

Vaughn, R.L., B. Würsig, D.E. Shelton, L.L. Timm and L.A. Watson (2008) Dusky dolphins influence prey accessibility for seabirds in Admiralty Bay, New Zealand. *Journal of Mammalogy* 89: 1051–1058.

Vaughn, R.L., B. Würsig and J. Packard (2009) Dolphin prey herding: prey ball mobility relative to dolphin and prey ball sizes, multispecies associates, and feeding duration. *Marine Mammal Science* (in press).

Vencl, F.V. (2004) Allometry and proximate mechanisms of sexual selection in *Photinus* fireflies and some other beetles. *Integrated Comparative Biology* 44: 242–249.

Verdolin, J.L. (2006) Meta-analysis of foraging and predation risk trade-offs in terrestrial systems. *Behavioral Ecology and Sociobiology* 60: 457–464.

Vidal, O. (1993) Aquatic mammal conservation in Latin America: problems and perspectives. *Conservation Biology* 7: 788–795.

Viddi, F.A. (2003) Ecology and conservation of the Chilean dolphin in southern Chile. *BP Conservation Project—Unpublished Final Report*. Available upon request from http://conservation.bp.com. 51 pp.

Viddi, F. and A.K. Lescrauwaet (2005) Insights on habitat selection and behavioural patterns of Peale's dolphins (*Lagenorhynchus australis*) in the Strait of Magellan, southern Chile. *Aquatic Mammals* 31: 176–183.

Villadsgaard, A., M. Wahlberg and J. Tougaard (2006) Echolocation signals of wild harbour porpoises, *Phocoena phocoena*. *Journal of Experimental Biology* 208: 56–64.

Visser, I.N. (1999a) Benthic foraging on sting rays by killer whales (*Orcinus orca*) in New Zealand waters. *Marine Mammal Science* 15: 220–227.

Visser, I.N. (1999b) A summary of interactions between orca (*Orcinus orca*) and other cetaceans in New Zealand waters. *New Zealand Natural Sciences* 24: 101–112.

Visser, I.N. (2000) Orca (*Orcinus orca*) in New Zealand waters. Ph.D. dissertation, University of Auckland, New Zealand. 193 pp.

Visser, I.N., D. Fertl, J. Berghan and R. Van Meurs (2000) Killer whale (*Orcinus orca*) predation on a shortfin mako shark (*Isurus oxyrinchus*), in New Zealand waters. *Aquatic Mammals* 26: 229–231.

Von Schreber, J.C.D. and J.A. Wagner (1846) *Die Säugetiere in Abbildungen nach der Natur mit Beschreibungen*. Volume 7. Wolfgang Walther, Erlangen.

Walker, R., O. Burger, J. Wagner and C.R. Von Rueden (2006) Evolution of brain size and juvenile periods in primates. *Journal of Human Evolution* 51: 480–489.

Walker, W.A., S. Leatherwood, K.R. Goodrich, W.F. Perrin and R.K. Stroud (1986) Geographical variation and biology of the Pacific white-sided dolphin, *Lagenorhynchus obliquidens,* in the northeastern Pacific. Pages 441–465 *in* M.M. Bryden and R.J. Harrison, eds. *Research on Dolphins*. Oxford University Press, Oxford.

Wang, D. (1993) Dolphin whistles: comparisons between populations and species. Ph.D. dissertation. The Institute of Hydrobiology, The Chinese Academy of Sciences, Wuhan, China. 247 pp.

Wang, D., B. Würsig and W. Evans (1995) Comparisons of whistles among seven odontocete species. Pages 299–323 *in* R.A. Kastelein, J.A. Thomas and P.E. Nachtigall, eds. *Sensory Systems of Aquatic Mammals*. De Spil Publishers, Woerden, The Netherlands.

Warrant, E. (2004) Vision in the dimmest habitats on Earth. *Journal of Comparative Physiology A-Neuroethology Sensory Neural and Behavioral Physiology* 190: 765–789.

Waterhouse, G. (1838a) Mammalia Part 2 No. 2. *The Zoology of the Voyage of H.M.S. Beagle*. Smith Elder and Co, London.

Waterhouse, G.R. (1838b) A new species of *Delphinus*. *Proceedings of the Zoological Society, London*: 23–24.

Watson-Capps, J.J. and J. Mann (2005) The effects of aquaculture on bottlenose dolphin (*Tursiops* sp.) ranging in Shark Bay, Western Australia. *Biological Conservation* 124: 519–526.

Watts, D.P. and A.E. Pusey (1993) Behavior of juvenile and adolescent great apes. Pages 148–167 *in* M.E. Pereira and L.A. Fairbanks, eds. *Juvenile Primates: Life history, development and behavior*. University of Chicago Press, Chicago, IL.

Watts, D.P. (2000) Grooming between male chimpanzees at Ngogo, Kibale National Park, I. Partner number and diversity and grooming reciprocity. *International Journal of Primatology* 21: 189–210.

Watts, D.P. (2002) Reciprocity and interchange in the social relationships of wild male chimpanzees. *Behaviour* 139: 343–370.

Watts, D.P. (2004) Intracommunity coalitionary killing of an adult male chimpanzee at Ngogo, Kibale National Park, Uganda. *International Journal of Primatology* 25: 507–521.

Weaver, A.C. (1987) An ethogram of naturally occurring behaviors of bottlenose dolphins (*Tursiops truncatus*) in southern California waters. M.Sc. thesis, San Diego State University, San Diego, CA. 93 pp.

Webb, B.F. (1973a) Cetaceans sighted off the west coast of the South Island New Zealand summer 1970. *New Zealand Journal of Marine and Freshwater Research* 7: 179–182.

Webb, B.F. (1973b) Dolphin sightings, Tasman Bay to Cook Strait, New Zealand, September 1968–June 1969. *New Zealand Journal of Marine and Freshwater Research* 7: 399–405.

Webber, M.A. (1987) A comparison of dusky and Pacific white-sided dolphins (Genus *Lagenorhynchus*): Morphology and Distribution. M.Sc. thesis, San Francisco State University, San Francisco, CA. 102 pp.

Weihs, D. (2004) The hydrodynamics of dolphin drafting. *Journal of Biology* 3: 8.

Weilgart, L.S. (2007) The impacts of anthropogenic ocean noise on cetaceans and implications for management. *Canadian Journal of Zoology* 85: 1091–1116.

Weimerskirch, H., P. Inchausti, C. Guinet and C. Barbraud (2003) Trends in bird and seal populations as indicators of a system shift in the Southern Ocean. *Antarctic Science* 15: 249–256.

Weir, B.S. and C.C. Cockerham (1984) Estimating F-statistics for the analysis of population-structure. *Evolution* 38: 1358–1370.

Weir, C.R. (2007) Occurrence and distribution of cetaceans off northern Angola, 2004/05. *Journal of Cetacean Research and Management* 9: 225–239.

Weir, J.S. (2007) Dusky dolphin nursery groups off Kaikoura, New Zealand. M.Sc. thesis, Texas A&M University, College Station, TX. 75 pp.

Weir, J.S., N.M.T. Duprey and B. Würsig (2008) Dusky dolphin (*Lagenorhynchus obscurus*) subgroup distribution: are shallow waters a refuge for nursery groups? *Canadian Journal of Zoology* 86: 1225–1234.

Wells, R.S. (1991) The role of long-term study in understanding the social structure of a bottlenose dolphin community. Pages 199–226 *in* K. Pryor and K.S. Norris, eds. *Dolphin Societies: Discoveries and puzzles*. University of California Press, Berkeley, CA.

Wells, R.S. (2003) Dolphin social complexity: lessons from long-term study and life history. Pages 32–56 *in* F.B.M. de Waal and P.L. Tyack, eds. *Animal Social Complexity: Intelligence, culture, and individualized societies*. Harvard University Press, Cambridge, MA.

Wells, R.S., M.D. Scott and A.B. Irvine (1987) The social structure of free-ranging bottlenose dolphins. Pages 247–305 *in* H.H. Genoways, ed. *Current Mammalogy*. Plenum Press, New York.

Wells, R.S. and K.S. Norris (1994) Patterns of reproduction. Pages 186–200 *in* K.S. Norris, B. Würsig, R.S. Wells and M. Würsig, eds. *The Hawaiian Spinner Dolphin*. University of California Press, Berkeley, CA.

Wells, R.S., D.J. Boness and G.B. Rathbun (1999) Behavior. Pages 324–422 *in* J.E. Reynolds III and S.A. Rommel, eds. *Biology of Marine Mammals*. Smithsonian Institution Press, Washington DC.

Weng, K.C., A.M. Boustany, P. Pyle, S.D. Anderson, A. Brown and B.A. Block (2007a) Migration and habitat of white sharks (*Carcharodon carcharias*) in the eastern Pacific Ocean. *Marine Biology* 152: 877–894.

Weng, K.C., J.B. O'Sullivan, C.G. Lowe, C.E. Winkler, H. Dewar and B.A. Block (2007b) Movements, behavior and habitat preferences of juvenile white sharks *Carcharodon carcharias* in the eastern Pacific. *Marine Ecology Progress Series* 338: 211–224.

Werner, E.E. and S.D. Peacor (2003) A review of trait-mediated indirect interactions in ecological communities. *Ecology* 84: 1083–1100.

Wessels, A. (2002) Onwaarskynlike ambassadeurs: vlagvertoonvaarte deur Suid-Afrikaanse oorlogskepe, 1922–2002. *Journal for Contemporary History* 27: 54–81.

Whitehead, H. (1996) Babysitting, dive synchrony, and indications of alloparental care in sperm whales. *Behavioral Ecology and Sociobiology* 38: 237–244.

Whitehead, H. (1997) Analysing animal social structure. *Animal Behaviour* 53: 1053–1067.

Whitehead, H. (1998) Cultural selection and genetic diversity in matrilineal whales. *Science* 282: 1708–1711.

Whitehead, H. (2002) Culture in whales and dolphins *in* W.F. Perrin, B. Würsig and J.G.M. Thewissen, eds. *Encyclopedia of Marine Mammals*. Academic Press, San Diego, CA.

Whitehead, H. (2007) Learning, climate and the evolution of cultural capacity. *Journal of Theoretical Biology* 245: 341–350.

Whitehead, H. (2009) Culture in whales and dolphins *in* W.F. Perrin, B. Würsig and J.G.M. Thewissen, eds. *Encyclopedia of Marine Mammals*, 2nd edn. Academic Press, San Diego, CA.

Whitehead, H. and S. Dufault (1999) Techniques for analyzing vertebrate social structure using identified individuals: review and recommendations. *Advances in the Study of Behavior* 28: 33–74.

Whitehead, H. and J. Mann (2000) Female reproductive strategies of cetaceans: life histories. Pages 219–246 *in* J. Mann, R.C. Connor, P.L. Tyack and H. Whitehead, eds. *Cetacean Societies*. University of Chicago Press, Chicago, IL.

Whitehead, H. and L. Weilgart (2000) The sperm whale: social females and roving males. Pages 154–172 *in* J. Mann, R.C. Connor, P.L. Tyack and H. Whitehead, eds. *Cetacean Societies*. University of Chicago Press, Chicago, IL.

Whitehead, P.J.P., G.J. Nelson and T. Wongratana (1988) Clupeoid fishes of the world, An annotated and illustrated catalogue of the herrings, sardines, pilchards, sprats, anchovies and wolf-herrings. Part 2. Engaulididae. FAO Fish. Synop. *FAO Species Catalogue* 7: 305–579.

Whitehead, H., R.R. Reeves and P.L. Tyack (2000) Science and the conservation, protection and management of wild cetaceans. Pages 308–332 *in* J. Mann, R.C. Connor, P.L. Tyack and H. Whitehead, eds. *Cetacean Societies*. University of Chicago Press, Chicago, IL.

Whitehead, H., L. Rendell, R.W. Osborne and B. Würsig (2004) Culture and conservation of non-humans with reference to whales and dolphins: review and new directions. *Biological Conservation* 120: 431–441.

Whitehead, H., B. Mcgill and B. Worm (2008) Diversity of deep-water ceataceans in relation to temperature: Implications for ocean warming. *Ecology Letters* 11: 1198–1207.

Whiten, A. and C.P. Van Schaik (2007) The evolution of animal 'cultures' and social intelligence. *Philosophical Transactions of the Royal Society of London. Series B, Biological Sciences* 362: 603–620.

Wiley, D.N., J.C. Moller, R.M. Pace III and C. Carlson (2008) Effectiveness of voluntary conservation agreements: case study of endangered whales and commercial whale watching. *Conservation Biology* 22: 450–457.

Williams, R., D.E. Bain, J.K.B. Ford and A.W. Trites (2002a) Behavioural responses of male killer whales to a 'leapfrogging' vessel. *Journal of Cetacean Research and Management* 4: 305–310.

Williams, R., A.W. Trites and D.E. Bain (2002b) Behavioural responses of killer whales (*Orcinus orca*) to whale-watching boats: opportunistic observations and experimental approaches. *Journal of Zoology* 256: 255–270.

Williams, R., D. Lusseau and P. Hammond (2006) Estimating relative energetic costs of human disturbance to killer whales (*Orcinus orca*). *Biological Conservation* 133: 301–311.

Williams, T.M., W.A. Friedl, M.L. Fong, R.M. Yamada, P. Sedivy and J.E. Haun (1992) Travel at low energetic cost by swimming and wave-riding bottlenose dolphins. *Nature* 355: 821–823.

Williams, T.M., J.E. Haun and W.A. Friedl (1999) The diving physiology of bottlenose dolphins (*Tursiops truncatus*)—I. Balancing the demands of exercise for energy conservation at depth. *Journal of Experimental Biology* 202: 2739–2748.

Willis, P., B.J. Crespi, L.M. Dill, R.W. Baird and M.B. Hanson (2004) Natural hybridization between Dall's porpoises (*Phocoenoides dalli*) and harbour porpoises (*Phocoena phocoena*). *Canadian Journal of Zoology* 82: 828–834.

Wilson, D.E. and D.M. Reeder (2005) *Mammal Species of the World. A taxonomic and geographic reference*, 3rd edn. Johns Hopkins University Press, Baltimore, MD.

Wilson, D.S. and E.O. Wilson (2007) Rethinking the theoretical foundation of sociobiology. *The Quarterly Review of Biology* 82: 327–348.

Wilson, E.O. (1975) *Sociobiology*. Harvard University Press, Cambridge, MA.

Wilson, R.P. and M.P.T. Wilson (1990) Foraging ecology of breeding *Spheniscus* penguins. Pages 181–206 *in* L.S. Davies and J.T. Darby, eds. *Penguin Biology*. Academic Press, New York.

Wilson, R.P., P.G. Ryan, A. James and M.P.T. Wilson (1987) Conspicuous coloration may enhance prey capture in some piscivores. *Animal Behaviour* 35: 1558–1560.

Wilson, R.P., Y. Ropert-Coudert and A. Kato (2002) Rush and grab strategies in foraging marine endotherms: the case for haste in penguins. *Animal Behaviour* 63: 85–95.

Wirsing, A.J., M.R. Heithaus and L.M. Dill (2006) Tiger shark (*Galeocerdo cuvier*) abundance and growth in a subtropical embayment: evidence from 7 years of standardized fishing effort. *Marine Biology* 149: 961–968.

Wirsing, A.J., M.R. Heithaus and L.M. Dill (2007) Fear factor: do dugongs (*Dugong dugon*) trade food for safety from tiger sharks (*Galeocerdo cuvier*)? *Oecologia* 153: 1031–1040.

Wirsing, A.J., M.R. Heithaus, A. Frid and L.M. Dill (2008) Seascapes of fear: evaluating sublethal predator effects experienced and generated by marine mammals. *Marine Mammal Science* 24: 1–15.

Wolf, J.B.W., G. Kauermann and F. Trillmich (2005) Males in the shade: habitat use and sexual segregation in the Galapagos sea lion (*Zalophus californianus wollebaeki*). *Behavioral Ecology and Sociobiology* 59: 293–302.

Wolff, J.O. and T. Van Horn (2003) Vigilance and foraging patterns of American elk during the rut in habitats with and without predators. *Canadian Journal of Zoology* 81: 266–271.

Woodward, B.L. and J.P. Winn (2006) Apparent lateralized behavior in gray whales feeding off the central British Columbia coast. *Marine Mammal Science* 22: 64–73.

Woottom, R.J. (1990) *Ecology of Teleost Fishes*. Fish and Fisheries Series, 1. Chapman and Hall, London.

Wrangham, R. (1982) Mutualism, kinship, and social evolution. Pages 269–289 *in* K.C.S. Group, ed. *Current Problems in Sociobiology*. Cambridge University Press, Cambridge.

Würsig, B. (1982) Radio tracking dusky porpoises in the South Atlantic. Pages 145–160 *in Mammals in the Seas*. FAO Fisheries Series No. 5, Volume IV. United Nations Food and Agriculture Organization, Rome.

Würsig, B. (1986) Delphinid foraging strategies. Pages 347–359 *in* R.J. Schusterman, J.A. Thomas and F.G. Wood, eds. *Dolphin Cognition and Behavior: A comparative approach*. Lawrence Erlbaum Associates, London.

Würsig, B. (1989) Cetaceans. *Science* 244: 1550–1557.

Würsig, B. (2002) Leaping behavior. Pages 689–692 *in* W.F. Perrin, B. Würsig and H. Thewissen, eds. *Encyclopedia of Marine Mammals*. Academic Press, San Diego, CA.

Würsig, B., R.R. Reeves and J.G. Ortega-Ortiz (2002) Global climate change and marine mammals. *In* P.G.H. Evans and J.A. Raga, eds. *Marine Mammals: Biology and Conservation*. Plenum Press, New York, NY.

Würsig, B. (2009) Intelligence and cognition. Pages 616–623 *in* W.F. Perrin, B. Würsig and J.G.M. Thewissen, eds. *Encyclopedia of Marine Mammals*, 2nd edn. Academic Press, San Diego, CA.

Würsig, B. and R. Bastida (1986) Long-range movement and individual associations of two dusky dolphins (*Lagenorhynchus obscurus*) off Argentina. *Journal of Mammalogy* 67: 773–774.

Würsig, B. and G.A. Gailey (2002) Marine mammals and aquaculture: conflicts and potential resolutions. Pages 45–59 *in* R.R. Stickney and J.P. McVey, eds. *Responsible Marine Agriculture*. CAB International Press, New York.

Würsig, B. and T. Jefferson (1990) Methods of photo-identification for small cetaceans. *Reports of the International Whaling Commission* 12: 43–52.

Würsig, B. and H. Whitehead (2009) Aerial behavior. Pages 5–10 *in* W.F. Perrin, B. Würsig and J.G.M. Thewissen, eds. *Encyclopedia of Marine Mammals*, 2nd edn. Academic Press, San Diego, CA.

Würsig, B. and M. Würsig (1979) Day and night of the dolphin. *Natural History* 88: 60–67.

Würsig, B. and M. Würsig (1980) Behavior and ecology of the dusky dolphin, *Lagenorhynchus obscurus*, in the South Atlantic. *Fisheries Bulletin* 77: 871–890.

Würsig, B., M. Würsig and F. Cipriano (1989) Dolphins in different worlds. *Oceanus* 32: 71–75.

Würsig, B., T.R. Kieckhefer and T.A. Jefferson (1990) Visual displays for communication in cetaceans. Pages 545–549 *in* J. Thomas and R. Kastelein, eds. *Sensory
Abilities of Cetaceans*. Plenum Press, New York.

Würsig, B., F. Cipriano and M. Würsig (1991) Dolphin movement patterns: informa-
tion from radio and theodolite tracking studies. Pages 79–112 *in* K. Pryor and
K.S. Norris, eds. *Dolphin Societies: Discoveries and puzzles*. University of
California Press, Berkeley, CA.

Würsig, B., F. Cipriano, E. Slooten, R. Constantine, K. Barr and S. Yin (1997) Dusky
dolphins (*Lagenorhynchus obscurus*) off New Zealand: status of present knowledge.
Report of the International Whaling Commission 47: 715–722.

Würsig, B., N. Duprey and J. Weir (2007) Dusky dolphins (*Lagenorhynchus obscurus*)
in New Zealand waters: present knowledge and research goals. DOC Research and
Development Series 270. Department of Conservation, Wellington. 28 pp.

Yazdi, P. (2002) A possible hybrid between a dusky dolphin (*Lagenorhynchus obscurus*)
and the southern right whale dolphin (*Lissodelphis peronii*). *Aquatic Mammals* 28:
211–217.

Ydenberg, R.C. and L.M. Dill (1986) The economics of fleeing from predators.
Advances in the Study of Behavior 16: 229–249.

Yin, S. (1999) Movement patterns, behaviors, and whistling sounds of dolphin groups
off Kaikoura, New Zealand. M.Sc. thesis. Texas A&M University, College Station,
TX. 107 pp.

Yin, S., B. Würsig and R. Constantine (2001) Acoustic behavior of dusky
(*Lagenorhynchus*) and common (*Delphinus delphis*) dolphin groups off Kaikoura,
New Zealand: Who's whistling, 14th biennial meeting on the Biology of Marine
Mammals, Vancouver, November.

Yodzis, P. (1988) The indeterminacy of ecological interactions as perceived through per-
turbation experiments. *Ecology* 69: 508–515.

Yodzis, P. (1998) Local trophodynamics and the interaction of marine mammals and
fisheries in the Benguela ecosystem. *Journal of Animal Ecology* 67: 635–658.

Yodzis, P. (2001) Must top predators be culled for the sake of fisheries? *Trends in
Ecology and Evolution* 16: 78–84.

Yurk, H., L. Barrett-Lennard, J.K.B. Ford and C.O. Matkin (2002) Cultural transmission
within maternal lineages: vocal clans in resident killer whales in southern Alaska.
Animal Behaviour 63: 1103–1119.

Zahavi, A. and A. Zahavi (1997) *The Handicap Principle*. Oxford University Press, NY.

Zornetzer, H.R. and D.A. Duffield (2003) Captive-born bottlenose dolphin × common
dolphin (*Tursiops truncatus* × *Delphinus capensis*) intergeneric hybrids. *Canadian
Journal of Zoology* 81: 1755–1762.

agonistic (behaviors) aggressive or defensive actions, such as fleeing or fighting, brought on by the interaction between individuals that are usually of the same species

allele an alternate form of a gene at a genetic locus

alloparental care behaviors by an adult that benefit young non-offspring; examples include food provisioning or protective behavior

allopatry populations or species occurring in separate, non-overlapping geographic areas

altruism a behavior that is detrimental to the actor and beneficial to another individual

anatomical of, or relating to, anatomy

anatomy the physical structure of living things

Antarctic Convergence a 32–48 km band of ocean surface water encircling Antarctica, where cold, northward-flowing Antarctic waters meet and sink beneath sub-Antarctic waters; sometimes referred to as the "Antarctic Polar Frontal Zone"

anthropomorphism attribution of human characteristics to non-human entities

aquaculture the culture or farming of aquatic species in marine and freshwaters, including natural waters and land-based facilities

artisanal fishery fisheries involving skilled but non-industrialized operators; typically a small-scale, decentralized operation; normally subsistence fisheries, although sometimes the catch may be sold

axillary the armpit, or an analogous part

backscatter the sound energy reflected back towards a sound source; echo

beak another term for the snout or rostrum of a dolphin, formed by the projecting tip of the upper and lower jaws

Bernoulli effect an increase in the speed of a fluid which occurs simultaneously with a decrease in pressure; thus, two objects close together may be forced towards each other due to the fluid between them flowing faster than in other areas

bioaccumulation the accumulation of a substance, such as a toxic chemical, in various tissues of a living organism

bottleneck/evolutionary bottleneck a drastic decline in the size of a population or entire species, due to an environmental catastrophe or human-associated mortality, often associated with a rapid loss of genetic diversity

broadband signals any acoustic signal with a bandwidth-to-center frequency ratio less than about 5

burst pulse a set of echolocation-like clicks which occur at high repetition rates of over 200 per second

by-catch non-targeted/unwanted marine creatures caught in nets or on hooks while fishing for another species

by-product mutualism an interaction in which individuals benefit from each others' selfish behavior

carapid a fish of the family Carapidae, such as a pearlfish

cephalopod a mollusk, such as an octopus or a nautilus, having a beaked head and prehensile tentacles

character a heritable attribute that provides information about the evolutionary relationships among organisms or groups of organisms

CITES the Convention on International Trade in Endangered Species of Wild Fauna and Flora (CITES) is an international agreement between governments; its aim is to ensure that international trade in specimens of wild animals and plants does not threaten their survival and it provides varying degrees of protection to more than 33 000 species of animals and plants

clean leap a type of leap in which the dolphin's entire body leaves the water then re-enters head-first, with little splash

click broadband signals with a rapid rise time

cognition brain functions associated with thinking, knowing, and remembering

commensal a situation where two different animal or plant species live in close association but are not interdependent; also used to describe individuals who live in this manner

competition someone or something working in opposition to the actor

condylar a rounded articulatory prominence at the end of a bone

congener refers to members (or species) of the same genus

continental shelf marks the outer limit of a continent and associated coastal plain and is covered by shallow seas and gulfs

crepuscular low light conditions that occur at sunrise and sunset

culture information or behavior which is shared by a population or subpopulation and acquired by learning from conspecifics

cyamids small crustaceans which are external parasites; sometimes referred to as whale lice, although they may also be found on dolphins and porpoises

deep scattering layer a horizontal, oceanic zone of zooplankton, fish, and squid which was first detected with echosounders as the swim-bladders of the fish reflect ("scatter") sound waves; this biological layer ascends to within tens of meters of the ocean surface at dusk and may descend to depths up to 1000 m during mid-day

delphinid a member of the family Delphinidae, or dolphins

demersal environment refers to the bottom of a body of water

density-mediated pathway an indirect effect resulting from changes in population density or growth rates of initiator and receiver species

dimorphism the existence of two distinct types of individual within a species, usually differing in one or more characteristics such as coloration, size, and shape. The most usual dimorphism is between the sexes, or sexual dimorphism

direct harvest hunting or harvesting of a target species

diurnal vertical migration the cycle of nightly ascent of the deep scattering layer to within tens of meters of the surface and subsequent descent before dawn

DNA deoxyribonucleic acid; a nucleic acid that stores the genetic information used in the development and functioning of all living organisms and some viruses

DVM a veterinarian doctor in the USA, "Doctor of Veterinary Medicine"

echelon position the orientation of a calf swimming parallel to its mother, just above her midline and by her dorsal fin

echolocation the use of sounds to determine presence of objects by the echoes produced from objects; in dolphins, this refers to high-frequency sound waves used to detect objects such as fish, obstructions, and other dolphins in the water

echosounder a type of sonar in which an acoustic beam is directed (usually) vertically down; composed of a transmitter that produces bursts of electrical energy at specific frequency and a transducer that converts the electrical energy into acoustic energy and directs the beam; the returning echoes (backscatter) are received by the transducer and converted again into electrical energy

ecosystem a system formed by the interaction between organisms and their environment

eco-tourism a form of tourism involving travel to areas of natural or ecological interest, typically under the guidance of a naturalist, for the purpose of observing wildlife and learning about conservation and the environment, and often involving active community participation (editors' note: only a small part of tourism that is labeled as "eco-tourism" actually qualifies as such)

El Niño a warming of the ocean surface off the coast of Peru and Ecuador that occurs every 4–12 years when upwelling of cold, nutrient-rich water does not occur due to a decrease in coastal winds that move surface waters offshore; this irregular cyclic swing of warm and cold phases in the tropical Pacific is sometimes referred to as ENSO (the El Niño Southern Oscillation). The term El Niño means "the child" in Spanish, and refers to the Christ-child, as it is usually first noticeable around Christmas

encephalization quotient a mathematic equation used to measure mammalian brain size in relation to body size

EPBC the Environment Protection and Biodiversity Conservation Act (EPBC) of 1999 is the Australian government's central piece of environmental legislation; it protects and manages nationally and internationally important flora, fauna, ecological communities, and heritage places

epipelagic the uppermost water layer where there is enough light for photosynthesis to occur

estrous cycle comprises the recurring physiologic changes that are induced by reproductive hormones in most placental mammalian females

estrus the reproductive phase when a female is sexually receptive

eusociality a type of sociality found in insects which is characterized by cooperative care of young, groups of sterile individuals specialized to perform specific tasks ("castes"), and overlapping generations

evolution the gradual change of heritable traits in a population over time, resulting in descendants that differ morphologically and physiologically from their ancestors

falcate curved like a scythe or sickle, in dolphins referring to the shape of the dorsal fin

fecund fertile; capable of producing offspring

fetal folds skin folds occurring in newborn dolphins as a result of their prenatal position in the mother's uterus

flank the side or lateral portion of a mammal between the end of the ribs and the hips; in a dolphin's body (which has no hindlimbs or pelvis), this refers to the side of the animal behind the dorsal fin

food web a complex network of interactions between producers and consumers within an ecosystem

foraging behavior which includes locating, containing/herding, and capturing prey

gambit of the group the assumption that individuals in a group are interacting with each other

genetic marker a DNA sequence that can be used to identify patterns of inheritance

gestation the developmental period of young in the uterus from conception until birth; pregnancy

gill net a curtain-like fishing net, suspended vertically in the water, with meshes of such a size as to catch fish by the gills

group focal follow a research sampling method used to record the behavior and associated data (e.g., time of day, location) for a group of animals

gyrification index a measure of neocortex surface area to total brain size

handicap principle refers to the hypothesis that if a trait is costly to produce, it is likely to be an honest or non-deceptive signal, as only an individual capable of paying the cost will adopt the signal

haplotype a set of genetic markers that are inherited together

harem a component of polygynous mating systems whereby a male defends a group of females with whom to mate

high-seas fisheries fisheries that are pelagic in nature, occurring on the open sea rather than in coastal locations

homologous similarities between structures or characteristics coming from a shared ancestry

horizontal culture a type of culture in which information is transmitted between members of the same generation

hydrophone a microphone or speaker that functions underwater. In biologists' parlance, this almost always is the underwater equivalent of a microphone

ichthyoplankton the eggs and larvae of fish found primarily in the upper 200 m of the water column

immunosuppression an act that reduces the activation or efficacy of the immune system

individual focal follow a research sampling method used to record behavior and associated data (e.g., time of day, location) for an individual animal

infant position the orientation of a calf swimming ventral and to one side of its mother, near her mammary slits

IUCN the International Union for Conservation of Nature, which is a democratic membership union that works towards solving global environmental and developmental challenges

IWC the International Whaling Commission, which is the intergovernmental body responsible for the conservation of whales and the management of whaling

jig fisheries a fishing technique commonly used in squid fisheries that employs "jig" fishing lures consisting of a lead sinker with hook which create a jerky, vertical motion

lactation the production and secretion of milk by the mammary glands; also refers to the period following birth during which milk is secreted

long-line fisheries commercial fisheries that use hundreds or even thousands of baited hooks hanging from a single line to catch target species such as swordfish, tuna, halibut and sablefish

Machiavellian intelligence intelligence used to serve an individual's own selfish interests; "political" intelligence

Māori the indigenous people of New Zealand, and their language

male alliance a long-term relationship formed between two or more males whereby individuals support one another during disputes with third parties

marine farming the farming of aquatic species in the marine environment

meiofauna small benthic invertebrates that live in both marine and freshwater environments; meiofauna is not a taxonomic grouping but loosely defines a group of organisms by their size, which is larger than microfauna but smaller than macrofauna

mesopelagic an oceanic layer that extends from 200 to 1000m in depth

microsatellite regions of DNA in which the sequence is short, containing 2–6 base pairs

mitochrondrial DNA a haploid genome (containing a single set of chromosomes) in the mitochondria that is passed on from mother to male and female offspring; often called mtDNA

MLRA the Marine Living Resources Act of 1998; a comprehensive act of the South African parliament outlining policy relating to all living marine resources in South Africa

MMPA the Marine Mammal Protection Act; enacted in the United States in 1972, this piece of legislation provides for the protection, conservation and management of marine mammals; similar legislation was enacted in New Zealand in 1978

MMPR the Marine Mammals Protection Regulations; regulations established in 1992 in New Zealand pursuant to section 28 of the Marine Mammal Protection Act of 1978

mobbing a type of antipredator behavior in which a group of animals approaches or aggressively interacts with a threatening or non-threatening predator

molecular consisting of molecules, in genetics referring to description of differences at the "molecule" or gene level

monogamy a mating pattern in which a male and female form a pair bond which may last for a portion of a breeding season, a whole breeding season, or a lifetime

monophyletic a taxonomic group comprising a single common ancestor and all the descendants of that ancestor

morphology the structure and form of an organism, excluding function

morphometrics the study of variation in shape and size of organisms

mtDNA see mitochondrial DNA

neocortex a portion of the cerebral cortex found only in mammals, which functions in tasks such as conscious thought, spatial reasoning, sensory perception, and language (in humans)

neritic zone the part of the ocean that extends from the low tide mark to the edge of the continental shelf and is usually characterized by relatively shallow depth of up to 200 m

(ecological) niche comprises the environment and the set of conditions under which an animal lives and reproduces given competition, predation, and the effects of its activities on those same factors

(trophic) niche the position or function of an organism in a food web

noisy leap a type of leap in which the dolphin lands on its side, back, or ventrum, making a noisy splash

nomenclature a set of rules used for naming organisms, i.e., taxa

nuclear DNA DNA in the nucleus of the cell

nuclear gene a gene located in the nucleus of a cell

nursery group a group consisting of mothers with mixed-age calves. At times, nursery groups may also have "other" attendant animals, usually females with variable relationships, or "aunts"

oblique culture a type of culture in which information is transmitted between unrelated members of different generations

odontocete a toothed cetacean, i.e., a toothed whale, dolphin, or porpoise

optimality models a quantitative evolutionary model that defines maximum fitness values; when applied to foraging, the model predicts that animals will forage in a way that maximizes their caloric intake per unit time

osteological characters characters derived from variable features of the bony skeleton

otolith a calcareous structure found in the inner ears of fishes

ovulation the release of an egg cell (ovum) from the ovary in female animals, regulated in mammals by hormones produced by the pituitary gland during the menstrual cycle

paraphyletic refers to a taxonomic group (a set of evolutionarily related species) that does not contain all of the descendants of its most recent common ancestor

parasitism a diseased condition resulting from parasitic infestation

parturition the process of giving birth

peduncle the base of a dolphin's tail, the narrow portion just forward of the flukes; sometimes called the tail stock

pelagic environment the environment of the open ocean or sea

pelagic species a species which occurs in open oceanic waters

pelagic zone the open ocean area away from the sea bed and beyond the tidal zone

phenotype the observable physical or biochemical characteristics of an organism, as determined by both genetic makeup and environmental influences

phylogenetics the study of the evolutionary relatedness between groups of organisms

phylogeny the sequence of events involved in the evolutionary development of a taxonomic group, often depicted as a branching diagram or "tree"

phylogeography the study of the historical processes that have influenced the contemporary geographic distributions of organisms, accomplished via the comparison of geographic structure of genetic variation within and between species

pigmentation coloration; in mammals, pigmentation is controlled by the type and amount of pigments (colored substances) inside particular skin cells called "melanocytes"

polyandry a mating pattern in which a female mates with more than one male in a single breeding season. Polyandry is quite rare in mammals

polygamy a general term used to describe a mating pattern in which an individual of either sex has more than one mate

polygynandry a mating pattern in which two or more males mate with two or more females, also "multi-mate"; this is often referred to as "promiscuity," but the latter implies absence of mate choice

polygyny a mating pattern in which a male mates with more than one female in a single breeding season

polymerase chain reaction a method of amplifying DNA in vitro by temperature-controlled repeated cycles of denaturing and synthesis; often called PCR

polyphyletic refers to a taxonomic group whose members share trait(s) but evolved separately in different places of the evolutionary tree

precocial an adjective used to describe offspring that are able to travel (somewhat) independently straight from birth

predation a biological interaction whereby one animal (the predator) feeds on another animal (the prey)

prey herding in dolphins, the act of swimming around prey to move it (e.g., closer to the surface), or to cause prey to bunch together more tightly

prey patchiness a parameter that describes the distribution of prey in space and time

primer a short, single-stranded polynucleotide of a particular sequence which anneals to complementary template DNA and "primes" extension (by allowing new deoxyribonucleotides to be added) during PCR amplification

progesterone a hormone that acts to prepare the uterus for implantation of the fertilized ovum, to maintain pregnancy, and to promote development of the mammary glands

promiscuous a type of mating pattern characterized by indiscriminate sexual interactions, with each male mating with several females and each female mating with several males; also referred to as "multi-mate" mating, and as "polygynandry" (see)

purse seine fisheries fisheries that use a net which closes at the base by pulling a rope through a series of rings, stopping schooling fish from descending to escape

radio tracking a method to document the movements of an animal by attaching a transmitter to it; the transmitter then emits a signal that can be picked up by satellite or Earth-bound receiver

reciprocal altruism a form of altruism in which one organism provides a benefit to another without expecting any immediate payment or compensation, but the "favor is returned" at a later time

Red List a list prepared by the International Union for Conservation of Nature (IUCN) that provides information on the taxonomy, distribution, and conservation status of plants and animals worldwide

rostrum see "beak"

scan sampling a method in which a "snapshot" or instantaneous sample is used to record the behavior and/or location of an individual in a group before moving to the next individual and doing the same

scouting a type of defensive behavior in which a few individuals may leave a group to assess a threat. "Scouting" is often claimed in popular articles/reports, but is very difficult to assess or prove

scramble competition in the context of mating, this refers to a form of competition which favors individuals who are able to locate mates rapidly and accurately

selfish herd effect a hypothesis that suggests that scattered individuals will aggregate tightly when exposed to danger, such that individuals at the edge of the group experience higher predation risk relative to individuals located near the center

semi-pelagic species species which live some portion of their lives in open oceanic waters

set net a gill net set across a current to catch fish; often used in small nearshore fisheries

sexual dimorphism the systematic difference in form between individuals of different sex in the same species

sexual selection the theory proposed by Charles Darwin that the frequency of traits can increase or decrease depending on attractiveness of the bearer

skin swab a tissue sample obtained by scraping the outer skin cells of an animal

spectrogram a plot of frequency versus time

sperm competition "competition" (which is usually a matter of volume and viability) among sperm of two or more males for the fertilization of an ovum

sponging a behavior exhibited primarily by female bottlenose dolphins in Shark Bay, Australia, whereby an individual breaks a marine sponge off the seafloor and subsequently places the sponge over its closed rostrum to apparently probe in the seafloor for fishes.

spyhop vertical lifting of the head above water by a cetacean so that the eye breaks the surface; potentially (but not proven) used for visual inspection of objects above the surface

stotting the springing gait of antelope and gazelle; also called pronging

stranding the act of beaching; running aground

sympatry the use of the same (overlapping) geographical areas by populations or species

synonym two different scientific names used for the same taxon

tandemly repeated units of DNA

taxa plural of taxon

taxon a group of biological organisms

taxonomy the science of describing, classifying, and naming organisms based on similarities of structure or origin, analysis of diagnostic characters, or genetic similarity

testosterone a steroid hormone from the androgen group which is the principal male sex hormone and an anabolic steroid; promotes the development of male sex organs, sperm, and secondary sexual characteristics

theodolite an instrument for measuring both horizontal and vertical angles, as used in triangulation networks; a key tool in surveying and engineering work, but has also been adapted for monitoring the movements of cetaceans

top-down control the process of an ecological community being regulated by events occurring at the top trophic level

trait a distinct variable attribute of an organism that is determined by the environment or the interaction of genes and the environment; may or may not be heritable

trait-mediated pathway an indirect effect resulting from a change in the trait of an intermediate or transmitter species present in a linked pathway between an initiator and a receiver species

trawling fisheries a method of fishing that involves pulling a large fishing net through the water behind one or more boats

Treaty of Waitangi signed on February 6, 1840, this treaty between the Māori of New Zealand and Great Britain is the founding document of the nation of New Zealand

trophic ecology the study of the feeding relationships between organisms within an ecosystem

trophic role the role of an organism in a food web

type specimen collected and catalogued remains of an individual that serves as the representative type for the description of a species

upwelling the process by which warm, less-dense surface water is drawn away from shore (or a subsurface structure) by currents and replaced by cold, denser water brought up from the subsurface

ventral pertaining to the front or lower surface of a structure; the belly

vertical culture a type of culture in which information is transmitted between parents and offspring

whistle dolphin sounds with tonal quality, often frequency modulated

Index

Printed and bound by CPI Group (UK) Ltd, Croydon, CR0 4YY

03/10/2024

01040412-0011